SECOND EDI

T0258964

FUNDAMENTALS OF GLACIER DYNAMICS

SECOND EDITION

FUNDAMENTALS OF GLACIER DYNAMICS

C.J. VAN DER VEEN

CRC Press
Taylor & Francis Group
Boca Raton   London   New York

CRC Press is an imprint of the
Taylor & Francis Group, an **informa** business

CRC Press
Taylor & Francis Group
6000 Broken Sound Parkway NW, Suite 300
Boca Raton, FL 33487-2742

First issued in paperback 2017

ISBN 13: 978-1-138-07721-8 (pbk)
ISBN 13: 978-1-4398-3566-1 (hbk)

Library of Congress Cataloging-in-Publication Data

Veen, C. J. van der (Cornelis J.), 1956-
    Fundamentals of glacier dynamics / author, C.J. van der Veen. -- Second edition.
    pages cm
    Includes bibliographical references and index.
    ISBN 978-1-4398-3566-1 (hardback)
    1. Glaciers. 2. Ice sheets. 3. Sea level. 4. Climatic changes. 5. Hydrology. 6. Glaciology. I. Title.

    GB2403.2.V44 2013
    551.31'2--dc23   2012050921

Visit the Taylor & Francis Web site at
http://www.taylorandfrancis.com

and the CRC Press Web site at
http://www.crcpress.com

Contents

Preface

In the 14 years since publication of *Fundamentals of Glacier Dynamics*, the field of glaciology has changed significantly. Satellite measurements and other remote-sensing techniques afford monitoring of glaciers and ice sheets on unprecedented spatial and temporal scales and have revealed large and rapid changes occurring in both Greenland and Antarctica. Identifying the processes responsible for these changes is paramount to understanding the interactions between ice sheets and the climate system and their future contribution to global sea level rise. The primary goal of this revised edition is to provide a theoretical framework for quantitatively interpreting observed glacier changes and for developing models of glacier flow. This framework is based on the partitioning of full stresses into a lithostatic component associated with gravitational action and resistive stresses that oppose this action. In my experience, this provides a more physically intuitive discussion of force budget than the more traditional approach involving deviatoric stresses.

In an effort to streamline the presentation and focus in this edition of *Fundamentals of Glacier Dynamics*, the more descriptive sections, including much of the last four chapters that appeared in the first edition, have been eliminated. This is not to imply that observations are of secondary importance—quite the contrary. However, many sources are available that provide excellent reviews of glacier observations, and a repeat would be redundant. The introductory Chapter 1 reviews the most important mathematical tools used throughout the remainder of the book. The new Chapter 8 discusses fracture mechanics and iceberg calving. Applications of the force-budget technique using measurements of surface velocity to locate mechanical controls on glacier flow are now consolidated into Chapter 11.

Many colleagues contributed directly and indirectly to the completion of this second edition. Special thanks are due to Leigh Stearns for reading through the entire draft manuscript and providing valuable comments and suggestions, and to Jeremy Bassis for his comments on several chapters. I am indebted to others who commented on individual chapters: Ed Bueler, Richard Hindmarsh, Faezeh Nick, Christian Schoof, and Dirk van As. Of course, I take full responsibility for the final text. Throughout my career as a glaciologist, my research has been funded mostly by the Office of Polar Programs of the National Science Foundation and by the Cryospheric Sciences Program of the National Aeronautics and Space Administration; this support is gratefully acknowledged.

C. J. van der Veen

Preface to the First Edition

This book is not intended to serve as a ski lift, transporting the reader effortlessly to the peaks of glaciological knowledge. Instead, it is designed to act as a trail guide, helping the reader achieve the basic level of understanding required to describe and model the flow and dynamics of glaciers. The emphasis is more on developing and outlining procedures than on providing a complete overview of all aspects of glacier dynamics. To this end, derivations leading to frequently used equations are presented step by step to allow the reader to grasp the mathematical details as well as physical approximations involved without having to consult the original works. While going through these derivations may be tedious, the reward is that in the end, the reader will have gained the understanding needed to apply similar techniques to somewhat different applications.

The choice of material presented in this book is, not surprisingly, based to a large extent on research that I have conducted over the past 15 years, first under the watchful eye of Hans Oerlemans at the University of Utrecht, and later in collaboration with Ian Whillans and others at the Byrd Polar Research Center. No claim is made here to present a complete and exhaustive review of the glaciological literature. In many instances, topics are discussed based on one or two key papers that I found very helpful and that may provide a good starting point for readers interested in a more in-depth review of the literature. While an effort was made to keep the entire text up to date, the fact that this book was written over a five-year period necessarily implies that some recent work and developments may have been overlooked. My apologies to those authors who feel slighted. The alternative would have been to continue updating the text, but in all likelihood, this would have postponed publication indefinitely.

The discussion of glacier dynamics is based on the force-budget technique. There is nothing magical or special about this technique. It is based on a different approach to solving Newton's second law of motion, which states that forces acting on a section of glacier must sum to zero (accelerations being negligible in glacier flow). Full stresses are partitioned into a lithostatic component associated with the action of gravity and resistive stresses that oppose this action. The main advantage of this approach is the distinction that is made between various physical processes controlling glacier flow. Consequently, the implications of omitting one or more terms in the balance equations becomes more clear than when the usual balance equations involving deviatoric stresses are considered. As should be the case, the final results are the same, whichever approach is used.

One of the most important steps in developing models to describe glacier flow is to determine which of the terms in the force-balance equations are important and should be retained. One procedure is to introduce scaling parameters and derive expressions for first- and higher-order solutions. Another approach, adopted here, is to use measurements of glacier speed and geometry to evaluate the role of various resistances to flow on a particular glacier. Based on these results, the balance equations can be simplified and a model for that particular glacier formulated. In either case, it is important for the modeler to keep the model limitations in mind when discussing predicted ice-sheet behavior. Equally important is to verify model results against available data. Because of the importance of these issues, evaluating the budget of forces using glacier measurements is discussed extensively in this book.

The organization of this book is fairly traditional, starting with developing the elementary tools, and ending with models of the Greenland and Antarctic ice sheets. Chapter 1 presents the obligatory overview of ice in the climate system. The essential theory needed to develop glacier models is discussed in Chapters 2–4. Simple analytical solutions are derived in Chapters 5 and 6, while Chapter 7 focuses on the thermodynamic aspects of glacier flow and the interactions between temperature and climate forcing. In most instances, numerical methods have to be used to solve the equations governing glacier flow. An overview of commonly used time-evolving models is given in Chapter 8. Some general aspects of glacier dynamics and discussion of important feedback mechanisms can be found in Chapter 9. The final three chapters discuss applications specific to smaller mountain glaciers, the Greenland Ice Sheet, and the Antarctic Ice Sheet, respectively. In the applications in later chapters, the important starting equations and underlying assumptions are summarized so that readers not interested in the details need not go through most of the earlier chapters.

The majority of my research over the past decade or so has been funded by the Office of Polar Programs of the National Science Foundation; this support is gratefully acknowledged. Further, I would like to thank my colleagues, and in particular Ian Whillans, for discussions and mutual criticisms, which have helped me in obtaining a better understanding of the topics discussed here.

C. J. van der Veen
May 13, 1999

About the Author

C. J. van der Veen is a glaciologist interested in the dynamics of fast-moving glaciers and ice streams. His research focuses on using measurements of ice velocity and glacier geometry to identify mechanical controls on glaciers and how changes in these controls affect glacier flow and stability. After completing his PhD at the University of Utrecht in the Netherlands, he spent 20 years at The Ohio State University as a research scientist with the Byrd Polar Research Center (1986–2006). He is currently a professor in the Geography Department and research scientist with the Center for Remote Sensing of Ice Sheets at the University of Kansas.

1 Mathematical Tools

1.1 VECTORS AND TENSORS

This book assumes that readers are familiar with the fundamental principles of vector calculus. This section briefly summarizes some of the results pertinent to applications discussed in later chapters. This overview is based on Chapter 1 in Joos (1934).

Quantities such temperature and mass that are characterized by the assignment of a single number (for example, 15°C and 243 kg) are called *scalars*. Other physical quantities, most notably displacement and velocity, cannot be completely defined by specifying a single number only. Consider, for example, the change in position of a stake placed in the ice surface. If the magnitude of the displacement of the stake is 100 m in a certain time interval, then without additional information, the new position of the stake may be anywhere on a circle with a radius of 100 m and centered on the original position. To completely describe the new position of the stake, the direction of displacement also needs to be given. In other words, displacement is defined by two numbers, namely, magnitude and direction. Quantities with this characteristic are called *vectors* and written as $\vec{u}$, with the arrow above indicating the vector quantity. Another frequently used convention is to use boldface (**u**) when referring to vectors. Graphically, vectors are shown as arrows, as in Figure 1.1.

Vectors can be subject to mathematical operations such as addition or subtraction of two or more vectors, or multiplication with a scalar. Given two vectors, $\vec{u}$ and $\vec{v}$, the sum is the vector that joins the initial point of $\vec{u}$ with the end point of $\vec{v}$ after putting the initial point of $\vec{v}$ at the end point of $\vec{u}$, as illustrated in Figure 1.1. From this figure it can be seen that vector addition is commutative, and

$$\vec{u} + \vec{v} = \vec{v} + \vec{u}. \tag{1.1}$$

Further, vector addition is associative and

$$(\vec{u} + \vec{v}) + \vec{w} = \vec{u} + (\vec{v} + \vec{w}) = \vec{u} + \vec{v} + \vec{w}. \tag{1.2}$$

Defining the vector $-\vec{v}$ as the vector with the same magnitude as $\vec{v}$ but with the opposite direction, vector subtraction, $\vec{u} - \vec{v}$, is equivalent to the addition, $\vec{u} + (-\vec{v})$. Finally, the product of a vector, $\vec{u}$, and a scalar, m, is a vector with the same direction of $\vec{u}$ (if m > 0) or pointing in the opposite direction (if m < 0), and with a magnitude that is equal to the absolute value of m times the magnitude of $\vec{u}$.

To define a vector, a coordinate system must be selected. Several such systems exist including the geographic coordinate system in which any point on the earth's surface is characterized by latitude and longitude, with sometimes a third coordinate included,

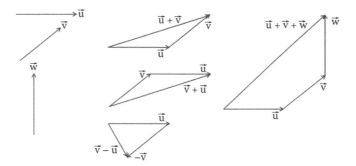

FIGURE 1.1 Illustrating addition and subtraction of three arbitrary vectors.

namely, height above sea level. In mathematics and physics, a commonly used system is the Cartesian coordinate system, which is used throughout most of this book. This system, shown in Figure 1.2, consists of three orthogonal (perpendicular) axes, referred to as the x-, y-, and z-axis. Any point in space is then uniquely defined by three coordinates, written as (x, y, z). Similarly, a vector is defined by three components. For example, velocity can be written as $\vec{u} = (u_x, u_y, u_z)$ or $\vec{u} = (u, v, w)$, with $= u_x = u$ the velocity component in the x-direction, and similar for the other two components. For the most part, this notation is used throughout this book. In a few instances, however, variable subscripts are used to indicate the three orthogonal directions as this allows for compact formulation of equations. For example, x_i with $i = 1$, 2, or 3 refers to the three orthogonal directions, with $x_1 = x$, $x_2 = y$, and $x_3 = z$. Similarly, $u_1 = u_x = u$, and so forth.

The coordinate system shown in Figure 1.2 is defined by three unit vectors, $\vec{i}$, $\vec{j}$, and $\vec{k}$, whose length equals unity and are oriented in the x-, y-, and z-direction, respectively. The velocity vector can then also be written as the sum of three vectors, each oriented in one of the three Cartesian directions. That is,

$$\vec{u} = u\vec{i} + v\vec{j} + w\vec{k}. \tag{1.3}$$

This notational convention invokes the additive properties of vectors, illustrated in Figure 1.1. Two or more vectors may be combined into a single vector by summing the corresponding components. Let $\vec{u}_1 = (u_1, v_1, w_1)$ and $\vec{u}_2 = (u_2, v_2, w_2)$ be

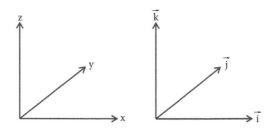

FIGURE 1.2 Cartesian coordinate system with the z-axis directed vertically upward (left panel). Each direction is defined by a unit vector (right panel).

the velocity of a point on a glacier measured over two consecutive years. The total velocity over the two-year interval is then

$$\vec{u}_t = \vec{u}_1 + \vec{u}_2 = (u_1 + u_2, v_1 + v_2, w_1 + w_2). \tag{1.4}$$

When discussing glacier flow, it is most convenient to take the z-axis in the vertically upward direction or, equivalently, perpendicular to the local geoid. Then the plane defined by the x- and y-axes corresponds to the (local) horizontal plane. The advantage of choosing this orientation is that the direction of the gravitational force coincides with the z-axis (but pointing in the opposite direction). This convention is opposite of that followed in many geophysics textbooks where the vertical axis points downward (see, for example, Turcotte and Schubert, 2002). The main difference is that gravity is negative when the vertical axis points upward and positive when this axis points downward.

In most glacier applications, the vertical velocity is small compared with the horizontal velocity components, and only the two-dimensional horizontal velocity vector, $\vec{u} = (u, v)$ needs to be considered. This vector is characterized by two quantities: magnitude, $|\vec{u}|$, and direction, ϕ, defined with respect to the x-axis. Referring to Figure 1.3, these quantities follow from trigonometry and

$$|\vec{u}| = \sqrt{u^2 + v^2}, \tag{1.5}$$

$$\tan\phi = \frac{v}{u}. \tag{1.6}$$

The velocity magnitude always has the same value, irrespective of the orientation of the (x, y) coordinate system. The direction, however, is defined as the angle between the velocity vector and the x-axis, and this angle will change if the orientation of the coordinate system changes.

Consider a Cartesian coordinate system (p, q) rotated over an angle α relative to the (x, y) system (both systems are in the horizontal plane so that the vertical z-axis is the rotation axis). To obtain the velocity components in the (p, q) system, the velocity components in the (x, y) system are rotated separately. From the geometry shown

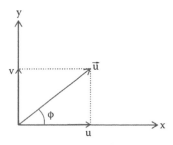

FIGURE 1.3   Components of the horizontal velocity vector, $\vec{u} = (u, v)$.

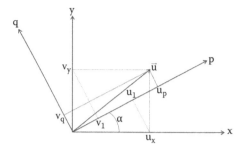

FIGURE 1.4 Illustrating vector rotation.

in Figure 1.4, it follows that $\cos \alpha = u_1/u$ and $\sin \alpha = v_1/v$. The total component of velocity in the p-direction is $u_p = u_1 + v_1$ and

$$u_p = u_x \cos \alpha + v_y \sin \alpha. \tag{1.7}$$

Similarly,

$$v_q = -u_x \sin \alpha + v_y \cos \alpha. \tag{1.8}$$

In geophysical applications, there are two ways that vectors can be multiplied, namely, the scalar and the vector products. The scalar product gives a number (scalar) that is equal to the product of the magnitudes of the two vectors multiplied by the cosine of the angle between them (Figure 1.5). That is,

$$\vec{F}\vec{d} = |\vec{F}|\,|\vec{d}|\cos \alpha, \tag{1.9}$$

where α represents the angle between the two vectors. It can be readily verified that this equation is equivalent to the sum of the corresponding components,

$$\vec{F}\vec{d} = F_x d_x + F_y d_y + F_z d_z. \tag{1.10}$$

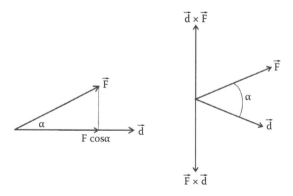

FIGURE 1.5 Illustrating the scalar product (left) and vector product (right) of two vectors.

Note that the scalar product of two vectors that are perpendicular is zero. To understand the physical meaning of the scalar product, let $\vec{F}$ be a constant force resulting in a displacement, $\vec{d}$. The scalar product then represents the work done by this force.

The vector product of two vectors $\vec{F}$ and $\vec{d}$ is itself a vector that is perpendicular to the plane formed by $\vec{F}$ and $\vec{d}$, and whose magnitude is equal to the area of the parallelogram formed by these two vectors. That is,

$$|\vec{F} \times \vec{d}| = |\vec{F}| |\vec{d}| \sin \alpha. \tag{1.11}$$

The resulting vector is given by

$$\vec{F} \times \vec{d} = (F_y d_z - F_z d_y)\vec{i} - (F_x d_z - F_z d_x)\vec{j} + (F_x d_y - F_y d_x)\vec{k}, \tag{1.12}$$

which can also be written in the form of a determinant

$$\vec{F} \times \vec{d} = \begin{vmatrix} \vec{i} & \vec{j} & \vec{k} \\ F_x & F_y & F_z \\ d_x & d_y & d_z \end{vmatrix}. \tag{1.13}$$

Note that the order in which the vectors appear in the vector product is significant and $\vec{F} \times \vec{d} = -\vec{d} \times \vec{F}$ (Figure 1.5). In physical applications, the vector product is used to calculate torque (vector product of radius and applied force) and magnetic force (vector product of magnetic field and velocity of a charge moving through this field).

An important mathematical operation used throughout this book is taking the derivative to evaluate how a quantity varies over time and in space. In the following, only space derivatives are considered. For a scalar quantity such as temperature, the derivative is a vector called the *gradient*, written as

$$\text{grad} T = \frac{\partial T}{\partial x}\vec{i} + \frac{\partial T}{\partial y}\vec{j} + \frac{\partial T}{\partial z}\vec{k}, \tag{1.14}$$

or, equivalently

$$\text{grad} T = \left(\frac{\partial T}{\partial x}, \frac{\partial T}{\partial y}, \frac{\partial T}{\partial z} \right). \tag{1.15}$$

Each component of this vector describes how, at a particular location, the temperature changes in each of the three orthogonal directions.

Inspection of these equations shows that the temperature gradient is the product of a scalar, T, with the symbolical vector ∇, defined as

$$\nabla = \frac{\partial}{\partial x}\vec{i} + \frac{\partial}{\partial y}\vec{j} + \frac{\partial}{\partial z}\vec{k}. \tag{1.16}$$

This vector, called *nabla,* was introduced by Sir William Hamilton in 1837 and named after an ancient Assyrian harp that has a similar shape as does the symbol. Note that while ∇ represents a vector, it usually does not have an arrow placed above it. This operator can also be applied to a vector. First, taking the scalar product of ∇ and the velocity $\vec{u} = (u, v, w)$ gives

$$\nabla \vec{u} = \frac{\partial u}{\partial x} + \frac{\partial v}{\partial y} + \frac{\partial w}{\partial z}. \tag{1.17}$$

This scalar quantity is called the *divergence* of the velocity, sometimes written as div $\vec{u}$. Taking the vector product of ∇ and $\vec{u}$ produces another vector, called the *curl* of the velocity, which essentially describes rotation present in the velocity field. Analogous to equation (1.13),

$$\text{curl } \vec{u} = \nabla \times \vec{u} = \begin{vmatrix} \vec{i} & \vec{j} & \vec{k} \\ \dfrac{\partial}{\partial x} & \dfrac{\partial}{\partial y} & \dfrac{\partial}{\partial z} \\ u & v & w \end{vmatrix}. \tag{1.18}$$

Next consider the space derivative of a vector. This involves calculation of the gradient of each of the three velocity components resulting in a set of three vectors defined by nine scalars. It can be shown that these nine components form a second-order *tensor.* Symbolically, a second-order tensor, $\vec{A}$, can be represented by a 3×3 matrix as

$$\vec{A} = \begin{pmatrix} A_{xx} & A_{xy} & A_{xz} \\ A_{yx} & A_{yy} & A_{yz} \\ A_{zx} & A_{zy} & A_{zz} \end{pmatrix}. \tag{1.19}$$

Two tensors that feature prominently in glacier dynamics are the stress tensor, describing the components of stress acting on an ice cube, and the strain-rate tensor, describing the rate at which the cube deforms as a result of these stresses. Both tensors are introduced in the next section, but to conclude this introductory discussion, two important tensor properties that will be used in later chapters need to be addressed, namely, invariants and tensor rotation.

As noted above, the magnitude of a vector (the length of the arrow) does not change if the coordinate system is rotated. A second-order tensor has three such scalar quantities whose values are independent of the orientation of the coordinate axes; these quantities are called *invariants* and represent the solutions of the so-called characteristic equation

$$\det (\vec{A} - \lambda \vec{E}) = \begin{vmatrix} A_{xx} - \lambda & A_{xy} & A_{xz} \\ A_{yx} & A_{yy} - \lambda & A_{yz} \\ A_{zx} & A_{zy} & A_{zz} - \lambda \end{vmatrix} = 0, \tag{1.20}$$

where $\vec{E}$ represents the identity tensor ($E_{xx} = E_{yy} = E_{zz} = 1$, and all other elements are equal to 0). Omitting the algebra, it can be shown that three solutions exist, namely,

$$\lambda_1 = A_{xx} + A_{yy} + A_{zz}, \tag{1.21}$$

$$\lambda_2 = A_{xx}^2 + A_{yy}^2 + A_{zz}^2 + A_{xy}^2 + A_{xz}^2 + A_{yx}^2 + A_{yz}^2 + A_{zx}^2 + A_{zz}^2, \tag{1.22}$$

$$\lambda_3 = \det(\vec{A}) =$$

$$= A_{xx}(A_{yy}A_{zz} - A_{yz}A_{zy}) - A_{xy}(A_{yx}A_{zz} - A_{yz}A_{zx}) + A_{xz}(A_{yx}A_{zy} - A_{yy}A_{zx}). \tag{1.23}$$

When considering the stress and strain-rate tensors, these invariants have a clear physical meaning, as discussed in the next section. A tensor for which the first invariant vanishes ($\lambda_1 = 0$) is called a *deviator* (although in glaciology, use of the term *deviator* is restricted to the stress tensor).

The value of the nine tensor components depends on the orientation of the coordinate system, and thus, formulas for rotating tensors are needed equivalent to equations (1.7) and (1.8) for vector rotation. For any arbitrary rotation in three-dimensional space, these formulas become rather lengthy, but for most glaciological applications it is sufficient to consider rotation in the horizontal plane only (that is, rotation around the vertical z-axis). Further, when considering the stress or strain-rate tensor, only six independent coefficients need to be considered because both tensors are symmetric, with $A_{ij} = A_{ji}$. Let $\vec{B}$ be the symmetric 2×2 tensor with components

$$\vec{B} = \begin{pmatrix} B_{xx} & B_{xy} \\ B_{xy} & B_{yy} \end{pmatrix}. \tag{1.24}$$

As in Figure 1.3 the angle between the two coordinate systems is α. The tensor components in the (p, q) coordinate system are then related to those in the (x, y) system as (Jaeger, 1969, p. 7)

$$B_{pp} = B_{xx} \cos^2 \alpha + B_{yy} \sin^2 \alpha + 2B_{xy} \sin \alpha \cos \alpha, \tag{1.25}$$

$$B_{qq} = B_{xx} \sin^2 \alpha + B_{yy} \cos^2 \alpha - 2B_{xy} \sin \alpha \cos \alpha, \tag{1.26}$$

$$B_{pq} = (B_{yy} - B_{xx}) \sin \alpha \cos \alpha + B_{xy} (\cos^2 \alpha - \sin^2 \alpha). \tag{1.27}$$

From equation (1.27) it follows that there is an angle ϕ for which $B_{pq} = 0$, namely, when

$$\tan 2\phi = \frac{2B_{xy}}{B_{xx} - B_{yy}}. \tag{1.28}$$

If $\vec{B}$ represents the stress tensor, this angle gives the direction of the two principal stresses in the horizontal plane.

1.2 STRESS AND STRAIN

Glacier flow is commonly described using concepts from continuum mechanics. The basic premise of continuum mechanics is that the material under consideration is continuous and that physical quantities such as mass and momentum associated with a small volume are distributed uniformly over that volume (Batchelor, 1967, p. 5). In that case, the response to applied stress can be described by a single constitutive relation or flow law. To model the flow of glaciers, such an approach works well and there is no need to consider deformation of each ice crystal individually. An obvious exception to this is when modeling the development of fabric patterns in ice under stress, as discussed in Section 2.4.

The goal of continuum mechanics is to describe how a material (in this case, glacier ice) deforms when subjected to force. This involves the constitutive relation, or flow law, which links the stress to the rate of deformation. Deformation of ice is discussed in Chapter 2, but first it may be instructive to have a closer look at the two central quantities involved, namely, stress and strain. A more extensive discussion on this topic can be found in many textbooks, for example, Means (1976) and Turcotte and Schubert (2002).

Stresses are forces per unit area. There are two types of stresses, namely, shear and normal stresses. A shear stress (also referred to as traction) acts along a surface (for example, friction generated by a block gliding over a rough surface); a normal stress acts perpendicular to the surface. In three-dimensional space, three perpendicular surfaces through a point are needed to describe all stresses acting at that point (Figure 1.6). For each surface, there are three stress components, namely, one normal stress acting perpendicular to the surface and two shear stresses acting along the two perpendicular directions in the surface. So, altogether there are nine stress components acting at each point. The stress components are written as σ_{ij}, with i, j = x, y, z, the three axes of the Cartesian coordinate system. The second subscript, j, denotes the direction normal to the surface on which the stress acts, while the first subscript, i, specifies the direction of the stress. For example, the shear stress σ_{xz} acts in the direction of the x-axis, along the plane normal to the z-axis

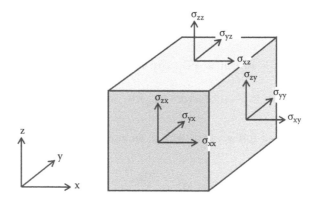

FIGURE 1.6 Stresses acting on the three perpendicular faces of a unit cube that is homogeneously stressed.

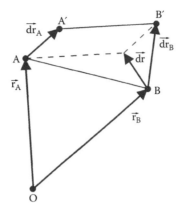

FIGURE 1.7 Deformation of a line segment.

(that is, the xy plane). It can be shown that balance of angular momentum requires that the stress tensor be symmetric and $\sigma_{ij} = \sigma_{ji}$ (for example, Jaeger, 1969, p. 6; Hutter, 1983, p.14; Greve and Blatter, 2009, Section 3.3.6). So, the distribution of stress is described completely by six independent components (three shear stresses and three normal stresses).

Under the combined actions of all stresses acting on an ice volume, deformation takes place. To see how this can be described, consider two neighboring points A and B. After deformation, their positions are A' and B', respectively (Figure 1.7). The deformation of the line segment AB is then A'B' − AB. Using the notation of Figure 1.7, the deformation vector is

$$\vec{dr} = \overline{A'B'} - \overline{AB} =$$

$$= \left(\vec{r_B} + \vec{dr_B}\right) - \left(\vec{r_A} + \vec{dr_A}\right) - \left(\vec{r_B} - \vec{r_A}\right) = \qquad (1.29)$$

$$= \vec{dr_B} - \vec{dr_A}.$$

For a Cartesian coordinate system x_j ($j = 1, 2, 3$), first-order Taylor expansion gives

$$\vec{dr} = \frac{\partial \vec{r}}{\partial x_j} dx_j, \qquad (1.30)$$

or, for each component of the displacement vector

$$dr_i = \frac{\partial r_i}{\partial x_j} dx_j, \qquad (1.31)$$

where summation over repeated indices is implied.

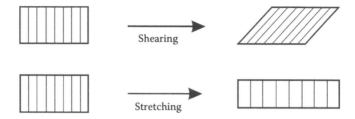

FIGURE 1.8 Illustrating the two modes of deformation.

The nine quantities, $\partial r_i/\partial x_j$, form a second-order tensor (a Jacobian), which can be written as the sum of a symmetric and an antisymmetric part

$$\frac{\partial r_i}{\partial x_j} = \frac{1}{2}\left(\frac{\partial r_i}{\partial x_j} + \frac{\partial r_j}{\partial x_i}\right) + \frac{1}{2}\left(\frac{\partial r_i}{\partial x_j} - \frac{\partial r_j}{\partial x_i}\right). \tag{1.32}$$

The second term on the right-hand side represents a rigid-body rotation, while the first term actually describes how the segment is deformed. This is called the strain tensor, with elements given by

$$\varepsilon_{ij} = \frac{1}{2}\left(\frac{\partial r_i}{\partial x_j} + \frac{\partial r_j}{\partial x_i}\right). \tag{1.33}$$

As illustrated in Figure 1.8, there are two types of deformation, namely, stretching (described by the diagonal elements of the strain tensor, with $i = j$) and shearing ($i \neq j$, the off-diagonal elements).

As ice deforms, it tries to reach a steady state in which individual ice crystals are still being deformed, but at a (locally) constant rate. In other words, a stationary stress- and velocity-field is established. Therefore, strain rates, rather than strains, are commonly used in glaciology. In the usual notation, strain rates are denoted by a dot above the elements of the strain tensor (referring to the time derivation). Then, by definition

$$\dot{\varepsilon}_{ij} = \frac{d}{dt}(\varepsilon_{ij}) =$$

$$= \frac{1}{2}\left[\frac{\partial}{\partial x_j}\left(\frac{dr_i}{dt}\right) + \frac{\partial}{\partial x_i}\left(\frac{dr_j}{dt}\right)\right] = \tag{1.34}$$

$$= \frac{1}{2}\left(\frac{\partial u_i}{\partial x_j} + \frac{\partial u_j}{\partial x_i}\right),$$

where the u_i represent the three components of ice velocity in the three orthogonal directions, x_i. There are six independent strain-rate components, and the strain-rate tensor exhibits the same symmetry as does the stress tensor.

The experiments of Rigsby (1958) suggest that the deformation of ice is practically independent of the hydrostatic pressure, provided that the difference between the ice temperature and the pressure-melting temperature is kept constant. Then the constitutive relation, linking the rate of deformation to applied stress, must be independent of the hydrostatic pressure. To achieve this, deviatoric stresses, rather than full stresses, are used to describe the rheological properties of glacier ice.

Let σ_{ij} denote the components of the full stress tensor (sometimes referred to as the Cauchy stress). The hydrostatic pressure is the sum of the three normal stresses, $P = (\sigma_{xx} + \sigma_{yy} + \sigma_{zz})/3$. The full stress tensor is now partitioned as follows

$$
\begin{pmatrix}
\sigma_{xx} & \sigma_{yx} & \sigma_{zx} \\
\sigma_{xy} & \sigma_{yy} & \sigma_{zy} \\
\sigma_{xz} & \sigma_{yz} & \sigma_{zz}
\end{pmatrix} =
$$

$$
= \begin{pmatrix}
(2\sigma_{xx} - \sigma_{yy} - \sigma_{zz})/3 & \sigma_{yx} & \sigma_{zx} \\
\sigma_{xy} & (2\sigma_{yy} - \sigma_{xx} - \sigma_{zz})/3 & \sigma_{zy} \\
\sigma_{xz} & \sigma_{yz} & (2\sigma_{zz} - \sigma_{xx} - \sigma_{yy})/3
\end{pmatrix} +
$$

$$
+ \begin{pmatrix}
(\sigma_{xx} + \sigma_{yy} + \sigma_{zz})/3 & 0 & 0 \\
0 & (\sigma_{xx} + \sigma_{yy} + \sigma_{zz})/3 & 0 \\
0 & 0 & (\sigma_{xx} - \sigma_{yy} - \sigma_{zz})/3
\end{pmatrix} \tag{1.35}
$$

The first tensor on the right-hand side is called the stress deviator. In short, its components are given by

$$
\tau_{ij} = \sigma_{ij} - \frac{1}{3}\sigma_{kk}\delta_{ij}, \tag{1.36}
$$

where δ_{ij} denotes the Kronecker delta ($\delta_{ij} = 1$ if $i = j$, and $\delta_{ij} = 0$ if $i \neq j$), and summation over repeat indexes is implied in equation (1.36). Thus, a deviatoric normal stress is defined as the full normal stress minus the hydrostatic pressure. The shear stresses are unaffected by this partitioning.

A few words on the notation commonly used in the glaciological literature may be helpful. In the literature, the components of stress are variably denoted by σ_{ij} or by τ_{ij}. Sometimes, σ_{ij} is reserved for normal stress ($i = j$) and τ_{ij} for shear stress ($i \neq j$). Deviatoric stresses are commonly denoted by a prime, for example, σ'_{ij}. In this book, in an effort to keep the notation simple and consistent, σ_{ij} refers to full stresses and τ_{ij} to deviatoric stresses.

As discussed in Section 1.1, a second-order tensor has three invariants. For the full stress tensor, the first invariant is the sum of the three normal stresses and equals three times the hydrostatic pressure, defined as the average pressure (or the average of the three normal stresses; hence the factor 1/3). For the deviatoric stress tensor, the first invariant is zero. The second invariant of this tensor gives the *effective stress*, τ_e, defined as

$$2\tau_e^2 = \tau_{xx}^2 + \tau_{yy}^2 + \tau_{zz}^2 + 2\left(\tau_{xy}^2 + \tau_{xz}^2 + \tau_{yz}^2\right). \qquad (1.37)$$

For the strain-rate tensor, the first invariant is

$$\lambda_1 = \dot{\varepsilon}_{xx} + \dot{\varepsilon}_{yy} + \dot{\varepsilon}_{zz}. \qquad (1.38)$$

For incompressible ice the density is constant and this first invariant is zero, and conservation of volume is expressed as

$$\dot{\varepsilon}_{xx} + \dot{\varepsilon}_{yy} + \dot{\varepsilon}_{zz} = 0. \qquad (1.39)$$

The effective strain rate is defined similar to the effective stress and related to the second invariant of the strain-rate tensor. The third invariants of the stress and strain-rate tensors are usually not considered in glacier modeling, with the exception of a few studies that considered normal stress effects or dilatancy (cf. Section 2.3).

1.3   ERROR ANALYSIS

The primary objective of this book is to provide the mathematical and theoretical framework necessary for modeling glacier flow and for quantitatively interpreting measurements on glaciers. One such application involves the force-budget technique, in which measured surface velocities are used to estimate forces acting on the glacier. Equally important as estimating the magnitude of the various resistive forces is to assign uncertainties or errors to these calculations. These uncertainties arise because of errors in the measurements. This section provides a brief overview of error analysis and how errors propagate through calculations. Taylor (1997) gives a more extensive introductory discussion of the topic.

In science, the word *error* refers to the uncertainty inherent when measuring physical quantities, and not to the colloquial meaning of *mistake*. How these errors are determined or estimated depends on the type of measurements. For the following discussion, these so-called input errors are assumed to be known and denoted by Δ. Now consider a function $F(\alpha_i)$ with variables α_i; the respective errors are denoted by ΔF and $\Delta \alpha_i$. If the errors in the input variables are independent, the error in the function is estimated from

$$(\Delta F)^2 = \sum_i \left(\frac{\partial F}{\partial \alpha_i} \Delta \alpha_i\right)^2. \qquad (1.40)$$

To illustrate how this formula is used, along-flow force balance is considered. The relevant equations are derived in Chapter 3 and readers unfamiliar with these equations may want to return to this section at a later stage.

If the x-axis is taken to coincide with the direction of flow, balance of forces is expressed by equation (3.22). Making the assumption that the resistive stresses, R_{xx} and R_{xy}, are constant throughout the ice thickness, this equation becomes

$$\tau_{dx} = \tau_{bx} - \frac{\partial}{\partial x}(HR_{xx}) - \frac{\partial}{\partial y}(HR_{xy}). \tag{1.41}$$

The second and third terms on the right-hand side are estimated from strain rates, which, in turn, are estimated from measured velocities.

Let U_1 and U_2 denote the velocities at two neighboring stations, located at x_1 and x_2, respectively. The average stretching rate between these stations is then

$$\dot{\varepsilon}_{xx} = \frac{U_2 - U_1}{x_2 - x_1}. \tag{1.42}$$

The error in the positions is neglected, and both velocity errors are equal to ΔU. The error in the strain rate is then

$$(\Delta\dot{\varepsilon}_{xx})^2 = \frac{(\Delta U_2)^2 + (\Delta U_1)^2}{(x_2 - x_1)^2} =$$

$$= \frac{2(\Delta U)^2}{(x_2 - x_1)^2}, \tag{1.43}$$

or

$$\Delta\dot{\varepsilon}_{xx} = \frac{\Delta U}{x_2 - x_1}\sqrt{2}. \tag{1.44}$$

This equation shows that the error in strain rate is inversely proportional to the distance over which the velocity gradient is calculated. This is important to keep in mind because it allows the strain-rate error to be minimized by increasing the differencing interval. Of course, increasing the differencing interval goes at the expense of spatial resolution, and strain-rate variations at a scale smaller than this interval will not be resolved. Depending on the application, careful consideration should be given to this tradeoff between minimizing errors and spatial resolution.

As an example, consider a glacier on which the average velocity increases by 1 m/yr for every km in the downflow direction. The average strain rate is then $\dot{\varepsilon}_{xx} = 10^{-3}$ yr^{-1}. If velocities are measured every 10 m, small-scale variations in strain rate can be calculated by differencing neighboring velocities. If the velocity error is 10 cm/yr, this would give an error in strain rate $\Delta\dot{\varepsilon}_{xx} = 0.014$ yr^{-1}, or an order of magnitude greater than the average strain rate itself. To reduce this error to, say, 10% of the average strain rate, the differencing interval should be increased to 1 km or greater. Thus, the velocity error essentially dictates the spatial scale over which strain rates can be meaningfully assessed. Because the strain rate equals the velocity gradient, the actual magnitude of the velocity is irrelevant and the preceding conclusion applies equally to glaciers that move at speeds of a few m/yr and those that move at speeds of several km/yr.

The next step is to estimate the error in the corresponding resistive stress, R_{xx}. This stress is related to the stretching rate through the flow law (3.48). To avoid getting lost in algebra and to keep the discussion clear, the simplifying assumption is made that $\dot{\varepsilon}_{xx}$ is the dominant strain rate contributing to the effective strain rate and that lateral spreading is zero. Then

$$R_{xx} = 2B\dot{\varepsilon}_{xx}^{1/n}. \qquad (1.45)$$

If the error in the viscosity parameter, B, and flow-law exponent, n, may be neglected, applying equation (1.40) gives

$$(\Delta R_{xx})^2 = \left(\frac{1}{n}2B\dot{\varepsilon}_{xx}^{1/n-1}\,\Delta\dot{\varepsilon}_{xx}\right)^2, \qquad (1.46)$$

which can be rewritten in terms of fractional uncertainty

$$\left|\frac{\Delta R_{xx}}{R_{xx}}\right| = \left|\frac{\Delta\dot{\varepsilon}_{xx}}{\dot{\varepsilon}_{xx}}\right|, \qquad (1.47)$$

where the vertical bars denote the absolute value. An intermediate value for the viscosity parameter is B = 500 kPa yr$^{1/3}$ for n = 3 (Figure 2.4). For the numerical example discussed above, the average resistive stress is 100 kPa. Assuming velocity gradients are calculated over a horizontal distance of 1 km, the error in strain rate is 1.4×10^{-4} yr^{-1} and, from equation (1.47), the error in the resistive stress, ΔR_{xx}, is 14 kPa.

The next step is to estimate the error in the longitudinal stress gradient term (the second term on the right-hand side of the balance equation (1.41)). Resistance to flow from gradients in longitudinal stress is written as

$$F_{lon} = \frac{\partial}{\partial x}(HR_{xx}). \qquad (1.48)$$

First consider the term in parenthesis on the right-hand side. This product of ice thickness, H, and resistive stress, R_{xx} has an error that can be estimated using equation (1.40). Similar to the derivation leading to equation (1.47) the fractional error is given by

$$\left(\frac{\Delta(HR_{xx})}{HR_{xx}}\right)^2 = \left(\frac{\Delta H}{H}\right)^2 + \left(\frac{\Delta R_{xx}}{R_{xx}}\right)^2. \qquad (1.49)$$

For an average ice thickness of 1500 m with measurement uncertainty of 50 m, and continuing with the numerical example, this gives $\Delta(HR_{xx}) = 2.2 \times 10^4$ kPa m. It can be readily verified that most of this uncertainty is due to the error in R_{xx} and that the error in ice thickness is relatively unimportant. Similar to the error in strain rate (equation (1.44)), the error in the longitudinal stress gradient is

$$\Delta F_{lon} = \frac{\Delta(HR_{xx})}{x_2 - x_1} \sqrt{2}. \qquad (1.50)$$

Again, taking a differencing distance of 1 km, this error is about 31 kPa. Typical values for the driving stress are in the range of 50 to 150 kPa, and consequently, the error in F_{lon} could be a significant fraction of the driving stress, rendering any conclusion about the role of longitudinal stress gradients in the balance of forces rather uncertain. The only way to reduce this uncertainty is by increasing the spatial interval to, say, 10 km.

This numerical example shows that small errors in measured velocities can result in large errors in the force balance terms. The reason for this is that the last two terms on the right-hand side of the balance equation (1.41) are proportional to the second derivative of the ice velocity or, equivalently, to the curvature of the velocity profile in the x- and y-directions, respectively. Small errors in the velocity are amplified by taking the derivative, and to reduce the resulting uncertainty, spatial derivatives must be calculated over greater distances. This may not always be the best option, however. For example, the width of the fast-moving part of Jakobshavn Isbræ in West Greenland is a little more than 3 km wide (Van der Veen et al., 2011), and to obtain an accurate estimating of the role of lateral drag from transverse velocity gradients, an approach involving linear regression is more appropriate. This procedure is discussed in Section 11.4.

1.4   PARAMETRIC UNCERTAINTY ANALYSIS

The analysis of errors discussed in Section 1.3 follows standard treatment of errors as can be found in, for example, introductory texts on experimental physics (Squires, 2001) or textbooks on error analysis (Taylor, 1997). Formula (1.40) for error propagation applies to any quantity that is a function of several variables, but its practical use may be limited if analytical expressions for the derivatives cannot be easily derived. In such cases, a parametric uncertainty analysis may provide an error estimate. In this procedure, input variables, α_i, are assigned values selected randomly from their probability density distributions, and the resulting value of F is calculated. This procedure is repeated many times to obtain a distribution for the dependent variable, F, from which the standard deviation or error can be estimated. For example, Van der Veen (2002b) applies this method to estimate the range of projected contributions from Greenland and Antarctica to global sea level change in 2100 AD, given uncertainties in parameter values and in warming scenarios. To illustrate this so-called Monte Carlo approach, consider the calculation of strain rates.

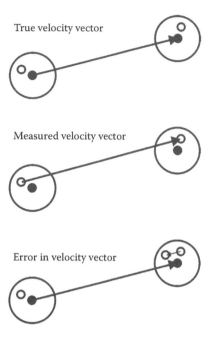

FIGURE 1.9 Illustrating the source of uncertainty in measured velocities. Filled black dots represent the true position of a location at two different times. Large circles indicate the uncertainty in measuring the true positions. Small open circles represent the actual measurements from which the measured velocity is derived (middle panel). The error in velocity is represented by the difference between the true velocity (upper panel) and the measured velocity and corresponds to the small vector connecting both measurement errors (lower panel).

Strain rates are linked to velocity gradients. If the x-direction is taken in the (approximate) flow direction, and U denotes the velocity in this direction, then the along-flow stretching rate is

$$\dot{\varepsilon}_{xx} = \frac{\partial U}{\partial x}. \tag{1.51}$$

Typically, the velocity is determined by measuring the displacement of a visible marker on the glacier surface. This could be a global positioning system (GPS) antenna anchored in the surface and surveyed repeatedly, or a visible feature such as a crevasse identified on successive satellite images. Referring to Figure 1.9 (upper panel), the true velocity vector is given by the change in position of the true positions over the interval of measurements. However, due to measurement uncertainties such as image pixel size, recognizing exactly the same feature on successive images, uncertainties in GPS positions, and other factors, positional errors are introduced and the measured velocity is the displacement of two points that are each displaced from their true positions (middle panel). The error in velocity is then given by the vector connecting the measured positions after subtracting the true displacement (lower panel). There is another error introduced because the measured velocity may be assigned to

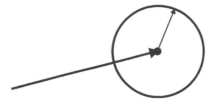

FIGURE 1.10 The error in measured velocity can be represented by the small vector starting at the end point of the velocity vector. The end point of this error vector can be located anywhere in the circle centered on the end point of the measured velocity vector. The radius of this circle is determined by the error in position and the time interval over which displacements are measured.

the wrong location, but that error is ignored here. Thus, the velocity can be represented by the measured velocity, plus some error that can be represented graphically as another vector to be added to the measured velocity (Figure 1.10). The magnitude of the error vector is determined by the error in locations. The angle can be anywhere between −180 and +180 degrees. Further, this error is independent of the magnitude and direction of the average measured velocity. It is essentially the difference between the errors in the measured positions after subtracting the average displacement.

Because interest is in the velocity error, the actual distance between the two true points does not matter and both are assigned the coordinates (0, 0). The two location errors can then be represented by two points (x_1, y_1) and (x_2, y_2). Assuming a standard deviation of 10 m for the position error, and assuming the position error is Gaussian distributed, the four coordinates are randomly selected and the two components of the velocity error vector calculated as

$$U = \frac{x_2 - x_1}{dt}, \tag{1.52}$$

$$V = \frac{y_2 - y_1}{dt}, \tag{1.53}$$

where dt = 1 year represents the time interval between successive position determinations. This process of randomly selecting the four coordinates of the two points is repeated 10,000 times to yield the probability distributions shown in Figure 1.11. As was to be expected, the average for both velocity components is zero, while the standard deviation is 14 m/yr, corresponding to the value predicted by equation (1.40).

The corresponding error in stretching rate can be estimated from a similar procedure. Consider two points at a nominal distance of 1 km apart. That is, the distance between the true locations is 1 km. At each location, the error velocity is determined as for Figure 1.11. The error in stretching rate is then estimated from

$$\dot{\varepsilon}_{xx} = \frac{U_2 - U_1}{dx}, \tag{1.54}$$

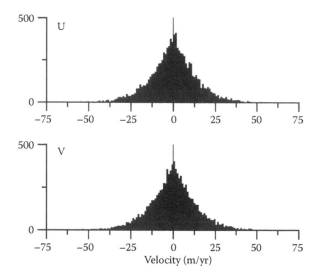

FIGURE 1.11 Probability distributions for the two components of the velocity error vector determined from a Monte Carlo simulation as explained in the text. For both distributions the mean (indicated by the vertical line) is zero and the standard deviation is 14 m/yr, corresponding to a positional error of 10 m and an interval of 1 year over which displacements are measured.

where dx represents the distance between the two velocity points. As noted, for this calculation, the nominal distance is 1 km, but the actual distance varies because of the positional error. Again, this procedure is repeated 10,000 times. The resulting probability distribution for the error in strain rate is shown in the upper panel of Figure 1.12 and corresponds to a Gaussian distribution with zero mean and standard deviation equal to 0.02 yr^{-1}. Interestingly, this standard deviation is nearly the same as that predicted by equation (1.42), in which the error in positions are neglected. The reason for this is that the relative error in distance, dx, between the two locations is much smaller than the relative error in the two velocities.

An alternative method for estimating strain rates is to use relative length changes. Consider a line segment connecting two points, A and B, with an initial length, L, aligned in the x-direction. After some time, δt, the length of this segment has changed by an amount δL. The corresponding strain is then

$$\varepsilon_{xx} = \frac{\delta L}{L}, \qquad (1.55)$$

and the strain rate is

$$\dot{\varepsilon}_{xx} = \frac{\delta L}{L \delta t}. \qquad (1.56)$$

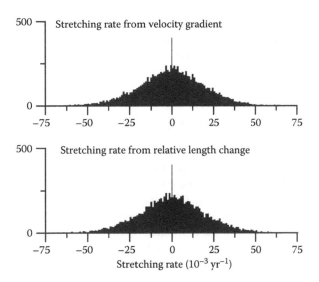

FIGURE 1.12 Probability distributions for error in along-flow stretching rate determined from a Monte Carlo simulation as explained in the text. The upper panel corresponds to stretching rate calculated from velocity gradients, and the lower panel to the stretching rate calculated from relative displacement. For both distributions the mean (indicated by the vertical line) is zero and the standard deviation is 0.02 yr^{-1}, corresponding to a positional error of 10 m and an interval of 1 year over which displacements are measured.

To demonstrate that this expression is similar to the definition of strain rate in terms of velocity gradient, note that the entire line segment is moving in the x-direction and any change in length must result from a velocity difference at the end points, A and B. That is

$$\frac{\delta L}{\delta t} = U_B - U_A, \tag{1.57}$$

and

$$\dot{\varepsilon}_{xx} = \frac{\delta L}{L \delta t} = \frac{U_B - U_A}{L} = \frac{\partial U}{\partial x}. \tag{1.58}$$

To allow for comparison with the error in strain rate determined from velocity errors, equation (1.56) is applied to two points separated by a nominal distance of 1 km and with a positional error with a standard deviation of 10 m, and the change in length is assumed to occur over a period of 1 year. Using the same Monte Carlo technique, the probability distribution of the error in stretching rate shown in the lower panel in Figure 1.12 is obtained. The mean and standard deviation are the same as for the distribution shown in the upper panel.

These examples illustrate how a parametric uncertainty analysis can be used to estimate the error in derived quantities. Admittedly, these examples are sufficiently trivial to allow for application of the error analysis discussed in Section 1.3. For dependent variables that are more complex functions of input parameters, the procedure for estimating the error is essentially the same as discussed in this section.

1.5   CALCULATING STRAIN RATES

Strain rates are related to velocity gradients, and most often, measured surface velocities are used to estimate surface strain rates. When maps of surface strain rates on a glacier are produced, the usual procedure is to use velocities interpolated to a regular grid with Cartesian axes (as in the example of Byrd Glacier, Antarctica, discussed in Section 11.2). This is an expedient procedure, but a concern is that artifacts could be introduced into the calculated strain rates. This is because velocity measurements usually are irregularly spaced, and the gridding may introduce some noise or irregularities, especially in regions where data density is low. Such noise may be inconsequential for the velocities themselves but becomes amplified when taking derivatives for strain rates. In calculating the resistive terms in the balance of forces, this noise may be further amplified because these resistive terms involve the spatial gradient of strain rates (or, equivalently, the second derivative of the velocity). No study has been conducted to evaluate whether this concern is warranted. Of course, if only gridded velocity data are made available, there is no other recourse than to use these products. Ideally, however, strain rates should be determined from irregularly spaced actual velocity measurements and then, if need be, interpolated to a regular grid.

A second instance where strain rates must be estimated from a sparse set of velocity determinations is when velocities are measured from a small number of GPS stations covering an area of interest. For example, Bassis et al. (2007) deployed 12 GPS stations around the tip of a rift on the Amery Ice Shelf, with the objective to estimate stresses acting on the rift. In such cases, calculating strain rates is slightly more involved than when gridded velocities are available. This section outlines how strain rates can be estimated using three stations whose position is tracked over time (either using GPS, feature tracking on repeat imagery, or measuring distances between the stations using electronic measurement devices).

In the horizontal (x, y) plane, there are three strain rates that describe the deformation. Following the convention adopted in this book, the x-axis is chosen to align with the average direction of ice flow. Then, $\dot{\varepsilon}_{xx}$ represents stretching in the flow direction, $\dot{\varepsilon}_{xy}$ represents lateral shearing, and $\dot{\varepsilon}_{yy}$ represents convergence or divergence in the transverse direction. To calculate these three components, three stations are required from which strain rates in three independent directions can be estimated.

The geometry for calculating strain rates is shown in Figure 1.13. Three survey stations are considered, A, B, and C. Recalling that strain rates are estimated from relative velocities or displacements, the position of the first station, A, may be considered fixed in time. The filled black circles in this figure represent the locations at some initial time, and the open circles labeled B′ and C′ represent the position of stations

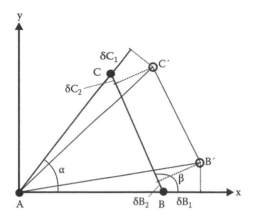

FIGURE 1.13 Geometry used to calculate strain rates from the relative displacements of three survey stations.

B and C (relative to station A) after some time, δt. Further, without loss of generality, the two stations A and B are positioned to align with the (initial) direction of flow, and thus with the x-axis (the results given below can be readily transformed to any arbitrary coordinate system using the tensor rotation formulas given in Section 1.1).

The initial survey gives the three distances, AB, AC, and BC. Repeating the survey after some time, δt, the position of stations B and C, relative to station A (which also has moved, but that does not enter into the calculation of strain rates), are given by B' and C', respectively. The displacements of these two stations can be projected onto the initial three directions corresponding to the sides of the triangle ABC. These displacements are denoted by δB_1 for the displacement of station B in the AB direction, δB_2 for the displacement of station B along the BC direction, and similarly for station C (Figure 1.13). Note that in this example, both δB_2 and δC_2 are negative, indicating shortening of the distance BC.

The three strain rates determined from the displacements are

$$\dot{\varepsilon}_{AB} = \frac{\delta B_1}{|AB|\,\delta t}, \tag{1.59}$$

$$\dot{\varepsilon}_{AC} = \frac{\delta C_1}{|AC|\,\delta t}, \tag{1.60}$$

$$\dot{\varepsilon}_{BC} = \frac{\delta B_2 + \delta C_2}{|BC|\,\delta t}. \tag{1.61}$$

Equivalently, these strain rates can be estimated from the gradients in velocity components in each of the three directions, AB, AC, and BC. This gives the same result as equations (1.59)–(1.61).

The three measured strain rates can be related to the strain rates defined in the (x, y) coordinate system using the tensor rotation formulas (1.25–1.27). The AB direction is aligned with the x-axis, and the angles between the AC and BC directions and the x-direction are α and β, respectively. Then

$$\dot{\varepsilon}_{AB} = \dot{\varepsilon}_{xx}, \tag{1.62}$$

$$\dot{\varepsilon}_{AC} = \dot{\varepsilon}_{xx} \cos^2 \alpha + \dot{\varepsilon}_{yy} \sin^2 \alpha + 2\dot{\varepsilon}_{xy} \sin\alpha \cos\alpha, \tag{1.63}$$

$$\dot{\varepsilon}_{BC} = \dot{\varepsilon}_{xx} \cos^2 \beta + \dot{\varepsilon}_{yy} \sin^2 \beta + 2\dot{\varepsilon}_{xy} \sin\beta \cos\beta. \tag{1.64}$$

After some tedious algebra, this set of equations can be solved for the three strain rate components (Turcotte and Schubert, 2002, p. 98):

$$\dot{\varepsilon}_{xx} = \dot{\varepsilon}_{AB}, \tag{1.65}$$

$$\dot{\varepsilon}_{yy} = \frac{1}{\tan\beta - \tan\alpha}[\dot{\varepsilon}_{AB}(\mathrm{ctn}\alpha - \mathrm{ctn}\beta) - \dot{\varepsilon}_{AC}(\mathrm{sea}\alpha\,\mathrm{csa}\alpha) + \dot{\varepsilon}_{BC}(\sec\beta\,\csc\beta)], \tag{1.66}$$

$$\dot{\varepsilon}_{xy} = \frac{1}{2(\mathrm{ctn}\beta - \mathrm{ctn}\alpha)}\left[\dot{\varepsilon}_{AB}(\mathrm{ctn}^2\alpha - \mathrm{ctn}^2\beta) - \dot{\varepsilon}_{AC}\csc^2\alpha + \dot{\varepsilon}_{BC}\csc^2\beta\right], \tag{1.67}$$

with $\sec\alpha = 1/\cos\alpha$, $\csc\alpha = 1/\sin\alpha$, and $\mathrm{ctn}\alpha = 1/\tan\alpha$.

As an example, consider the displacements of three survey stations on Helheim Glacier, east Greenland, whose positions were continuously tracked over a 20-day period in June 2007, using GPS (data provided by Leigh Stearns). GPS positions are in latitude and longitude and elevation above the geoid. Here, only the horizontal positions are considered. The first step is to convert the horizontal positions to a local Cartesian coordinate system. The most often used Cartesian coordinate system for regional studies is the Universal Transverse Mercator (UTM) system developed by the U.S. Army Corps of Engineers in the 1940s. Figure 1.14 shows the displacements of the three stations in this coordinate system; Table 1.1 gives the start and end positions of the three stations, both in absolute UTM coordinates and relative to the initial position of station A in the UTM coordinate system.

As in Figure 1.13, a more convenient Cartesian coordinate system is introduced with the x-axis in the direction of the line connecting the initial positions of stations A and B. This requires coordinate transformation using formulas equivalent to equations (1.7) and (1.8). The rotation angle follows from the coordinates of station B relative to station A and equals –22°. This coordinate transformation gives the positions relative to the initial position of station A listed in Table 1.2. As noted

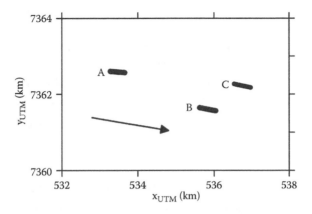

FIGURE 1.14 Displacement of three survey stations on Helheim Glacier, east Greenland, in the UTM (zone 24) coordinate system. Positions were measured over a 20-day period in June 2007. The arrow indicates the direction of the principal tensile strain. (Data provided by Leigh Stearns.)

above, strain rates are estimated from relative displacements and the position of station A may be considered fixed in time. Keeping this station stationary gives the relative end positions of stations B and C listed in Table 1.2 under the second set of coordinates.

The next step is to determine the four displacements, δB_1, δB_2, δC_1, and δC_2, in the initial three directions of the sides of the triangle ABC (Figure 1.13). This

TABLE 1.1
Start and End Positions of Three Survey Stations on Helheim Glacier, East Greenland, in the UTM Coordinate System

Station	Start Position		End Position	
	x_{UTM} (m)	y_{UTM} (m)	x_{UTM} (m)	y_{UTM} (m)
A	533,287.61	7,362,596.78	533,667.96	7,362,564.84
B	535,618.03	7,361,641.36	536,046.75	7,361,561.02
C	536,530.00	7,362,265.00	536,974.00	7,362,172.00
	Relative Start Position		**Relative End Position**	
A	0	0	380.35	−31.94
B	2,330.42	−955.42	2,759.13	−1,035.76
C	3,242.39	−331.78	3,686.39	−424.78

Source: Data provided by Leigh Stearns.

Note: The first set of coordinates gives positions in absolute coordinates, and the second set of coordinates gives positions relative to the initial position of survey station A.

TABLE 1.2
Start and End Positions of Three Survey Stations on Helheim Glacier, East Greenland, in the Local Coordinate System with the X-Axis Directed along the Line Connecting the Initial Positions of Stations A and B

Station	Relative Start Position		Relative End Position	
	x (m)	y (m)	x (m)	y (m)
A	0	0	364.04	114.72
B	2,518.66	0	2,945.82	88.29
C	3,125.90	922.97	3,571.99	1,005.34
	Relative Start Position		**Relative End Position**	
A	0	0	0	0
B	2,518.66	0	2,581.78	−26.44
C	3,125.90	922.97	3,207.96	890.62

Note: The first set of coordinates gives positions in relative coordinates, and the second set of coordinates gives relative end positions assuming that survey station A remained stationary.

is a rather tedious procedure involving determining many relative angles. For the example under consideration, equations (1.59)–(1.61) for the three strain rates may be approximated as

$$\dot{\varepsilon}_{AB} = \frac{|AB'| - |AB|}{|AB|\,\delta t},\tag{1.68}$$

$$\dot{\varepsilon}_{AC} = \frac{|AB'| - |AC|}{|AC|\,\delta t},\tag{1.69}$$

$$\dot{\varepsilon}_{BC} = \frac{|B'C'| - |BC|}{|BC|\,\delta t}.\tag{1.70}$$

It is left to the reader to check that, for this example, using these approximations results in calculated strain rates differing by a few percent from the values calculated using the exact formulas.

Changes in the length of line segments are readily calculated from the relative coordinates given in Table 1.2. Applying equations (1.68)–(1.70) gives the following values for the three strain rates: $\dot{\varepsilon}_{AB} = 0.458$ yr^{-1}, $\dot{\varepsilon}_{AC} = 0.392$ yr^{-1}, and $\dot{\varepsilon}_{BC} = 0.093$ yr^{-1}. Using the transformation equations (1.65)–(1.67), strain rates in the orthogonal (x, y) coordinate system are found to be $\dot{\varepsilon}_{xx} = 0.458$ yr^{-1}, $\dot{\varepsilon}_{yy} = -1.060$ yr^{-1}, and $\dot{\varepsilon}_{xy} = 0.267$ yr^{-1}. As was to be expected, the angle of the principal tensile strain rate is 9.7° (using equation (1.28)), which corresponds to the flow direction of the three stations.

2 Ice Deformation

2.1 CREEP OF GLACIER ICE

Many laboratory experiments have been conducted to study the creep (or deformation) of ice. These include measurements on the creep of single ice crystals by Glen and Perutz (1954), Steinemann (1954), Rigsby (1958), Readey and Kingery (1964), and Montagnat and Duval (2004); papers discussing the deformation of polycrystalline ice include Glen (1952), Steinemann (1958), Jonas and Müller (1969), Duval (1981), Mellor and Cole (1982), and Jacka and Maccagnan (1984). Extensive reviews on this subject can be found in Weertman (1973a), Hobbs (1974), Glen (1975), Hooke (1981), Budd and Jacka (1989), and Schulson and Duval (2009). Below, a short summary of the most important results is given. For most applications discussed in this book, this elementary review is sufficient; those readers who are interested in a more comprehensive discussion are referred to the in-depth treatment of this topic by Schulson and Duval (2009).

The most common form of ice is Ice Ih with a crystal structure as shown in Figure 2.1. Each molecule consists of an oxygen atom strongly bonded to two hydrogen atoms forming a V shape. These bonds between oxygen and hydrogen atoms are called *covalent bonds*. The resulting H_2O molecule has a slightly positive charge on the side of the two hydrogen atoms and a slightly negative charge on the side of the oxygen atom. As a result, the partially positive hydrogen atom on one water molecule is electrostatically attracted to the partially negative oxygen of a neighboring molecule, forming a weak hydrogen bond (Figure 2.1a). In an ice crystal, each oxygen atom is surrounded by four nearest neighbors arranged near the vertices of a regular tetrahedron centered around the molecule under consideration (Pounder, 1965; Schulson, 1999). Referring to Figure 2.1b, three of the atoms surrounding the central oxygen atom O form an equilateral triangle in a plane called the basal plane; these atoms are marked as O^2. The fourth oxygen atom, O^1 is positioned such that the line O^1O is perpendicular to the basal plane. This direction is referred to as the c-axis (Pounder, 1965).

Next consider the four oxygen atoms surrounding atom O^1. One of these four neighboring atoms is atom O, while the other three atoms, marked O^3 also lie in a plane perpendicular to O^1O, forming an inverted tetrahedron centered on O^1. Figure 2.1b shows only how O and O^1 are surrounded by four oxygen atoms. The other oxygen atoms, O^2 and O^3, are of course also surrounded by four oxygen atoms but extending the illustration to depict the full three-dimensional structure would make the figure too confusing. Following Pounder (1965), consider therefore the two projections shown in Figure 2.1c and d.

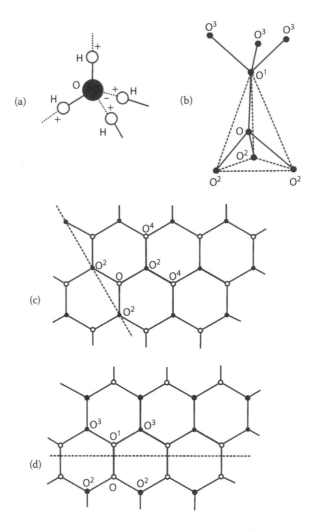

FIGURE 2.1   Crystal structure of ice. (a) placement of oxygen and hydrogen atoms; solid lines represent covalent bonds and dashed lines represent weaker hydrogen bonds. (b) structure of an ice lattice showing the tetrahedral bond arrangement; solid lines represent hydrogen bonds, and dashed lines outline the tetrahedron. (c) projection of the ice lattice on a basal plane; open and filled circles represent oxygen atoms on different planes, and the dashed line gives the direction of the c-axis. (d) projection of the ice lattice on a plane through the c-axis (the plane perpendicular to the dashed line in panel d); the dashed line represents the glide plane. (Panels b–d modified from Pounder, E. R., *Physics of Ice*, Pergamon Press, London, 1965.)

Figure 2.1c shows the arrangement of oxygen atoms projected on the basal plane. Two sets of atoms are shown and distinguished by open and filled circles. The filled circles correspond to atoms lying in the basal plane passing through the three O^2 atoms. The open circles correspond to the next basal plane containing atom O. Note that this figure shows only the slanting bonds of the O^2O type and not the bonds in the direction of the c-axis (O^1O). Atom O has three downward slanting bonds to the three atoms O^2. Each of these atoms must have three upward slanting bonds, one of which is to atom O. The other two slanting bonds are to two atoms, marked O^4, that must lie in the same basal plane as atom O. What emerges is a pattern of hexagons that zigzag back and forth between two basal planes (Pounder, 1965).

Now consider the arrangement of oxygen atoms projected on a plane parallel to the c-axis and passing through a pair of O^2 atoms (Figure 2.1d). In this case, open and filled circles represent atoms that are in different vertical planes (with vertical being the direction of the c-axis). The upward bond from atom O must be vertical so that the positions of the atoms O^1 are fixed. Further, each O^3 atom is directly above an O^2 atom. The other allowed geometry would be if the O^3 atoms occupied the vacant centers of the hexagons shown in Figure 2.1d; this is the arrangement found in, for example, diamond (Pounder, 1965). The important consequence of this arrangement is that ice has only one important axis of symmetry, namely, the c-axis; and properties of a single ice crystal may be assumed isotropic in all directions perpendicular to the c-axis, but anisotropy can be expected between properties measured parallel and perpendicular to the c-axis (Pounder, 1965).

To recapitulate, the ice crystal structure is such that oxygen atoms are arranged in approximately parallel planes, called the basal planes. The direction perpendicular to the basal planes is referred to as the c-axis. Deformation of a single crystal occurs readily if there is a component of shear stress acting on the basal plane. In that case, the layers of molecules glide over one another, similar to a deck of cards. Nonbasal glide is much more difficult to initiate and requires much larger stress; for that reason, nonbasal glide is also referred to as hard glide. For the majority of crystal orientations, there will be a component of shear stress in the basal plane; thus it seems reasonable to presume that deformation of ice is achieved mainly by basal glide. An example in which basal glide cannot occur is when a compressive or tensile stress is applied perpendicular to the basal plane. For practical purposes, a crystal in this situation can be considered nondeforming.

Hobbs (1974) argues that polycrystalline ice can deform into any arbitrary shape without changing its volume only if there are at least five independent slip directions. In the basal plane, there are only two (perpendicular) slip directions, and therefore Hobbs (1974) concludes that deformation of polycrystalline ice is probably controlled by nonbasal glide. As noted above, this hard glide is much more difficult to induce than basal glide, so that, under similar conditions, a sample of ice may be expected to deform at a slower rate than a single crystal deforming by basal slip. Experiments suggest that this is indeed the case (c.f. Hobbs, 1974), but there may be an alternative explanation.

The argument that there must be at least five independent slip directions is credited to Taylor (1938), who assumed that deformation occurs uniformly throughout a polycrystalline sample. In that case, the strain in each crystal or volume element must be the same as the bulk deformation of the sample. Because the sample is able to deform in any direction, Taylor's assumption of homogeneous strain implies that each individual crystal must be able to deform into any shape. As shown by Von Mises (1928), this requires at least five independent slip directions (c.f. Hutchinson, 1976). Hutchinson (1977) showed that relaxing the assumption of uniform strain and adopting a self-consistent method reduces the number of required slip directions to four and deformation can be accommodated by basal slip plus slip of basal

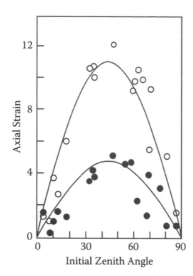

FIGURE 2.2   Axial strain of individual crystals as a function of their initial zenith angle (the angle between the crystal c-axis and the direction of compression) after 3% bulk strain (dots) and after 7% bulk strain (open circles). The two curves represent the theoretical distribution $\varepsilon_c = \pi \varepsilon_b \cos \phi \sin \phi$, with ε_c the crystal strain, ε_b the bulk strain of the aggregate, and ϕ the initial zenith angle of the crystal. (From Azuma, N., and A. Higashi, *Ann. Glaciol.,* 6, 130–134, 1985. Reprinted from the *Annals of Glaciology* with permission of the International Glaciological Society and the authors.)

dislocations on prismatic planes (parallel to the c-axis). The latter slip mode requires larger stresses than basal slip. An alternative possibility is that deformation of a polycrystalline aggregate is achieved by straining only those crystals that are aligned favorably, that is, crystals with a component of shear stress in the basal plane. In a sample of randomly oriented crystals, this is always possible. Because unfavorably oriented crystals deform much more slowly than those that are oriented optimally, the rate of deformation of the sample (which is simply the average of the deformation rates of all crystals in the sample) may be expected to be (much) smaller than the rate of deformation of a single crystal that is favorably aligned. The difficulty with this model is that it is not entirely clear how individual crystals can be strained at different rates without creating voids or overlaps in the aggregate.

Experimental evidence supports the hypothesis that basal glide is the dominant mechanism by which deformation occurs. Azuma and Higashi (1985) subjected a sample with an initially random fabric to compression and measured the uniaxial strain of individual crystals as a function of the initial angle between the compressive axis and the crystal c-axis. The results, shown in Figure 2.2, show that crystals with their c-axis along the compressive axis, or perpendicular to it, undergo very little deformation. This is because these crystals have little or no component of shear stress in their basal plane, and for single crystals, the rate of deformation is proportional to this shear stress. The resolved basal shear stress is largest if the angle between c-axis and direction of compression is 45° and these crystals

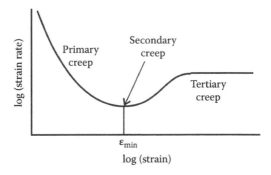

FIGURE 2.3   Typical creep curve for glacier ice as measured in laboratory experiments.

undergo maximum deformation, as indicated by the experiments of Azuma and Higashi (1985).

For single ice crystals subjected to a constant stress and oriented for basal glide, the strain rate initially increases with time until a steady state is reached. Weertman (1973a) argues that the accelerating creep can be attributed to the initially low density of dislocations. Dislocations are line defects bounding areas in a crystal where slip has already taken place. Outward propagation of these dislocations increases the area where slip occurs, resulting in deformation by basal slip (Poirier, 1985). As creep continues, more dislocations are produced by multiplication processes, allowing the creep rate to accelerate. In single ice crystals, this effect appears to be more important than the interaction of dislocations, which causes a decrease in creep rate as observed in metal single crystals.

For a specimen of polycrystalline ice, on the other hand, the rate of deformation decreases rapidly with time after a (constant) stress is applied; this stage is called primary or transient creep (Figure 2.3). The mechanism responsible for the decrease in creep rate remains unknown (Weertman, 1973a; Schulson and Duval, 2009), although it is often suggested that the decrease in creep rate is caused by the formation of dislocation tangles, which inhibit dislocation glide. An alternative explanation is proposed by Van der Veen and Whillans (1994), who argue that the bulk strain rate of a specimen decreases as the sample deforms as a result of progressive rotation of crystal c-axes toward orientations unfavorable for further deformation by basal glide (c.f. Section 2.4).

According to the conventional view, the period of primary creep is followed by a period of secondary creep, during which the strain rate remains approximately constant as deformation continues. Paterson (1994) argues that during secondary creep, the ice may be recrystallizing at grain boundaries where (elastic) stresses are particularly high. The observed constant strain rate probably results from a temporary balance between softening at these boundaries and hardening elsewhere. A different view is expressed by Budd and Jacka (1989), who note that the concept of steady-state secondary creep for ice is invalid because of its inherently transitory nature. The model simulations of Van der Veen and Whillans (1994) support this view and suggest that the minimum creep rate reached during deformation simply marks the onset of recrystallization, which softens the ice by aligning crystals in favorable orientations (c.f. Section 2.4).

After passing through the minimum, the strain rate increases again to, perhaps, a constant value, during what is referred to as tertiary creep. This increase is attributed to recrystallization (Hooke, 1981; Li et al., 1996) and may occur in discrete steps, as suggested by Hooke and Hudleston (1980).

A typical creep curve obtained in the laboratory is shown in Figure 2.3. The common view for modeling glacier flow is that secondary and tertiary creep are most important. Primary creep, if related to long-range interactions between dislocations (Schulson and Duval, 2009), does not occur within the main body of a glacier because this ice has been deforming under a given stress regime for a sufficiently long time to reach the stage of secondary or tertiary creep. Schulson and Duval (2009) suggest that tertiary creep can be described by a similar relation as that used for secondary creep, but with enhancement factors ranging from about 3 in compression or pure shear to between 8 and 10 for simple shear. Secondary creep of anisotropic ice is described by Glen's flow law discussed in Section 2.2. Note that this interpretation is contradictory to the view that primary creep reflects hardening of ice as crystals are oriented and a distinct fabric develops (Van der Veen and Whillans, 1994).

2.2   CONSTITUTIVE RELATION

To arrive at the constitutive relation most often used in glaciology, the analysis of Nye (1953) is followed; a mathematically more rigorous approach can be found in Hutter (1983). The first assumption is that ice is incompressible and that it remains isotropic throughout the flow, with crystal c-axes randomly oriented. In that case, the principal axes of the strain-rate tensor must be parallel to those of the stress tensor. The second assumption is that ice deformation is independent of hydrostatic pressure, and the components of strain rate are proportional to the components of deviatoric stress (Section 1.2). That is,

$$\dot{\varepsilon}_{ij} = \chi \tau_{ij}, \tag{2.1}$$

where χ is a positive scalar function that may depend on the entire stress state, temperature of the ice, and perhaps other factors as well.

The next step is to determine the form of the function χ. Because the constitutive response of glacier ice must be independent of the particular coordinate system chosen, χ must be a function of the invariants of the strain rate and stress deviator tensors. Invariants of a tensor are scalar quantities whose values are not affected by a transformation of the coordinate system (Section 1.1). Both the stress and strain-rate tensors are of second order and therefore have three invariants each. For the deviatoric stress tensor, the invariants are (equations (1.21)–(1.23))

$$I_1 = \tau_{xx} + \tau_{yy} + \tau_{zz}, \tag{2.2}$$

$$I_2 = \tau_{xx}^2 + \tau_{yy}^2 + \tau_{zz}^2 + 2\left(\tau_{xy}^2 + \tau_{xz}^2 + \tau_{yz}^2\right), \tag{2.3}$$

$$I_3 = \det(\tau_{ij}). \tag{2.4}$$

From the definition of deviatoric stresses (equation (1.36)), it follows immediately that the first invariant is zero. By adopting equation (2.1), it also follows that the first invariant of the strain-rate tensor is zero (which is true for any incompressible material). It is now postulated that the rheology of glacier ice is independent of the third invariants of the strain-rate and deviatoric stress tensor. Thus

$$\chi = \chi[I_2(\tau), I_2(\dot{\varepsilon})]. \tag{2.5}$$

Rather than using the second invariants, the quantities called effective strain rate, $\dot{\varepsilon}_e$, and effective stress, τ_e, are used. These are defined through

$$2\dot{\varepsilon}_e^2 = \dot{\varepsilon}_{xx}^2 + \dot{\varepsilon}_{yy}^2 + \dot{\varepsilon}_{zz}^2 + 2\left(\dot{\varepsilon}_{xy}^2 + \dot{\varepsilon}_{xz}^2 + \dot{\varepsilon}_{yz}^2\right), \tag{2.6}$$

$$2\tau_e^2 = \tau_{xx}^2 + \tau_{yy}^2 + \tau_{zz}^2 + 2\left(\tau_{xy}^2 + \tau_{xz}^2 + \tau_{yz}^2\right). \tag{2.7}$$

Based on the laboratory experiments of Glen (1952), Nye (1953) suggested the following relation between the effective stress and effective strain rate

$$\dot{\varepsilon}_e = A\,\tau_e^n, \tag{2.8}$$

where A and n are the flow parameters. From (2.1) and the definitions of effective stress and strain rate, it follows that

$$\chi = A\,\tau_e^{n-1}, \tag{2.9}$$

and hence

$$\dot{\varepsilon}_{ij} = A\,\tau_e^{n-1}\,\tau_{ij}. \tag{2.10}$$

Equation (2.10) is called Nye's generalization of Glen's law, or Glen's law for short. Virtually all modeling of the flow of glaciers is based on this constitutive relation.

For some applications, the inverse formulation of equation (2.10) is needed, for example when estimating stresses from measured strain rates. Combining (2.10) and (2.8) gives

$$\dot{\varepsilon}_{ij} = A\left(\frac{\dot{\varepsilon}_e}{A}\right)^{1-1/n}\tau_{ij}, \tag{2.11}$$

and

$$\tau_{ij} = B\dot{\varepsilon}_e^{1/n-1}\,\dot{\varepsilon}_{ij}, \tag{2.12}$$

with

$$B = A^{-1/n}. \tag{2.13}$$

The flow parameters, A and B, are dependent on many factors, most notably the temperature of the ice.

The above derivation is based solely on mathematical considerations of permissible relations between the stress and strain-rate tensors and does not consider the physical processes by which ice deforms. Deformation of crystals results from defects existing in the crystal lattice where the periodicity in which atoms are arranged is locally interrupted. As these defects move around under an applied stress, some deformation of the crystal takes place. Thus, if this deformation can be estimated, and if the concentration of defects and their migration velocity are known, the flow law can be derived on theoretical grounds, at least in principle (Poirier, 1985). While glide of dislocation (line defects) is the primary agent of deformation of polycrystalline ice, other microscopic processes may contribute to observed macroscopic deformation. For example, material may be transported by diffusion (diffusion creep) or by shear along grain boundaries (grain boundary sliding). These two processes are strongly coupled and mutually accommodating. Grain boundary sliding creates voids or overlaps that are accommodated by diffusion creep, and vice versa (Poirier, 1985). Which microscopic process is most important to bulk deformation depends on a number of factors including ice temperature and magnitude of the applied stress. The impossibility of observing microscopic processes in situ on glaciers makes it near impossible to evaluate which process is most important or relevant to glacier flow and under what conditions. For this reason, there is no generally accepted flow law for glacier ice derived from consideration of crystal-scale processes (c.f. Schulson and Duval, 2009). Some alternatives to Glen's law have been proposed, and these are discussed in the next section.

The constitutive relation (2.10), or the inverse formulation (2.12), contains two parameters, namely, the factor n in the exponent and the rate factor A or viscosity parameter B (the latter two are related through equation (2.13)). Many laboratory and field measurements have been conducted to determine the values of these parameters. It should be kept in mind, however, that none of these experiments has unambiguously confirmed the validity of the general form of the constitutive relation, although Schulson and Duval (2009) note that equation (2.11) was verified with tests performed in shear and compression. To determine the form of the constitutive relation, combined shear and compression experiments are needed (Morland, 1979), but most laboratory experiments are uniaxial stress experiments. Interpretation of field data has not led to an unequivocal constitutive relation, either. The major problem is that field measurements yield only strain rates, and to find the relation between strain rate and stress, the stresses must be inferred indirectly. In most cases, the stresses are calculated from a simplified theory, which may greatly affect the resulting correlation. Van der Veen and Whillans (1990) used a full-stress model to simulate flow along the Dye-3, Greenland, strain network and compared field data with model predictions. Adopting various forms of the flow law, their results failed to convincingly verify the applicability of Glen's flow law.

Laboratory experiments generally support values of n around 3 for effective stresses in the range of 200 to 2000 kPa (Hooke, 1981; Budd and Jacka, 1989), in agreement with theoretical considerations (Weertman, 1973a). At stresses more common in glaciers (<200 kPa), the exponent may be less than 3 (including n = 1, describing a Newtonian viscous fluid), but these experiments must be interpreted with caution because of the great difficulty in conducting such tests and because in many of the experiments, the ice is deformed for only a limited time to strains of a few percent (Weertman, 1973a; Budd and Jacka, 1989; Schulson and Duval, 2009). Nevertheless, analysis of field data and laboratory experiments supports a transition from deformation with n ≈ 3 at high stresses and n ≤ 2 at low stresses (Montagnat and Duval, 2004). This issue is discussed further in Section 2.4.

The rate factor determined from laboratory and field experiments applies to (steady-state) secondary creep, linking the minimum strain rate (Figure 2.3) to deviatoric stress. It is widely accepted that the rate factor is an exponential function of temperature. Hooke (1981) evaluated available data and found the following best fit

$$A = A_o \exp\left(-\frac{Q}{RT} + \frac{3C}{(T_r - T)^k} \right), \qquad (2.14)$$

or

$$B = B_o \exp\left(\frac{T_o}{T} - \frac{C}{(T_r - T)^k} \right), \qquad (2.15)$$

where $A_o = 9.302 \cdot 10^7$ kPa^{-3} yr^{-1}, $Q = 78.8$ kJ/mol (the activation energy for creep), $R = 8.321$ J/(mol K) (the gas constant), $C = 0.16612$ K^k, $T_r = 273.39$ K, $k = 1.17$, $B_o = 2.207 \cdot 10^{-3}$ kPa $\cdot$ yr$^{1/3}$, $T_o = 3155$ K, and the ice temperature, T, is expressed in K (Kelvin). In Figure 2.4 these two relations are shown (dashed lines). Also shown

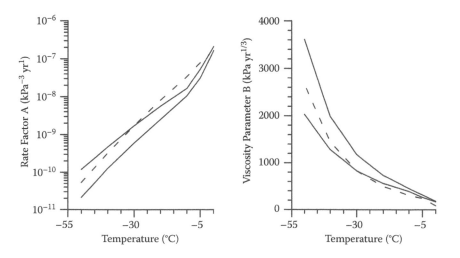

FIGURE 2.4 Rate factor, A, and viscosity parameter, B, as a function of ice temperature. The solid curves represent the upper and lower values recommended by Paterson and Budd (1982), and the dashed curve the best-fit relations (2.14) and (2.15) determined by Hooke (1981).

in this figure are the minimum and maximum values recommended by Paterson and Budd (1982). The uncertainty in the value of the rate factor implies that, for a given stress and temperature, the strain rate may vary by as much as a factor of 5. Hooke (1981) discusses possible explanations for why the different experiments yield such wide scatter. Factors that may play a role are melting effects in the ice samples, variations in grain size, impurity content, and ice density. However, perhaps most important is the effect that ice fabric has on the creep rate. Before discussing this issue, the next section delves into alternative (isotropic) flow laws that have been proposed for glacier modeling.

2.3   MORE ABOUT THE CONSTITUTIVE RELATION

Nye's generalization of Glen's law, discussed in the previous section, is commonly used to describe the rheological properties of glacier ice. However, the form of this relation has not been confirmed unambiguously. On the contrary, it can be argued that many laboratory experiments have disproven the applicability of Glen's law. As early as 1958, Glen recognized that the few experimental results available at that time disagreed with the simple flow law (2.10). Based on theoretical arguments, summarized below, Glen (1958) presented a more general constitutive relation.

Making the assumption that ice is isotropic implies that the strain-rate tensor must possess the same symmetry as the full stress tensor. This means that the strain-rate tensor must be a polynomial function of the stress tensor. That is

$$\dot{\varepsilon} = \sum_{k=0}^{K} F_k \, \sigma^k, \tag{2.16}$$

where the F_k represents arbitrary functions of the three invariants of the stress tensor. A particular strain-rate component is then given by

$$\dot{\varepsilon}_{ij} = F_1 \, \delta_{ij} + F_2 \, \sigma_{ij} + F_3 \, \sigma_{ik} \, \sigma_{kj} + F_4 \, \sigma_{ik} \, \sigma_{kl} \, \sigma_{lj} + \cdots \cdots . \tag{2.17}$$

The fourth and further terms on the right-hand side can be written in terms of the first three terms using the Hamilton–Cayley theorem (Truesdell and Noll, 1965, p. 26)

$$\sigma_{ij}^3 - I_1 \, \sigma_{ij}^2 - I_2 \, \sigma_{ij} - I_3 \, \delta_{ij} = 0, \tag{2.18}$$

where the I_i represent the three invariants of the full stress tensor, and $\delta_{ij} = 1$ if $i = j$ and $\delta_{ij} = 0$ if $i \neq j$ (the Kronecker delta).

To eliminate the effect of hydrostatic pressure on the rate of deformation, the strain rates are linked to deviatoric stresses, τ_{ij}, defined in Section 1.2. From the above it follows immediately that the most general constitutive relation is

$$\dot{\varepsilon}_{ij} = F_1 \, \delta_{ij} + F_2 \, \tau_{ij} + F_3 \, \tau_{ik} \, \tau_{kj}. \tag{2.19}$$

A further simplification is possible if the ice is taken to be incompressible; that is, the density of the ice remains constant during deformation. Because the change in density is directly related to the first invariant of the strain-rate tensor, incompressibility requires that (c.f. Section 1.2)

$$\dot{\varepsilon}_{xx} + \dot{\varepsilon}_{yy} + \dot{\varepsilon}_{zz} = 0. \tag{2.20}$$

Using equation (2.19) to determine the strain rates, and noting that the three normal deviatoric stresses sum to zero, gives the following relation:

$$F_1 = -\frac{2}{3} I_2 F_3. \tag{2.21}$$

The flow law can now be written in the general form

$$\dot{\varepsilon}_{ij} = F_2(I_2, I_3)\tau_{ij} + F_3(I_2, I_3)\left(\tau_{ik}\tau_{kj} - \frac{2}{3}I_2\delta_{ij}\right), \tag{2.22}$$

where I_2 and I_3 are the second and third invariants of the deviatoric stress tensor, as defined in Section 1.1. Only if the second term on the right-hand side is neglected and also the dependency of the first term on the third invariant is ignored does this expression reduce to the commonly used constitutive relation.

By neglecting the second term on the right-hand side of equation (2.22), normal stress effects are tacitly excluded. However, Reiner (1945) predicted on theoretical grounds that normal stress effects (or dilatancy, a term coined by Reynolds in 1885) should be present in any viscous material whose rheological properties are described by a nonlinear functional relationship between deviatoric stress and strain rate. Most non-Newtonian fluids for which suitable measurements have been made exhibit non-zero and unequal normal stress differences in shearing flow (Schowalter, 1978). To illustrate the manifestation of normal stress effects, consider the situation in which only a shear stress, τ_{xz}, is applied. From the constitutive relation (2.22) the strain-rate tensor is found to be

$$\begin{pmatrix} \dot{\varepsilon}_{xx} & \dot{\varepsilon}_{xy} & \dot{\varepsilon}_{xz} \\ \dot{\varepsilon}_{xy} & \dot{\varepsilon}_{yy} & \dot{\varepsilon}_{yz} \\ \dot{\varepsilon}_{xz} & \dot{\varepsilon}_{yz} & \dot{\varepsilon}_{zz} \end{pmatrix} = F_2\begin{pmatrix} 0 & 0 & \tau_{xz} \\ 0 & 0 & 0 \\ \tau_{xz} & 0 & 0 \end{pmatrix} + \frac{1}{3}F_3\begin{pmatrix} \tau_{xz}^2 & 0 & 0 \\ 0 & -2\tau_{xz}^2 & 0 \\ 0 & 0 & \tau_{xz}^2 \end{pmatrix}. \tag{2.23}$$

Thus, there is a swelling or contraction perpendicular to the plane of shear, given by

$$\dot{\varepsilon}_{yy} = -\frac{2}{3} F_3 \tau_{xz}^2, \tag{2.24}$$

and a corresponding contraction or swelling in the other two directions (that is, in the plane of shear), such that the volume does not change. If the material is laterally constrained, for example by rigid walls, an extra pressure is exerted upon these walls.

Although Glen (1958) discussed these normal stress effects, this phenomena has received very little attention in glaciology. Two papers in which these effects are addressed are those of McTigue et al. (1985) and Man and Sun (1987). From these analyses it appears that the manifestation of normal stress effects on glaciers may be small. McTigue et al. (1985) calculate that these effects may cause a depression of several meters in the free surface of a glacier in an open channel. However, Man and Sun (1987) argue that this surface depression (or heave) is considerably smaller. Despite the apparent smallness of these effects, Van der Veen and Whillans (1990) find that inclusion of normal stress effects into the constitutive relation may be important in explaining the small-scale flow features observed near Dye-3, Greenland, but whether such a modified flow law has much impact on the large-scale dynamics of ice sheets remains to be demonstrated.

While Glen's law with exponent n = 3 is widely adopted for glacier modeling, many laboratory experiments indicate that this value may be smaller (n ≤ 2) at low deviatoric stresses (< 100 kPa) common in glaciers. Earlier studies may be suspect because the slow deformation rates involved make it unclear whether steady-state creep was achieved (Weertman, 1973a). More recent experiments, however, have convincingly demonstrated the existence of a transitional stress value where the exponent changes as different physical processes become more or less dominant (for example, Durham et al., 2001; Goldsby and Kohlstedt, 2001; Goldsby, 2006). Three different creep regimes can be identified. At low stresses, deformation proceeds primarily by grain boundary sliding with n = 1.8 (termed *superplastic flow*), while at higher stresses dislocation creep becomes dominant and n = 4. Contrary to dislocation creep, grain boundary sliding is strongly dependent on grain size. At very low stresses and for the smallest grain sizes, basal slip accommodated by grain boundary sliding is most important, with n = 2.4 (Goldsby, 2006; Goldsby and Kohlstedt, 2001). A fourth mechanism was proposed by Goldsby and Kohlstedt (2001), namely, diffusional flow, but could not be observed at practical laboratory strain rates.

Based on experimental results, Goldsby and Kohlstedt (2001) propose the following constitutive relation

$$\dot{\varepsilon} = \dot{\varepsilon}_{\text{diff}} + \left(\frac{1}{\dot{\varepsilon}_{\text{basal}}} + \frac{1}{\dot{\varepsilon}_{\text{gbs}}} \right)^{-1} + \dot{\varepsilon}_{\text{disl}}, \tag{2.25}$$

where the subscripts refer to diffusional flow (diff), basal or easy slip (basal), grain boundary sliding (gbs), and dislocation creep (disl). Each strain rate on the right-hand side is described by a power law of the form

$$\dot{\varepsilon} = A \frac{\sigma^n}{d^p} \exp\left(-\frac{Q + PV}{RT} \right), \tag{2.26}$$

where A and n are flow parameters, σ is the differential stress, d is grain size with p the corresponding exponent, Q is the activation energy for creep, P is the hydrostatic pressure, V is the activation volume for creep, R is the gas constant, and T is absolute temperature. Values for the various parameters for each creep mechanism are given in Goldsby and Kohlstedt (2001).

Goldsby (2006) considers the flow at Byrd Station, West Antarctica, and compares grain size at depth measured in the deep ice core with shear stress at depth and concludes that the ice probably deforms by grain boundary slip over nearly the entire ice thickness. Only near the base of the ice sheet is the ice expected to deform in a transitional regime between creep limited by grain boundary slide and dislocation creep. This conclusion may be questioned because the only stress considered is the shear stress, τ_{xz}, assumed to increase linearly with depth (as in the lamellar flow model discussed in Section 4.2). However, other stress components, in particular the along-flow stretching stress, τ_{xx}, are believed to be important (Whillans and Johnsen, 1983), and hence Goldsby (2006) may have underestimated the differential stress. To adequately test the applicability of the flow law (2.25) would require a full-stress model and comparison with model predictions against observations, as was done by Van der Veen and Whillans (1990) for flow along the Dye-3, Greenland, strain network.

An interesting consequence of the Goldsby–Kohlstedt flow law is that this law seems to eliminate the need to account for crystal anisotropy. The role of fabric development on glacier deformation and flow is discussed more in the next two sections, but in short, ice near the base of ice sheets often exhibits a single-orientation fabric with the crystal c-axes oriented vertically upward. This anisotropy renders the ice relatively soft with respect to shear stresses, and to accommodate this effect, the rate factor in Glen's flow law is multiplied with an *enhancement factor* where crystal anisotropy is expected. For grain boundary sliding, the strain rate is proportional to $1/d^{1.4}$ where d represents grain size. In deeper and older ice, a reduction in grain size compared with that of younger ice is often observed, possibly associated with the Holocene—Last Glacial Maximum climatic transition (Weiss et al., 2002). Thus, creep rates associated with grain boundary sliding will increase in the lower ice layers, thereby eliminating the need for introducing an enhancement factor.

Peltier et al. (2000) adopt the Goldsby–Kohlstedt flow law (2.25) to model the shape of the Laurentide Ice Sheet during the Last Glacial Maximum using a coupled thermomechanical model tuned to the present-day Greenland Ice Sheet. Pettit and Waddington (2003) apply a similar two-term flow law to model ice flow around ice divides where deviatoric stresses are small. Peltier et al. (2000) claim that the model based on the flow law (2.25) gives a height-to-width ratio in better agreement with the inferred geometry during the Last Glacial Maximum than does the model based on Glen's flow law (2.10). The comparison is somewhat qualitative, and the new rheology essentially replaces one unknown parameter, the enhancement factor at depth to account for anisotropy in crystal fabric, with another unknown parameter, grain size. Without additional observations and more stringent comparison protocols, it remains unclear whether the Goldsby–Kohlstedt law represents an improvement over Glen's law when it comes to large-scale ice-sheet modeling.

Glen's flow law (and variations on it; c.f. Hutter, 1983) applies to isotropic ice. Where the ice has developed a strong anisotropy (for example, a vertical clustering of the crystal c-axes), none of these relations adequately describes the rheological properties. Therefore, the next two sections discuss how fabric affects the deformation of glacier ice.

2.4 FABRIC EFFECTS IN GLACIER ICE

The typical creep curve shown in Figure 2.3 may reflect changes in the fabric pattern of polycrystalline ice as it is being deformed. Azuma and Higashi (1985) found that compression of an initially random fabric results in a clustering of c-axes around the compressional axis; tension causes the c-axes to rotate away from the tensile axis (Fujita et al., 1987). Similar results have been found for metallic crystals (c.f. Nicolas and Poirier, 1976). Based on geometrical considerations, Boas and Schmid (1931) showed why such crystal rotation occurs.

Referring to Figure 2.5 for the case of crystals deforming by basal glide only, lateral constraints imposed by neighboring crystals allow only deformation as shown in the figure. That is, there is no rotation of lines perpendicular to the compressive axis or parallel to the tensile axis (the basis for this assumption is not entirely clear). Under these conditions, geometry shows that the c-axis rotates toward the compressional axis and away from the tensile axis. This fabric development has important consequences for the rate of deformation.

Consider as an example an aggregate of polycrystalline ice with a random fabric; that is, all orientations of crystal c-axes are equally probable. If the sample is subjected to uniaxial compression, deformation causes the c-axes of individual crystals to rotate toward the compressive axis. The rate of rotation is dependent on the strain of the crystal, so that crystals that are oriented optimally for basal glide (at an angle of 45° to the compressive axis) are rotated the most because these crystals have the largest shear stress acting on their basal planes. However, because the rotation causes the crystals to be oriented less favorably for deformation by basal glide, the resolved basal shear stress decreases and their rate of deformation decreases. Ultimately, when all crystals have rotated toward the compressive axis and the resolved shear stress on their basal planes is almost zero, very little deformation occurs.

Van der Veen and Whillans (1994) use a numerical model to simulate the development of fabric in an ice sample subjected to stress. Their model is based on the assumptions that crystals deform by basal glide only, and that the stress is uniform throughout the aggregate. Rotation of the crystal c-axes is calculated somewhat differently than shown in Figure 2.5, but this difference is not crucial to the results. They start their model simulations with 400 crystals whose c-axes are distributed randomly. Calculated creep curves are shown in Figure 2.6.

In the simulations in which recrystallization is not included (curves labeled "No Re-X"), the bulk strain rate decreases as the sample deforms, due to the progressive rotation of crystal c-axes toward orientations unfavorable for further deformation by basal glide. This rotation causes the aggregate to become harder so that the rate of deformation of the aggregate (which is simply the sum of all crystal deformations) decreases with increasing strain. Thus, primary creep (the stage of decreasing strain

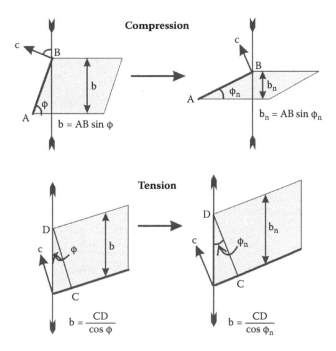

FIGURE 2.5 Geometrical basis for the Taylor–Bishop–Hill model, illustrating the rotation of a crystal c-axis. The heavy side of the crystal represents the basal plane, with the c-axis perpendicular to it. Under compression, the c-axis rotates toward the compressive axis and the new zenith angle and crystal strain are related as $\varepsilon = (b_n - b)/b = \sin\phi_n/\sin\phi - 1$. For deformation under tension, the crystal c-axis rotates away from the tensile axis, and $\varepsilon = (b_n - b)/b = \cos\phi_n/\cos\phi - 1$. (Reprinted from Van der Veen, C. J., and I. M. Whillans, *Cold Reg. Sci. Techn.*, 22, 171–195, 1994. With permission from Elsevier.)

rate; Figure 2.3) may be caused by crystal rotation rather than by the formation of dislocation tangles, or other microscopic processes suggested by Weertman (1973a) and Schulson and Duval (2009). It may be noted that after a few percent bulk strain, the c-axes distribution is almost the same as the initial random fabric. Nevertheless, the rotation of crystals since the start of deformation is sufficient to cause an appreciable decrease in strain rate.

If the sample is deformed for a sufficiently long time, the rate of deformation approaches zero. In Nature this does not happen because of recrystallization. This is an important softening process in which old, strained crystals are replaced by new and strain-free crystals. The experiments of Kamb (1972) suggest that the new crystals are formed at directions that are optimal for deformation by basal glide. In the case of uniaxial compression, this direction is at an angle of 45° to the compressive axis, and the recrystallization fabric is a girdle with a half-angle of 45° around this axis. Because the new crystals are oriented optimally for basal glide, the rate of deformation increases again after the onset of recrystallization.

Van der Veen and Whillans (1994) make the assumption that a crystal recrystallizes after it has accumulated a certain threshold strain. The resulting creep curves

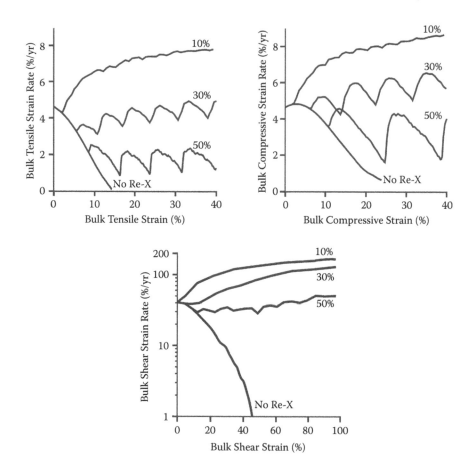

FIGURE 2.6 Modeled change in bulk strain for an aggregate of 400 crystals for uniaxial extension, uniaxial compression, and simple shear. The labels indicate the value of the threshold strain for recrystallization used in each simulation. The curves labeled "No Re-X" correspond to the simulations that do not include recrystalization and clearly show how the rate of deformation decreases with continued deformation, reaching almost zero after all crystals have rotated toward the "stiff" orientations. Note that the strain-rate scale for simple shear is logarithmic. (Reprinted from Van der Veen, C. J., and I. M. Whillans, *Cold Reg. Sci. Techn.*, 22, 171–195, 1994. With permission from Elsevier.)

(Figure 2.6) are strikingly similar to those measured in laboratory experiments. The stage of decreasing strain rate (primary creep) is followed by tertiary creep, in which recrystallization makes the ice softer and the bulk strain rate increases and continues to do so until all crystals are aligned optimally. The increase in strain rate may be approximately continuous if the value for the threshold strain for recrystallization is small, or the increase in bulk strain rate may occur in discrete steps (for larger values of the threshold strain), as suggested by Hooke and Hudleston (1980). In the latter case, periods of recrystallization alternate with periods of crystal rotation and strain hardening. Unfortunately, no laboratory experiments have been carried out to sufficiently large strains to determine whether such quasi-cyclic behavior exists or not.

The creep curves shown in Figure 2.6 support the suggestion by Budd and Jacka (1989) that the concept of steady-state secondary creep is misleading. The curves suggest that only two creep stages exist, namely, primary creep (decreasing bulk strain rate) and tertiary creep (increasing strain rate). The minimum creep rate reached during the deformation ("secondary creep") simply marks the onset of recrystallization.

There thus exists an important feedback between ice deformation and fabric development. Recrystallization causes the aggregate to become softer with respect to the applied stress, and continued deformation makes the ice much softer than ice with randomly oriented crystals. This may be the case for the basal ice in Greenland, characterized by vertical clustering of the directions of crystallographic c-axes. The dominant stress at depth is the vertical shear stress, and crystals are oriented optimally for deformation by basal glide. As a result, the shear strain rates are about three times as large as those that would occur if the ice had a random fabric (Dahl-Jensen, 1985). In the common terminology, the enhancement factor is about 3 for this ice.

Van der Veen and Whillans (1994) calculate a maximum softening factor of 2.5 for ice under uniaxial compression, and 4.3 for ice subjected to simple shear (using a nonlinear single-crystal flow law). These values are referenced to the initial strain rate. Common practice in laboratory experiments is to use the minimum strain rate as reference. Russell-Head and Budd (1979) argue that at such low strains, significant recrystallization has not yet occurred and the sample crystallography can be considered that of the initial sample. However, during the initial phase of deformation, rotation of crystals causes the ice to become gradually stiffer with respect to the applied stress. The minimum strain rate just before the onset of recrystallization does not correspond to the rate of deformation of a randomly oriented aggregate. Depending on the threshold strain for recrystallization, the difference between initial and minimum strain rate may be as much as a factor of 1.5 (Figure 2.6). Because the value of the minimum strain rate depends on the unknown value of the threshold strain, Van der Veen and Whillans (1994) chose the initial strain rate as reference. Consequently, their enhancement factors may be expected to be smaller than those given in the literature.

The softening effect of developing fabric is larger for deformation under simple shear than for uniaxial compression. This is in agreement with other studies summarized in Budd and Jacka (1989). These studies indicate that for uniaxial deformation, the total softening from the minimum strain rate to the steady-state value is about 3, while for simple shear the enhancement factor is about 8.

2.5   CREEP IN AXIALLY SYMMETRIC ICE

Continued deformation and recrystallization leads to special fabric patterns in which the crystal c-axes are clustered around a limited number of orientations. Such a fabric makes the ice anisotropic; that is, the ice may be softer with respect to certain stresses and harder with respect to others than ice with randomly oriented crystal c-axes. A common example is the single-maximum fabric that develops in ice subjected to simple shear, characterized by all crystal c-axes oriented in the direction of the vertical axis. Such ice is said to be axially symmetric, or transversely isotropic. This means that with the axis of symmetry in the z-direction, say, the ice has a different stiffness with respect to the shear stresses τ_{xz} and τ_{yz} than with respect to stresses in the other directions (τ_{xx}, τ_{yy}, or τ_{xy}). The stiffness is the same for both normal stresses, τ_{xx} and τ_{yy}.

The softening of ice with a single maximum fabric is commonly described by an enhancement factor. That is, the rate factor, A, in the constitutive relation (2.10) is multiplied by a depth-dependent enhancement factor, whose value is larger than 1. This is an expedient technique, but strictly speaking, it does not account for anisotropy of the ice. The softening of the ice must be described by a tensor. For example, an aggregate in which all c-axes are aligned vertically may be expected to be stiffer with respect to longitudinal stresses, and softer with respect to shear stresses, than isotropic ice with a random fabric. This effect cannot be described by a single enhancement factor.

A number of anisotropic constitutive relations for polar ice have been suggested. Most of these are based on results from laboratory experiments and are not yet fully supported by a mathematical theory (for example, Lile, 1978; Baker, 1981, 1982; Jacka and Budd, 1989). A theoretically more sound constitutive relation was formulated by Johnson (1977). In somewhat modified form, this relation was applied to glacier ice by Lliboutry and Andermann (1982), Pimienta and others (1987), and Van der Veen and Whillans (1990). The derivation of the last authors (based on Johnson, 1977) is summarized below.

Van der Veen and Whillans (1990) show that the isotropic constitutive relation (2.10) can also be written as

$$\dot{\varepsilon}_{ij} = \frac{\partial W}{\partial \sigma_{ij}}, \tag{2.27}$$

with

$$W = \frac{2}{n+2} A \, \tau_e^{n+1}, \tag{2.28}$$

the dissipation potential. Because the effective stress, τ_e, is a function of the second invariant of the deviatoric stress tensor (equations (2.3) and (2.7)), the generalized flow law is based on the assumption that the dissipation potential is a function of the second invariant only.

Ericksen and Rivlin (1954) discuss the requirements for the dissipation potential that applies to the case of axially symmetric ice with the axis of symmetry in the z-direction. Noting that the first invariant of the stress deviator is zero by definition, and neglecting any effects of the third invariant (as in the isotropic case), these authors show that W can only be a function of the following three quantities

$$\tau_{xx}^2 + \tau_{yy}^2 + \tau_{zz}^2 + 2\tau_{xz}^2 + 2\tau_{yz}^2 + 2\tau_{xy}^2, \tag{2.29}$$

$$\tau_{zz}, \tag{2.30}$$

and

$$\tau_{xz}^2 + \tau_{yz}^2 + \tau_{zz}^2. \tag{2.31}$$

Based on this result, Van der Veen and Whillans (1990) introduce the following three invariants

$$T_1^2 = \tau_{zz}^2, \tag{2.32}$$

$$T_2^2 = \tau_{xx}^2 + \tau_{yy}^2 + 2\tau_{xy}^2, \tag{2.33}$$

and

$$T_3^2 = \tau_{xz}^2 + \tau_{yz}^2. \tag{2.34}$$

Any combination of these three quantities can be used to construct the dissipation potential. The simplest form that is reducible to expression (2.28) for the isotropic case, is

$$W = \frac{1}{m} A G^m, \tag{2.35}$$

with

$$G = \frac{\alpha}{2} T_1^2 + \frac{\beta}{2} T_2^2 + \frac{\gamma}{2} T_3^2 = \tag{2.36}$$

$$= \frac{1}{2}\left(\alpha\tau_{zz}^2 + \beta\tau_{xx}^2 + \beta\tau_{yy}^2\right) + \beta\tau_{xy}^2 + \frac{\gamma}{2}\left(\tau_{xz}^2 + \tau_{yz}^2\right).$$

After some tedious tensor arithmetic, the following expressions for the strain-rate components can be derived:

$$\dot{\varepsilon}_{xx} = A G^{m-1}\left(\beta\tau_{xx} + \frac{1}{3}(\beta-\alpha)\tau_{zz}\right), \tag{2.37}$$

$$\dot{\varepsilon}_{yy} = A G^{m-1}\left(\beta\tau_{yy} + \frac{1}{3}(\beta-\alpha)\tau_{zz}\right), \tag{2.38}$$

$$\dot{\varepsilon}_{zz} = A G^{m-1}\left(\frac{2}{3}\alpha + \frac{1}{3}\beta\right)\tau_{zz}, \tag{2.39}$$

$$\dot{\varepsilon}_{xz} = A G^{m-1}\left(\frac{\gamma}{2}\tau_{xz}\right), \tag{2.40}$$

$$\dot{\varepsilon}_{yz} = A G^{m-1}\left(\frac{\gamma}{2}\tau_{yz}\right), \tag{2.41}$$

$$\dot{\varepsilon}_{xy} = A G^{m-1}\left(\beta\tau_{xy}\right). \tag{2.42}$$

Comparison of these expressions with the isotropic flow law (2.11) shows that the anisotropic constitutive relation reduces to Glen's law for $\alpha = \beta = 1$ and $\gamma = 2$, and the exponent $m = (n + 1)/2$.

It should be noted that the anisotropic constitutive relation proposed by Van der Veen and Whillans (1990) is only applicable to glacier ice in which the c-axes are clustered around the vertical direction. A more general theory should also include an equation describing the change with time of the axis of symmetry (i.e., the c-axes direction). More general theories have been developed by Green (1964a,b) and in a series of papers by Ericksen (Truesdell and Noll, 1965, p. 523, discuss Ericksen's theory and provide the references to the original papers). Few attempts have been made as of yet to apply these theories to glacier ice. Because vertical clustering of c-axes has been observed at many sites where the flow is dominated by vertical shear, it seems appropriate to first investigate theories for anisotropy that can explain observed patterns of flow with a given configuration of c-axes. If such theories fail to explain the observations, there is little point in taking the next step and developing complex theories that also include explicit calculation of the time-evolution of the orientation of the axis of symmetry.

3 Mechanics of Glacier Flow

3.1 FORCE BALANCE

Glaciers and ice sheets generally move very slowly, with speeds up to a few tens of meters per day at most. This suggests that in good approximation accelerations may be neglected so that Newton's second law reduces to an equilibrium of forces. Forces applied to the surface of a volume element must balance the body force (gravity) that acts on the entire volume. To arrive at the balance equations, consider a small volume element in the glacier, centered at (x, y, z) with dimensions ∂x, ∂y, and ∂z as shown in Figure 3.1. It is convenient to choose the Cartesian coordinate system such that the z-axis is vertical and positive upward.

There are three components of stress acting in the x-direction, namely the normal stress, σ_{xx}, on the (y, z) faces, and the two shear stresses, σ_{zx} and σ_{yx}, on the (x, y) and (x, z) faces, respectively. The normal stress acting on the right face of the volume element is approximately

$$\sigma_{xx}(x) + \frac{\partial \sigma_{xx}}{\partial x} \frac{\partial x}{2}, \tag{3.1}$$

and that acting on the left face is

$$\sigma_{xx}(x) - \frac{\partial \sigma_{xx}}{\partial x} \frac{\partial x}{2}. \tag{3.2}$$

The net normal force in the x-direction acting on the volume element is the difference between the force on the right face and on the left face, multiplied by the area of each face:

$$\left(\sigma_{xx}(x) + \frac{\partial \sigma_{xx}}{\partial x} \frac{\partial x}{2}\right) \partial y \, \partial z - \left(\sigma_{xx}(x) - \frac{\partial \sigma_{xx}}{\partial x} \frac{\partial x}{2}\right) \partial y \, \partial z =$$

$$= \frac{\partial \sigma_{xx}}{\partial x} \partial x \, \partial y \, \partial z. \tag{3.3}$$

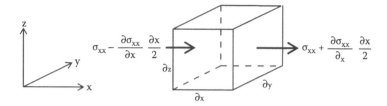

FIGURE 3.1 Normal stress acting on two opposing sides of a small volume element.

The shear stresses acting on the top and bottom faces are, respectively,

$$\sigma_{xz}(z) + \frac{\partial \sigma_{xz}}{\partial z}\frac{\partial z}{2}, \tag{3.4}$$

$$\sigma_{xz}(z) - \frac{\partial \sigma_{xz}}{\partial z}\frac{\partial z}{2}, \tag{3.5}$$

leading to a net force in the x-direction

$$\frac{\partial \sigma_{zx}}{\partial z}\partial x\,\partial y\,\partial z. \tag{3.6}$$

Similarly, the net force associated with shear stresses acting on the front and back face is given by

$$\frac{\partial \sigma_{yx}}{\partial y}\partial x\,\partial y\,\partial z. \tag{3.7}$$

With the x-axis chosen in the horizontal direction, there is no component of gravity acting in this direction. Summing the three net forces given by (3.3), (3.6), and (3.7) and dividing by the volume of the element $\partial x \cdot \partial y \cdot \partial z$, balance of forces in the x-direction is expressed by

$$\frac{\partial \sigma_{xx}}{\partial x} + \frac{\partial \sigma_{yx}}{\partial y} + \frac{\partial \sigma_{zx}}{\partial z} = 0. \tag{3.8}$$

For the other horizontal direction (the y-direction) the balance equation is similar:

$$\frac{\partial \sigma_{yx}}{\partial x} + \frac{\partial \sigma_{yy}}{\partial y} + \frac{\partial \sigma_{zy}}{\partial z} = 0. \tag{3.9}$$

The z-axis is chosen vertically, and positive upward, so that the gravitational body force acting per unit volume equals $-\rho g$, where ρ denotes the density of ice and g the acceleration due to gravity. Force balance for the vertical direction is then given by

$$\frac{\partial \sigma_{xz}}{\partial x} + \frac{\partial \sigma_{yz}}{\partial y} + \frac{\partial \sigma_{zz}}{\partial z} = \rho g. \tag{3.10}$$

The stress tensor is symmetric, and there are six independent components of stress (Section 1.2). Because there are only three force-balance equations, the stress distribution cannot be determined without additional equations or simplifying assumptions. By invoking the constitutive relation, the number of unknowns can be reduced to three, namely, the three components of velocity. In principle, this would allow the velocity distribution, and hence the stresses, to be calculated. However, because the flow law for glacier ice is nonlinear, analytical solutions can be derived for only a limited number of very simple cases. These solutions are discussed in Chapter 4.

The force-balance equations derived above describe local conditions that apply to small volumes at any location within the glacier. From the perspective of modeling glacier flow, it is often necessary to extend the analysis to apply to the entire thickness to allow identification of (potential) sites of resistance to glacier flow, such as drag at the glacier bed or lateral drag at fjord walls. Budget of forces in a section of glacier has been discussed many times before (for example, Budd, 1970a, b; Echelmeyer and Kamb, 1986; Hutter, 1983, p. 265; Kamb, 1986; Kamb and Echelmeyer, 1986a, b; Nye, 1957, 1969b; Paterson, 1994, p. 263). The usual procedure is to assume plane flow in the x-direction (so that derivatives with respect to the other horizontal y-direction are zero) and integrate the balance equations (3.8) and (3.9) over the ice thickness, subject to the boundary conditions that the upper surface must be stress free. The resulting equation is

$$\tau_b \approx -\rho g H \left| \frac{\partial h}{\partial x} \right| + 2G - T, \tag{3.11}$$

where τ_b represents the shear stress at the base of the glacier, H the ice thickness, and h the elevation of the surface. The second and third terms on the right-hand side are defined respectively as

$$G = \frac{\partial}{\partial x} \int_{h-H}^{h} \tau_{xx} \, dz, \tag{3.12}$$

and

$$T = \int_{h-H}^{h} \int_{h-H}^{z} \frac{\partial^2 \tau_{xz}}{\partial x^2} \, d\bar{z} \, dz. \tag{3.13}$$

In equation (3.12), $\tau_{xx} = (\sigma_{xx} - \sigma_{zz})/2$ represents the longitudinal stress deviator for the case of plane strain.

The balance equation (3.11) is difficult to interpret because the physical meanings of G and T are not immediately clear. It is often argued that when averaged over horizontal distances 20 times the ice thickness or more, both terms may be neglected, while on the intermediate scale, with averages taken over about four ice thicknesses, T is negligible but G may be important. On smaller scales, all terms in equation (3.11) must be taken into account. This averaging scheme is based on observations and scaling arguments (for example, Kamb and Echelmeyer, 1986b; Greve and Blatter, 2009). However, the physical significance of neglecting G and/or T is not obvious. This confusion can be avoided if resistive stresses, rather than deviatoric stresses, are used to formulate the balance equations. By using resistive stresses, the physical significance of G and T becomes more clear and the implications of neglecting either or both become more evident.

The partitioning of full stresses into lithostatic and resistive components is based on the notion that glacier flow is driven by gravity and opposed by resistive forces such as basal drag (Whillans, 1987). The lithostatic stress at some depth in the ice is equal to the weight of the ice above that depth, and horizontal gradients in this stress drive the glacier flow. The resistive stresses are the difference between full stress and the lithostatic component, and they usually impede the flow of the glacier. Thus, a separation is made between action, or gravitational forces, and reaction, or resistive forces. In geophysical applications, the resistive stress is often referred to as the *tectonic stress* (Turcotte and Schubert, 2002, p. 77). Engelder (1993, p. 10) defines tectonic stresses as "components of the in situ stress field which are a deviation from a reference state." A convenient reference state is the lithostatic stress (Turcotte and Schubert, 2002, p. 77).

The lithostatic stress is defined as the weight of ice above a level:

$$L = -\rho g (h - z).$$ (3.14)

Full stresses are now written as

$$\sigma_{ij} = R_{ij} + \delta_{ij} L,$$ (3.15)

where δ_{ij} represents the Kronecker delta ($\delta_{ij} = 1$ for $i = j$; $\delta_{ij} = 0$ for $i \neq j$) and R_{ij} denotes the components of the resistive stress tensor. As with the partitioning of full stresses in stress deviator and hydrostatic pressure (equation (1.36)), only the normal components of the full stress tensor are affected by the partitioning (3.15). To arrive at the force-balance equations in terms of resistive stresses, the analysis of Van der Veen and Whillans (1989a) is followed.

Using the separation of full stresses into resistive and lithostatic components, equation (3.8) expressing force balance in the x-direction becomes

$$\frac{\partial [R_{xx} - \rho g (h-z)]}{\partial x} + \frac{\partial R_{xy}}{\partial y} + \frac{\partial R_{xz}}{\partial z} = 0.$$ (3.16)

Integrating from the base of an ice column ($z = h - H$) to the surface ($z = h$) gives

$$\int_{h-H}^{h} \frac{\partial R_{xx}}{\partial x} \, dz - \rho g H \frac{\partial h}{\partial x} + \int_{h-H}^{h} \frac{\partial R_{xy}}{\partial y} \, dz + R_{xz}(h) - R_{xz}(h-H) = 0. \quad (3.17)$$

The order of integration and differentiation can be switched by applying Leibnitz's rule:

$$\frac{\partial}{\partial x} \int_{h-H}^{h} R_{xx} dz - R_{xx}(h) \frac{\partial h}{\partial x} + R_{xx}(h-H) \frac{\partial(h-H)}{\partial x} +$$

$$+ \frac{\partial}{\partial y} \int_{h-H}^{h} R_{xy} \, dz - R_{xy}(h) \frac{\partial h}{\partial y} + R_{xy}(h-H) \frac{\partial(h-H)}{\partial y} + \quad (3.18)$$

$$-\rho g H \frac{\partial h}{\partial x} + R_{xz}(h) - R_{xz}(h-H) = 0.$$

This equation can be greatly simplified by applying the surface boundary condition. The upper surface must be stress free; that is, the shear stress parallel to the surface must be zero. Using standard tensor transformation (for example, Jaeger, 1969), this condition reads, in terms of resistive stresses

$$R_{xx}(h) \frac{\partial h}{\partial x} + R_{xy}(h) \frac{\partial h}{\partial y} - R_{xz}(h) = 0. \quad (3.19)$$

Next, basal drag is defined to include all basal resistance

$$\tau_{bx} = R_{xz}(h-H) - R_{xx}(h-H) \frac{\partial(h-H)}{\partial x} - R_{xy}(h-H) \frac{\partial(h-H)}{\partial y}. \quad (3.20)$$

The driving stress is given by the familiar formula involving ice thickness and surface slope

$$\tau_{dx} = -\rho g H \frac{\partial h}{\partial x}. \quad (3.21)$$

The balance equation (3.18) then reduces to

$$\tau_{dx} = \tau_{bx} - \frac{\partial}{\partial x} \int_{h-H}^{h} R_{xx} \, dz - \frac{\partial}{\partial y} \int_{h-H}^{h} R_{xy} \, dz. \quad (3.22)$$

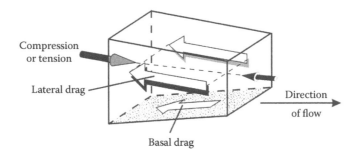

Compression
or tension

Lateral drag

Direction
of flow

Basal drag

FIGURE 3.2 Resistive stresses opposing the driving stress (not shown), directed in the direction of decreasing surface slope. Flow resistance is associated with gradients in longitudinal stress (when compression from upglacier is larger or smaller than compression from downglacier), with friction generated at the sides of the glacier, and with drag at the bed.

Similarly, force balance for the second horizontal y-direction is given by

$$\tau_{dy} = \tau_{by} - \frac{\partial}{\partial x} \int_{h-H}^{h} R_{xy}\, dz - \frac{\partial}{\partial y} \int_{h-H}^{h} R_{yy}\, dz. \tag{3.23}$$

The balance equation for the vertical direction is discussed later in Section 3.4.

Each of the terms in equations (3.22) and (3.23) has a clear physical meaning as illustrated in Figure 3.2. Considering the balance equation for the x-direction, the driving stress represents the action, making the glacier flow. In most cases, this action is resisted by drag at the glacier base (first term on the right-hand side), by the difference between the normal forces acting on the right and left faces of an ice column (second term), and by lateral drag (third term). In some cases, however, normal stresses and lateral drag may act in accordance with the driving stress. Because each term is associated with a physical process, the implications of omitting one or more terms become more clear than when the balance equation (3.11) is used.

The main advantage of using the balance equations (3.22) and (3.23) is that these equations make a clear distinction between the various sources of flow resistance. Thus, they provide a useful tool for studying the mechanics of glaciers. Where velocity measurements are available, the importance of each of the resistive terms can be assessed, using the constitutive relation to link resistive stresses to strain rates. Determining the location and magnitude of forces opposing the flow of a glacier is the first step in developing models to describe the flow of glaciers.

In retrospect, the term "resistive" stress is a somewhat unfortunate choice because, as noted, these stresses do not always offer resistance to flow. Gradients in longitudinal stress (the second term on the right-hand side of equation (3.22)) can act in cooperation with the driving stress and pull the ice forward. Similarly, slower-moving ice outboard of fast-moving ice streams and outlet glaciers can be dragged in the downflow direction (Van der Veen et al., 2009). Perhaps a more appropriate term would be *flow stress* or, following geophysical terminology, *tectonic stress*. The R_{ij}

represent the stresses acting on the glacier that are associated with glacier flow and deformation as opposed to the driving stress, which describes the action of gravity. Nevertheless, the terminology appears to have made its way into the glaciological literature (for example, Cuffey and Paterson, 2010, Section 8.2.2), and a name change at this stage would likely introduce more confusion.

3.2 INTERPRETING FORCE BALANCE

Glacier flow is driven by the downslope component of gravity. This action is described by horizontal gradients in the lithostatic stress or, equivalently, by the driving stress as given by equation (3.11) and a similar expression for the second horizontal direction. The driving stress is balanced by friction between the moving ice and the bed, by lateral drag where the glacier is laterally bounded by rock or slower-moving ice, and by gradients in longitudinal tension and/or compression. The formal derivation of the balance equations given in the preceding section may obscure the physical meaning of the various terms. Therefore, this section provides a more intuitive derivation of the depth-integrated force-balance equation based on the geometric approach of Van der Veen and Payne (2003).

Figure 3.3 shows the lithostatic stresses acting on a section of glacier extending from the bed to the surface. Summing the gravitational forces acting on the column gives the component of driving stress in the x-direction. At each depth in the column, the lithostatic stress equals the weight of the ice above and, neglecting low-density firn near the surface, the lithostatic stress increases linearly with depth:

$$L(z) = -\rho g (h - z), \tag{3.24}$$

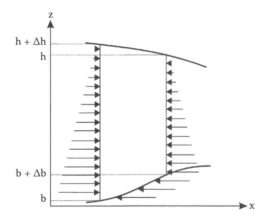

FIGURE 3.3 Vertical column through a glacier showing lithostatic stresses. Only those acting on the column are shown and summed to give the driving stress. (Reprinted from Van der Veen, C. J., and A. J. Payne, in *Mass Balance of the Cryosphere: Observations and Modeling of Contemporary and Future Changes*, Cambridge University Press, Cambridge, 2003. With permission of Cambridge University Press.)

where $z = h$ represents the surface elevation; $z = b$ denotes the elevation of the glacier bed. Integrating over the full depth gives the total lithostatic force acting on the left-hand face of the column:

$$\int_{b}^{h} \rho g(h - z)\, dz = \frac{1}{2} \rho g(h - b)^2. \tag{3.25}$$

At the right-hand side of the column, the surface and bed elevations are $(h + \Delta h)$ and $(b + \Delta b)$, respectively, and the force acting on that face is

$$\int_{b+\Delta b}^{h+\Delta h} \rho g(h - z)\, dz = \frac{1}{2} \rho g[(h + \Delta h) - (b + \Delta b)]^2. \tag{3.26}$$

Note that forces acting in the x-direction are taken positive so that the force acting on the right side is negative. To evaluate the lithostatic stress at the bed, the average column thickness may be used. This stress acts on a vertical step in the glacier sole with height Δb, giving the total lithostatic force at the bed:

$$-\rho g\left[\left(h + \frac{\Delta h}{2}\right) - \left(b + \frac{\Delta b}{2}\right)\right] \Delta b. \tag{3.27}$$

Summing the three forces acting on the column gives the net lithostatic force:

$$-\frac{1}{2} \rho g\left[\left(h + \frac{\Delta h}{2}\right) - \left(b + \frac{\Delta b}{2}\right)\right] \Delta h. \tag{3.28}$$

Denoting the average thickness by $\bar{H}$ and dividing by the length of the column, Δx, gives the driving force per unit map area,

$$-\rho g\bar{H} \frac{\Delta h}{\Delta x}. \tag{3.29}$$

Taking the limit $\Delta x \to 0$ yields the driving stress

$$\tau_{dx} = -\rho g H \frac{\partial h}{\partial x}. \tag{3.30}$$

A similar expression can be derived for the driving stress corresponding to the other horizontal direction.

The derivation of driving stress above differs from the conventional approach in which stresses in a plane sloping slab are considered (for example, Paterson, 1994, p. 240; Greve and Blatter, 2009, p. 83) and lamellar flow is assumed in which the driving stress is balanced by drag at the glacier base. From these definitions, it is not clear how the definition of driving stress can be extended to more complex geometries or to ice shelves on which basal drag is nonexistent. In the present approach, the term *driving stress* is reserved for the action due to gravity and it is given by equation (3.30) for any glacier geometry, irrespective of which resistive stress opposes this gravitational action.

Note that the slope of the base of the column is irrelevant for calculating the driving stress, and the lithostatic stress acting at the bed cancels out when adding the three forces acting on the column. While this may appear to be counterintuitive, why the bed slope does not enter into the definition of driving stress can be understood by considering a fluid of constant density confined in a cup. There is no tendency of the fluid to flow to the middle of the cup, despite the slope of the bottom of the cup. A nonzero surface slope is needed to drive the flow. The implication is that glaciers can flow "uphill," that is, against a reversed bed slope, provided the surface elevation decreases in the downflow direction. One could argue that for mountain glaciers on a steep slope, it is the bed slope that determines the driving stress. However, for such glaciers, the ice thickness generally is more or less constant in the flow direction and the surface slope is approximately equal to the bed slope. Equation (3.30), which involves only the surface slope, remains valid.

Resistive stresses are associated with glacier flow and, generally, act to oppose the driving stress. Consider first the horizontal normal stress, R_{xx}, acting on the vertical column, as shown in Figure 3.4. How this stress varies with depth is generally not known. Note that the contribution of the normal stress acting on the sloping bed is included in the definition of basal drag (equation (3.20)) and need not be considered

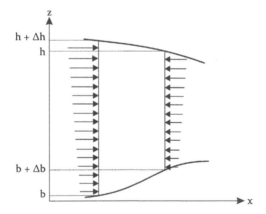

FIGURE 3.4 Horizontal resistive stresses acting on a vertical column through a glacier. The difference between the net force acting on the right and left faces gives the resistance to flow associated with longitudinal tension or compression. Note that the depth variation of these horizontal stresses is generally unknown. (Reprinted from Van der Veen, C. J., and A. J. Payne, in *Mass Balance of the Cryosphere: Observations and Modeling of Contemporary and Future Changes*, Cambridge University Press, Cambridge, 2003. With permission of Cambridge University Press.)

at this point. The longitudinal resistive force acting on the left face is obtained by integrating the resistive stress over the ice thickness:

$$\int_b^h R_{xx}(z)\,dz = H\bar{R}_{xx},$$
(3.31)

where the overbar denotes the depth-averaged resistive stress. This force acts in the x-direction and is taken positive; the longitudinal force on the right face is oriented in the negative x-direction and therefore negative. Assuming the resistive stress on the right face differs from that on the left face by an amount ΔR_{xx}, integrating over the ice thickness gives

$$-\int_{b+\Delta b}^{h+\Delta h}\left[R_{xx}(z)+\Delta R_{xx}(z)\right]dz =$$

$$= -H\bar{R}_{xx} - \Delta H\,\bar{R}_{xx} - H\,\Delta\bar{R}_{xx} - \Delta H\,\Delta\bar{R}_{xx}.$$
(3.32)

Here $\Delta H = \Delta h - \Delta b$ is the difference in ice thickness at the two faces. Summing equations (3.31) and (3.32) and neglecting the higher-order term in the latter equation gives the net longitudinal force acting on the column

$$-\left[\frac{\Delta H}{\Delta x}\bar{R}_{xx} + H\frac{\Delta\bar{R}_{xx}}{\Delta x}\right]\Delta x.$$
(3.33)

Dividing by the length of the column gives the force per unit map area, and after taking the limit $\Delta x \to 0$, this net force is

$$-\frac{\partial H}{\partial x}\bar{R}_{xx} - H\frac{\partial\bar{R}_{xx}}{\partial x} = -\frac{\partial}{\partial x}(H\bar{R}_{xx}).$$
(3.34)

Expression (3.34) shows that resistance to flow is offered by *gradients* in resistive stress, rather than by the *magnitude* of the stress. This is readily understood by considering two people pushing on opposite ends of a large object; if both push equally hard, the object will not move because the net force is zero. Similarly, on glaciers, the longitudinal stress itself may be large compared with the driving stress, but the effect on force balance may be small when the gradients are much smaller than the driving stress. Note that gradients in longitudinal stress can oppose the driving stress but can also act in cooperation with the driving stress, where the ice is being pushed from upglacier or pulled from downglacier.

Where a glacier or ice stream is bounded by rock walls or slower-moving ice, lateral drag provides resistance to flow. This frictional resistance acts on vertical faces parallel to the x-direction and, following a derivation similar to that for the longitudinal force, is given by

$$-\frac{\partial}{\partial y}(H\bar{R}_{xy}),$$
(3.35)

where the horizontal y-axis is perpendicular to the x-axis.

The third source of resistance to flow is drag at the glacier bed, as expressed by basal drag, τ_{bx}. This drag may be due to frictional forces (skin drag) between the moving ice and the bed, and resistive stresses acting on a sloping bed (form drag). Formally, basal drag is defined as

$$\tau_{bx} = R_{xz}(b) - R_{xx}(b)\frac{\partial b}{\partial x} - R_{xy}(b)\frac{\partial b}{\partial y}. \tag{3.36}$$

The first term on the right-hand side describes skin friction and is associated with vertical shear in velocity at the glacier base.

Balance of forces requires the driving stress to be balanced by resistance to flow offered by gradients in longitudinal stress, lateral drag, and basal drag. Thus

$$\tau_{dx} = \tau_{bx} - \frac{\partial}{\partial x}(H\bar{R}_{xx}) - \frac{\partial}{\partial y}(H\bar{R}_{xy}). \tag{3.37}$$

A similar equation can be derived for the balance of forces in the other horizontal direction. These balance equations form the basis of the force-budget technique discussed in the next section.

3.3   THE FORCE-BUDGET TECHNIQUE

Neither the formal derivation in Section 3.1 nor the more intuitive derivation in Section 3.2 makes any assumption about glacier geometry or relative importance of resistance to flow, and the resulting force-balance equations apply to any glacier geometry and flow regime. A common practice is to apply scaling methods to these equations to determine the dominant terms in the balance equations (for example, Greve and Blatter, 2009). An alternative procedure is to determine the budget of forces where measurements of glacier geometry and (surface) velocity are available. The driving stress is calculated from the geometry (thickness and surface slope), while lateral drag and resistance associated with gradients in longitudinal stress can be estimated from measured velocities. Basal drag cannot be calculated directly; rather, this quantity is estimated from the force-balance equation as the value needed for all forces to sum up to zero.

Where surface velocity measurements are available, surface strain rates can be evaluated from the horizontal gradients in velocity. To relate these strain rates to stresses, the constitutive relation needs to be invoked. As discussed in Section 2.2, the constitutive relation links strain rates to deviatoric stresses. This means that to apply the constitutive relation to link resistive stresses to strain rates, it is necessary to write the resistive stresses in terms of stress deviators first. Comparing the two schemes for separating full stresses

$$\sigma_{ij} = \tau_{ij} + \delta_{ij}P, \tag{3.38}$$

and

$$\sigma_{ij} = R_{ij} + \delta_{ij}L, \tag{3.39}$$

where $P = (\sigma_{xx} + \sigma_{yy} + \sigma_{zz})/3$ represents the spherical stress. Eliminating the full stress from these equations gives

$$\tau_{ij} + \delta_{ij}(P - L) = R_{ij}. \tag{3.40}$$

Taking $i = j = z$ yields

$$P - L = -\tau_{zz} + R_{zz} =$$
$$= \tau_{xx} + \tau_{yy} + R_{zz}, \tag{3.41}$$

because, by definition, the three normal deviatoric stresses sum to zero. The resistive stresses can now be written in terms of deviatoric stresses as follows:

$$R_{xx} = 2\tau_{xx} + \tau_{yy} + R_{zz}, \tag{3.42}$$

$$R_{yy} = 2\tau_{yy} + \tau_{xx} + R_{zz}, \tag{3.43}$$

$$R_{xy} = \tau_{yy}, \tag{3.44}$$

$$R_{xz} = \tau_{xz}, \tag{3.45}$$

$$R_{yz} = \tau_{yz}. \tag{3.46}$$

Applying the constitutive relation (2.12), the horizontal resistive stresses can be expressed in terms of strain rates:

$$R_{ij} = B\dot{\varepsilon}_e^{1/n-1}\dot{\varepsilon}_{ij}, \qquad i \neq j = x, y, z, \tag{3.47}$$

$$R_{ii} = B\dot{\varepsilon}_e^{1/n-1}(2\dot{\varepsilon}_{ii} + \dot{\varepsilon}_{jj}) + R_{zz}, \quad i = j = x, y. \tag{3.48}$$

These expressions contain the vertical resistive stress, R_{zz}. This stress can be calculated from consideration of the vertical force balance. This is discussed further in Section 3.4, but for now, it is sufficient to note that this stress is zero in most applications.

For the case in which the surface strain rates can be taken representative of those at depth, the theory for the force-budget technique is now complete. Measured surface velocities are used to determine strain rates, from which the resistive stresses at the surface are calculated (using equations (3.47) and (3.48), and neglecting the vertical resistive stress, R_{zz}). Taking these stresses constant through the ice thickness, the last two terms in the horizontal balance equations (3.22) and (3.23) can be estimated and, with the driving stress calculated from the glacier geometry, the remaining term, basal drag, follows from the requirement that the sum of all forces must be zero.

This so-called isothermal block-flow model has the great merit of being simple to carry out. Its drawback is that the viscous terms in the balance equations (the last two terms in (3.22) and (3.23)) tend to be overestimated because surface values are applied to the entire glacier thickness. Thus, the inferred value for basal drag represents a limiting value. The other extreme value for basal drag can be found by setting the viscous terms equal to zero; that is, by equating basal drag to the driving stress. The actual value for basal drag may be expected to fall in between these two extremes. An example illustrating the steps involved in the force-budget technique is discussed in Section 11.2.

The assumption of isothermal block flow need not be made, and Van der Veen and Whillans (1989a) describe a solution scheme that solves for the three components of velocity throughout a section of glacier (c.f. Van der Veen, 1989; Van der Veen, 1999b, Section 3.4). Calculations start at the glacier surface, where measured velocities are prescribed, and steadily progress downward by solving force balance for successive ice layers. The main difficulty with this downward calculation is that it is an inverse technique. That is, observed surface effects (surface velocities and surface slope) are used to determine the basal conditions that are causing these effects. However, as discussed in Section 4.6, the ice acts as a filter and the amplitude of horizontal variations decreases upward. This means that surface features reflect basal conditions, but the variations at the surface are much smaller than those at the glacier bed. Also, the extent of glacial filtering is strongly dependent on the wavelength of the feature and a basal feature with a horizontal scale of, say, one ice thickness may be difficult to detect at the surface, if at all. By using the downward calculation scheme, the filter works in the opposite way, and surface effects are amplified as the calculation proceeds downward. In particular, the small-scale surface features are amplified most. Typically, measured surface velocities contain relatively small errors that are amplified in the downward force-budget calculation. A more practical approach is to use a full-stress solver and find a solution that minimizes the difference between calculated and measured surface velocities in a least-squares sense.

3.4 BRIDGING EFFECTS

Where the weight of an ice column is fully supported by the bed underneath, the vertical normal stress at a depth $(z - h)$ below the ice surface is equal to the weight of the ice above. That is

$$\sigma_{zz} = -\rho g(h - z), \tag{3.49}$$

and the vertical resistive stress, R_{zz}, is zero (equations (3.14) and (3.15)). For most applications, this is a valid approximation. Over horizontal distances that are short compared with the ice thickness, differences from lithostatic may become important. For example, in the lee of a subglacial hill, the basal ice may become separated from the bed if cavitation occurs. The cavity does not support the weight of the ice, leading to shear-stress gradients that effectively transfer the weight to surrounding areas where the ice is in contact with the bed (Section 7.2). This is similar to a bridge (hence the term *bridging effects*). The span of the bridge is not supported from below so that the full vertical normal stress is zero there and R_{zz} equals minus the lithostatic stress. The abutments carry the weight of the entire bridge, and the vertical normal stress is larger than the weight of the material directly above (Figure 3.5).

FIGURE 3.5 Illustrating bridging effects. The weight of the span of the bridge is not supported from below, but transferred to the abutments through variations in the shear stress; the abutments adjacent to the bridge support more than the weight of material directly above. (From Van der Veen, C. J., and I. M. Whillans, *J. Glaciol.,* 35, 53–60, 1989a. Reprinted from the *Journal of Glaciology* with permission of the International Glaciological Society and the authors.)

Force balance in the vertical direction is described by equation (3.10). Substituting the partitioning of full stresses into lithostatic and resistive components, this equation becomes

$$\frac{\partial R_{xz}}{\partial x} + \frac{\partial R_{yz}}{\partial y} + \frac{\partial R_{zz}}{\partial z} = 0. \tag{3.50}$$

Neglecting bridging effects ($R_{zz} \approx 0$) is thus equivalent to neglecting horizontal gradients in the vertical shear stresses, R_{xz} and R_{yz}. To estimate whether this approximation is reasonable, the balance equation is integrated over the full ice thickness to give

$$R_{zz}(b) = \int_b^h \frac{\partial R_{xz}}{\partial x} \, dz = \frac{\partial}{\partial x} \int_b^h R_{xz} \, dz + R_{xz}(b)\frac{\partial b}{\partial x}, \tag{3.51}$$

where $R_{zz}(h)$ and $R_{xz}(h)$ have been set to zero. The second term in equation (3.50) involving the transverse direction is omitted for brevity, but this has no effect on the following order-of-magnitude estimate.

In the first approximation, the vertical shear stress increases linearly with depth from zero at the surface to a maximum value equal to basal drag at the glacier base (c.f. Section 4.2). That is

$$R_{xz}(z) = \frac{h - z}{H} \, \tau_{bx}. \tag{3.52}$$

Substituting in equation (3.51) gives

$$R_{zz}(b) = \frac{1}{2}H\frac{\partial \tau_{bx}}{\partial x} + \tau_{bx}\frac{\partial b}{\partial x}. \tag{3.53}$$

The first term on the right-hand side may become important in the vicinity of *sticky spots* or subglacial lakes, while the second term may be significant where pronounced basal relief exists. Consider first a sticky spot with a length of 5 km and basal drag equal to 150 kPa (admittedly a high value; see Figure 4.3), surrounded by near-frictionless sediments. For a 1 km thick glacier, this gives

$$\frac{1}{2}H\frac{\partial \tau_{bx}}{\partial x} \approx \frac{1000}{2}\frac{150}{5000} = 15 \text{ kPa}. \tag{3.54}$$

Compared to the lithostatic stress at the bed, $L(b) = \rho gH \approx 9000$ kPa, the vertical resistive stress associated with spatial variations in basal drag is insignificant. Similarly, even if the bed slope is as large as 100 m/km, the associated vertical resistive stress is only 15 kPa for a basal drag equal to 150 kPa. This estimate of the magnitude of R_{zz} indicates that this stress generally will be small and may be set to zero when considering force balance or modeling glacier flow.

3.5   STOKES EQUATION APPLIED TO GLACIER FLOW

In fluid dynamics, flow of a viscous fluid is described by the Navier–Stokes equations (for example, Chorin and Marsden, 1992, p. 33; Pedlosky, 1982, p. 173). These equations express balance of forces in terms of velocity gradients and form the basis for many finite-element solvers. Given the increased application of finite-element routines to glaciological problems, this section illustrates how the formulation commonly used in fluid mechanics is equivalent to the balance equations (3.8)–(3.10).

For an incompressible fluid, the Navier–Stokes equations can be written in vector form as

$$-\rho\frac{\partial \vec{u}}{\partial t} - \rho(\vec{u}\cdot\nabla)\vec{u} = \nabla P - \nabla\cdot[\eta(\nabla\vec{u} + (\nabla\vec{u})^T)] - \vec{F}, \tag{3.55}$$

where ρ represents density, $\vec{u} = (u, v, w)$ the velocity vector, P the pressure, η the dynamic viscosity, and $\vec{F}$ the volume or body force. The superscript T indicates the transpose of the velocity gradient tensor. For a 2×2 tensor with components a_{ij}, the transpose is defined as the tensor with components a_{ji}. Derivations of the Navier–Stokes equations for a viscous incompressible fluid can be found in Brown (1991, Section 6.3.3) and Greve and Blatter (2009, Section 5.1.1).

The terms on the left-hand side of equation (3.55) represent acceleration

$$\frac{d\vec{u}}{dt} = \frac{\partial \vec{u}}{\partial t} + \vec{u}\cdot\nabla\vec{u}. \tag{3.56}$$

For glacier flow, accelerations may be neglected and the basic equation becomes

$$\nabla P - \nabla \cdot [\eta(\nabla \vec{u} + (\nabla \vec{u})^T] = \vec{F}. \tag{3.57}$$

This simplification of the Navier–Stokes equation is referred to as the Stokes equation for an incompressible fluid.

The body force for glacier flow is gravity acting in the vertical z-direction. Then

$$F_z = -\rho g, \tag{3.58}$$

while the components in the x- and y-directions are zero. In this expression, $\rho = 920 \, \text{kg/m}^3$ represents the density of ice, and $g = 9.8 \, \text{m/s}^2$ the gravitational acceleration.

The above form (3.57) of the Stokes equation is in compact vector notation. To make the correspondence to the equations commonly used in glaciology more clear, the three corresponding equations (for each of the three orthogonal directions) are considered:

$$\frac{\partial P}{\partial x} - \frac{\partial}{\partial x}\left(2\eta\frac{\partial u}{\partial x}\right) - \frac{\partial}{\partial y}\left(\eta\left(\frac{\partial u}{\partial y} + \frac{\partial v}{\partial x}\right)\right) - \frac{\partial}{\partial z}\left(\eta\left(\frac{\partial u}{\partial z} + \frac{\partial w}{\partial x}\right)\right) = 0, \tag{3.59}$$

$$\frac{\partial P}{\partial y} - \frac{\partial}{\partial x}\left(\eta\left(\frac{\partial v}{\partial x} + \frac{\partial u}{\partial y}\right)\right) - \frac{\partial}{\partial y}\left(2\eta\frac{\partial v}{\partial y}\right) - \frac{\partial}{\partial z}\left(\eta\left(\frac{\partial v}{\partial z} + \frac{\partial w}{\partial z}\right)\right) = 0, \tag{3.60}$$

$$\frac{\partial P}{\partial z} - \frac{\partial}{\partial x}\left(\eta\left(\frac{\partial w}{\partial x} + \frac{\partial u}{\partial z}\right)\right) - \frac{\partial}{\partial y}\left(\eta\left(\frac{\partial w}{\partial y} + \frac{\partial v}{\partial z}\right)\right) - \frac{\partial}{\partial z}\left(2\eta\frac{\partial w}{\partial z}\right) = -\rho g. \tag{3.61}$$

To rewrite the Stokes equation applied to glacier flow in the more common force-balance form, the flow law for glacier ice needs to be considered. This relation links deviatoric stresses to strain rates (or velocity gradients) as (Section 2.2)

$$\tau_{ij} = B\dot{\varepsilon}_e^{1/n-1} \dot{\varepsilon}_{ij}. \tag{3.62}$$

In this expression, the effective strain rate is

$$\dot{\varepsilon}_e = \left[\frac{1}{2}\left(\dot{\varepsilon}_{xx}^2 + \dot{\varepsilon}_{yy}^2 + \dot{\varepsilon}_{zz}^2\right) + \left(\dot{\varepsilon}_{xy}^2 + \dot{\varepsilon}_{xz}^2 + \dot{\varepsilon}_{yz}^2\right)\right]^{1/2}. \tag{3.63}$$

Defining the effective viscosity as

$$\eta = \frac{1}{2}B\dot{\varepsilon}_e^{1/n-1}, \tag{3.64}$$

the flow law can be simplified to

$$\tau_{ij} = 2\eta\,\dot{\varepsilon}_{ij}. \tag{3.65}$$

In particular,

$$\tau_{xx} = 2\eta \frac{\partial u}{\partial x}, \qquad (3.66)$$

$$\tau_{xy} = \eta \left(\frac{\partial u}{\partial y} + \frac{\partial v}{\partial x} \right), \qquad (3.67)$$

$$\tau_{xz} = \eta \left(\frac{\partial u}{\partial z} + \frac{\partial w}{\partial x} \right). \qquad (3.68)$$

The Stokes equation (3.59) for the x-direction can then be written as

$$\frac{\partial P}{\partial x} - \frac{\partial}{\partial x}(\tau_{xx}) - \frac{\partial}{\partial y}(\tau_{xy}) - \frac{\partial}{\partial z}(\tau_{xz}) = 0. \qquad (3.69)$$

In fluid dynamics, the pressure, P, is defined as

$$P = - (\sigma_{xx} + \sigma_{yy} + \sigma_{zz})/3. \qquad (3.70)$$

Note that this definition has the opposite sign from that commonly used in glaciology! Then, deviatoric stresses, τ_{ij}, are linked to full stresses, σ_{ij}, as

$$\tau_{ij} = \sigma_{ij} + \delta_{ij} P, \qquad (3.71)$$

and equation (3.69) becomes

$$- \frac{\partial}{\partial x}(\sigma_{xx}) - \frac{\partial}{\partial y}(\sigma_{xy}) - \frac{\partial}{\partial z}(\sigma_{xz}) = 0. \qquad (3.72)$$

This equation is the same as the force-balance equation (3.8) that serves as the starting equation for modeling glacier flow,

$$\frac{\partial}{\partial x}(\sigma_{xx}) + \frac{\partial}{\partial y}(\sigma_{xy}) + \frac{\partial}{\partial z}(\sigma_{xz}) = 0. \qquad (3.73)$$

The Stokes equations do not involve assumptions about relative magnitudes and can be solved using iterative procedures after supplementing with appropriate boundary conditions. Examples of studies applying this method include Johnson and Staiger (2007), Gagliardini et al. (2007), Gagliardini and Zwinger (2007), and Durand et al. (2009b). In these studies, finite-element codes are used to solve equation (3.55) by numerical iteration. Typically, plane flow along a glacier flowline is considered. Isothermal conditions may be assumed or the temperature throughout the glacier is calculated by simultaneously solving the thermodynamic equation. Anyone who has tried this knows that key to finding a correct or physically plausible solution is to impose appropriate boundary conditions at the four boundaries of the model domain (ice surface, ice/bed interface, and the upglacier and downglacier ends of the flowline).

The upper surface must be stress free, and both the shear stress parallel to the surface and the normal stress perpendicular to the surface must be zero. Equation (3.19) expresses the first of these conditions. The second condition enters when considering balance of forces in the vertical direction and is approximated as $R_{zz}(h) = 0$ in Section 3.4. In vector notation, the requirement for a stress-free surface is written as

$$\vec{n}_s \cdot (\vec{\sigma}_s \cdot \vec{n}_s) = 0, \tag{3.74}$$

(Durand et al., 2009b), or, equivalently,

$$P_s \vec{I} - \left[\eta \left(\nabla \vec{u}_s + (\nabla \vec{u}_s)^T \right) \right] = 0, \tag{3.75}$$

(Johnson and Staiger, 2007). In these equations, the subscript s refers to surface values and $\vec{n}_s$ represents the unit vector outward and perpendicular to the ice surface; $\vec{\sigma}$ represents the full stress.

The basal boundary condition depends on what type of glacier is being modeled. For a grounded glacier frozen to the bed, the following no-slip condition is prescribed (Johnson and Staiger, 2007):

$$\vec{u}_b = 0. \tag{3.76}$$

On a grounded glacier subject to basal sliding, a sliding relation linking basal velocity to basal drag must be imposed. For example, Durand and others (2009b) use

$$\tau_b = \vec{t}_b \cdot (\vec{\sigma}_b \cdot \vec{n}_b) = CU_b^m, \tag{3.77}$$

and

$$\vec{u}_b \cdot \vec{n}_b = 0, \tag{3.78}$$

where $\vec{t}_b$ is the unit vector tangent to the bed, such that $\vec{n}_b \times \vec{t}_b = 0$. The second basal requirement states that the sliding velocity must be parallel to the bed. For the case of floating ice shelves, basal drag must vanish, and

$$\tau_b = \vec{t}_b \cdot (\vec{\sigma}_b \cdot \vec{n}_b) = 0. \tag{3.79}$$

Ignoring the effect of long-period ocean waves (c.f. Sergienko, 2010), the normal stress at the base of the ice shelf equals the water pressure, and

$$\vec{\sigma}_b \cdot \vec{n}_b = -\vec{n}_b (\rho g H). \tag{3.80}$$

For a flowline originating at the ice divide, the upglacier boundary conditions follow from the requirement of zero mass flux,

$$\vec{u}_d \cdot \vec{n}_d = 0, \tag{3.81}$$

and vanishing shear stress

$$\vec{t}_d \left[P_d \vec{I} - \left[\eta (\nabla \vec{u}_d + (\nabla \vec{u}_d)^T) \right] \right] = 0, \tag{3.82}$$

(Johnson and Staiger, 2007). In these expressions, the subscript d refers to the (usually) vertical ice divide boundary and $\vec{n}_d$ is the unit vector perpendicular to this boundary, while $\vec{t}_d$ is the unit vector tangent to the divide boundary.

Where the lower end of the model domain coincides with the floating calving front, the normal stress is prescribed in terms of sea-water pressure as

$$\vec{\sigma}_c \cdot \vec{n}_c = -\vec{n}_{cb} P_w, \tag{3.83}$$

where

$$P_w = \begin{cases} 0 & \text{if} \quad \dfrac{\rho}{\rho_w} H \leq z < H \\[3mm] \rho_w g \left(\dfrac{\rho}{\rho_w} H - z \right) & \text{if} \quad 0 \leq z < \dfrac{\rho}{\rho_w} H \end{cases} \tag{3.84}$$

with $z = 0$ corresponding to sea level (Sergienko, 2010).

For applications in which the modeled flowline does not extend from the divide to the calving front, formulating boundary conditions for the up- and downglacier boundaries is less straightforward. Because these boundary conditions will depend on the particular geometry being modeled, no general expressions can be given here.

3.6   CREEP CLOSURE OF ENGLACIAL TUNNELS

The balance equations discussed in this chapter, together with the constitutive relation discussed in Chapter 2, form the basis for models of glacier flow. In most instances, simplifying assumptions are invoked to allow an analytical solution to be found. Examples are discussed in following chapters, and in particular Chapter 5. For most of these applications, a Cartesian coordinate system is used with the three coordinate axes mutually perpendicular. However, in some situations it is more convenient to use cylindrical coordinates (r, θ, z), with the coordinates of a point defined by the radius, r, the azimuth angle, θ, and elevation, z (Figure 3.6). This system is most suitable if the problem being modeled possesses rotational symmetry around the z-axis, such as an axi-symmetric ice sheet (discussed in Section 5.4). In this section, another example of the use of cylindrical coordinates in solving for the stress distribution is considered, namely, creep closure of an englacial drainage tunnel. Such tunnels are maintained by a balance between the rate at which the tunnels close due to creep of the overlying ice, and melting of the tunnel walls by

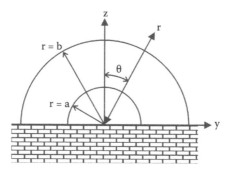

FIGURE 3.6 Geometry used in the determination of the creep closure rate of a semicircular tunnel. The radius of the tunnel is a; at sufficiently large distances from the tunnel (r = b >> a), the flow of the glacier is unperturbed by the tunnel and, in the x-direction, perpendicular to the plane of drawing.

the water flowing through them (c.f. Section 7.5). Tunnel closure was first discussed by Nye (1953) in connection with estimating the flow-law parameters from the observed rate of borehole closure. The analysis given below is somewhat different from that of Nye (1953) and demonstrates that even for a comparatively well-defined and simple problem, deriving an analytical solution may become rather involved.

To allow an analytical solution to be found, the shape of the englacial tunnel, incised in the basal ice, is assumed to be semicircular, with radius a as shown in Figure 3.6. The axis of the tunnel is taken to follow the main direction of ice flow and extending infinitely far in the x-direction, so that the problem becomes essentially two-dimensional. Additionally, there is no ice flow in the transverse y-direction, other than localized creep into the tunnel.

The Cartesian and cylindrical coordinates are related as

$$y = r \cos\theta,$$
$$z = r \sin\theta. \tag{3.85}$$

The components of the stress tensor in the polar coordinates are found by using standard tensor transformation formulas

$$\sigma_{rr} = \frac{\sigma_{zz} + \sigma_{yy}}{2} + \frac{\sigma_{zz} - \sigma_{yy}}{2}\cos 2\theta + \sigma_{yz}\sin 2\theta, \tag{3.86}$$

$$\sigma_{\theta\theta} = \frac{\sigma_{zz} + \sigma_{yy}}{2} - \frac{\sigma_{zz} - \sigma_{yy}}{2}\cos 2\theta - \sigma_{yz}\sin 2\theta, \tag{3.87}$$

$$\sigma_{r\theta} = \sigma_{yz}\cos 2\theta - \frac{\sigma_{zz} - \sigma_{yy}}{2}\sin 2\theta. \tag{3.88}$$

The equations describing force balance in polar coordinates can be derived by considering the forces acting on a small element within the glacier (for example, Xu, 1992), similar to the derivation in Section 3.1. Summing all forces acting in the radial direction and setting the result equal to zero (because accelerations may be neglected) gives

$$\frac{\partial \sigma_{rr}}{\partial r} + \frac{1}{r}\frac{\partial \sigma_{r\theta}}{\partial \theta} + \frac{\sigma_{rr} - \sigma_{\theta\theta}}{r} + F_r = 0. \tag{3.89}$$

Similarly, for the tangential direction, force balance is given by

$$\frac{1}{r}\frac{\partial \sigma_{\theta\theta}}{\partial \theta} + \frac{\partial \sigma_{r\theta}}{\partial \theta} + \frac{2\sigma_{r\theta}}{r} + F_\theta = 0. \tag{3.90}$$

In these expressions, F_r and F_θ represent the components of the body force in the radial and tangential directions, respectively. The only body force acting on the glacier is gravity, $-\rho g$, directed vertically downward. Thus,

$$F_r = -\rho g \cos\theta,$$
$$F_\theta = +\rho g \sin\theta. \tag{3.91}$$

The stress solution to be derived must satisfy boundary conditions. At the wall of the tunnel ($r = a$) the radial normal stress must balance the water pressure within the tunnel, P_w, while the shear stress along the tunnel wall may be set to zero. Thus

$$\sigma_{rr}(a) = -P_w, \tag{3.92}$$

$$\sigma_{r\theta}(a) = 0. \tag{3.93}$$

The tunnel is assumed to be small compared with the dimensions of the glacier, and it may be expected that the tunnel only affects stresses within a circular area surrounding its walls. In other words, at radial distances much larger than the radius of the tunnel ($r \gg a$), the stress solution should be unaltered by the presence of the tunnel.

The undisturbed stress solution (referenced to the Cartesian coordinate system) can be found as follows. From the assumption that the flow is independent of the along-flow x-direction, it follows that the stress deviator in this direction must be zero ($\tau_{xx} = 0$). Similarly, the stress deviator in the transverse direction must be zero ($\tau_{yy} = 0$). Applying these conditions, and using the definition of stress deviator, the two normal-stress components in both horizontal directions are found to be

$$\sigma_{xx} = \sigma_{zz},$$
$$\sigma_{yy} = \sigma_{zz}. \tag{3.94}$$

For the present discussion, σ_{xx} need not be considered. Neglecting bridging effects, the vertical normal stress is simply the weight of the ice above; thus,

$$\sigma_{zz} = -\rho g(H - z), \tag{3.95}$$

if $z = 0$ is chosen at the base of the glacier, and H represents the thickness of the overlying ice. The undisturbed shear stress, σ_{yz}, is zero and applying the transformation formulas (3.86)–(3.88) gives the undisturbed stresses referenced to the cylindrical coordinates

$$\sigma_{rr}(r) = -P_i + \rho g r \cos\theta, \tag{3.96}$$

$$\sigma_{\theta\theta}(r) = -P_i + \rho g r \cos\theta, \tag{3.97}$$

$$\sigma_{r\theta}(r) = 0, \tag{3.98}$$

with $P_i = \rho g H$ the ice overburden pressure.

The next step is to reformulate the flow law for strain rates and stresses referenced to cylindrical coordinates. In Cartesian coordinates, strain rates are linked to deviatoric stresses using the conventional flow law (Section 2.2)

$$\dot{\varepsilon}_{ij} = A\tau_e^{n-2}\tau_{ij}, \tag{3.99}$$

with A the rate factor and τ_e the effective stress. For the geometry under consideration, the effective stress is

$$\tau_e^2 = \tau_{zz}^2 + \tau_{yz}^2 =$$

$$= \frac{1}{4}(\sigma_{zz} - \sigma_{yy})^2 + \sigma_{yz}^2, \tag{3.100}$$

or, in terms of stresses referred to the cylindrical coordinates

$$\tau_e^2 = \frac{1}{4}(\sigma_{rr} - \sigma_{\theta\theta})^2 + \sigma_{r\theta}^2. \tag{3.101}$$

The strain-rate tensor is subject to the same transformation formulas (3.86)–(3.88) as the stress tensor. Noting that $\dot{\varepsilon}_{xx}$ is zero so incompressibility requires $\dot{\varepsilon}_{yy} + \dot{\varepsilon}_{zz} = 0$, the strain-rate components in cylindrical coordinates are

$$\dot{\varepsilon}_{rr} = \frac{1}{2}(\dot{\varepsilon}_{zz} - \dot{\varepsilon}_{yy})\cos 2\theta + \dot{\varepsilon}_{yz}\sin 2\theta, \tag{3.102}$$

$$\dot{\varepsilon}_{\theta\theta} = -\dot{\varepsilon}_{rr}, \tag{3.103}$$

$$\dot{\varepsilon}_{r\theta} = \dot{\varepsilon}_{yz}\cos 2\theta - \frac{1}{2}(\dot{\varepsilon}_{zz} - \dot{\varepsilon}_{yy})\sin 2\theta. \tag{3.104}$$

Equation (3.103) can be interpreted as the incompressibility condition. Invoking the flow law (3.99) to express the Cartesian strain-rate components in terms of Cartesian stress deviators and applying the stress-transformation formulas gives the flow law for the cylindrical coordinate system:

$$\dot{\varepsilon}_{rr} = \frac{A}{2^n} \left[\left(\sigma_{rr} - \sigma_{\theta\theta} \right)^2 + 4\sigma_{r\theta}^2 \right]^{(n-1)/2} (\sigma_{rr} - \sigma_{\theta\theta}), \tag{3.105}$$

$$\dot{\varepsilon}_{r\theta} = \frac{A}{2^{n-1}} \left[\left(\sigma_{rr} - \sigma_{\theta\theta} \right)^2 + 4\sigma_{r\theta}^2 \right]^{(n-1)/2} \sigma_{r\theta}. \tag{3.106}$$

As in the Cartesian coordinate system, strain rates in the cylindrical coordinate system are linked to velocity gradients. The two components of velocity are u_r in the radial direction and u_θ in the tangential direction (in addition, there may be a component in the along-flow x-direction, but this velocity need not be considered in the present analysis). The strain-rate components are (for example, Xu, 1992)

$$\dot{\varepsilon}_{rr} = \frac{\partial u_r}{\partial r}, \tag{3.107}$$

$$\dot{\varepsilon}_{\theta\theta} = \frac{u_r}{r} + \frac{1}{r}\frac{\partial u_\theta}{\partial \theta}, \tag{3.108}$$

$$\dot{\varepsilon}_{r\theta} = \frac{1}{r}\frac{\partial u_r}{\partial \theta} + \frac{\partial u_\theta}{\partial r} - \frac{u_\theta}{r}. \tag{3.109}$$

Applying equation (3.103) gives

$$\frac{\partial u_r}{\partial r} + \frac{u_r}{r} + \frac{1}{r}\frac{\partial u_\theta}{\partial \theta} = 0, \tag{3.110}$$

as the incompressibility condition expressed in terms of velocity gradients.

With the basic equations in place, the stress solution in the vicinity of the tunnel can be derived. According to the boundary conditions (3.93) and (3.98), the shear stress, $\sigma_{r\theta}$ is zero on both boundaries of the area affected by the tunnel ($a \leq r \leq b$). It may therefore be expected that this stress is small, if not zero, in the entire domain. Neglecting the shear stress, the force-balance equations (3.89) and (3.90) reduce to

$$\frac{\partial \sigma_{rr}}{\partial r} + \frac{\sigma_{rr} - \sigma_{\theta\theta}}{r} - \rho g \cos\theta = 0, \tag{3.111}$$

$$\frac{1}{r}\frac{\partial \sigma_{\theta\theta}}{\partial \theta} + \rho g \sin\theta = 0. \tag{3.112}$$

The second equation contains as the only unknown the tangential normal stress, $\sigma_{\theta\theta}$, and can thus be solved, subject to the boundary condition (3.97). Substituting the solution in the first balance equation (3.111) allows the other normal stress component to be found.

Integrating equation (3.112) with respect to the radial distance, r, gives

$$\sigma_{\theta\theta} = \rho g r \cos\theta + C(r),$$ (3.113)

in which the integration constant, C(r), may be a function of the radial distance, r. Applying the boundary condition (3.97) gives

$$C(r) = -P_i,$$ (3.114)

for r >> a. Any function of the radial distance that reduces to (3.114) for large values of r can be chosen for C. The simplest possible form, namely, $C = -P_i$ for all values of r, is selected here.

Substituting the solution (3.113) in the balance equation (3.111) gives

$$\frac{\partial \sigma_{rr}}{\partial r} + \frac{\sigma_{rr}}{r} - \frac{C}{r} - 2\rho g \cos\theta = 0.$$ (3.115)

The solution of the homogeneous part is found by setting C = 0 and is

$$\sigma_{rr}^{(h)} = \frac{B}{r},$$ (3.116)

with B a constant. To find the solution of the full equation, assume that B = B(r). Substituting in equation (3.115) gives

$$\frac{\partial B}{\partial r} = -C + 2\rho g r \cos\theta.$$ (3.117)

This expression can be integrated to give

$$B(r) = -P_i r + \rho g r^2 \cos\theta + G,$$ (3.118)

in which G represents yet another integration constant that may be a function of the angle, θ.

The solution for the radial normal stress now becomes

$$\sigma_{rr}(r) = -P_i + \rho g r \cos\theta + \frac{G}{r}.$$ (3.119)

For large values of the radial distance, r, this expression reduces to the undisturbed solution (3.96). At the wall of the tunnel, the radial normal stress is

$$\sigma_{rr}(a) = -P_i + \rho g a \cos\theta + \frac{G}{a},$$ (3.120)

and should balance the water pressure, P_w. This yields the integration constant

$$G = a P_e - \rho g a^2 \cos \theta, \tag{3.121}$$

in which $P_e = P_i - P_w$ represents the effective basal pressure. The radial normal stress thus becomes

$$\sigma_{rr}(r) = -P_i + \frac{a}{r} P_e + \rho g r \left(1 - \frac{a^2}{r^2} \right) \cos \theta. \tag{3.122}$$

An expression for the radial velocity, u_r, can be found by integrating expression (3.105) for the radial strain rate with respect to r. From the stress solutions (3.113) and (3.122), it follows that

$$\sigma_{rr} - \sigma_{\theta\theta} = \frac{a}{r} P_e - \rho g \frac{a^2}{r} \cos \theta, \tag{3.123}$$

giving the radial strain rate

$$\dot{\varepsilon}_{rr}(r) = \frac{A}{2^n} \left(a P_e - \rho g a^2 \cos \theta \right)^n \frac{1}{r^n}. \tag{3.124}$$

Integrating with respect to r yields

$$u_r(r) = \frac{A}{2^n} \left(a P_e - \rho g a^2 \cos \theta \right)^n \frac{-1}{(n-1) r^{n-1}}. \tag{3.125}$$

The integration constant has been set to zero to satisfy the condition that the radial velocity must vanish for large values of the radial distance.

The rate at which the tunnel closes is given by $-u_r(a)/a$, or, using equation (3.125),

$$S = \frac{A}{(n-1) 2^n} \left(P_e - \rho g a \cos \theta \right)^n. \tag{3.126}$$

The effective pressure, P_e, is defined as the difference between the ice overburden pressure and the water pressure. Compared to the weight of the ice above ($\rho g H$), the second term between the brackets may be neglected and

$$S = \frac{A P_e^n}{(n-1) 2^n}. \tag{3.127}$$

This result is essentially the same as that derived by Nye (1953), except for the constants in the denominator. His derivation, however, is somewhat different from the one given above and, admittedly, much shorter.

Nye (1953) makes the assumption that the flow is purely radial, so that the tangential component of velocity vanishes. The incompressibility condition then reduces to

$$\frac{\partial u_r}{\partial r} + \frac{u_r}{r} = 0,$$

(3.128)

which can be integrated to give

$$u_r(r) = -\frac{C}{r},$$

(3.129)

with C a positive constant. Differentiating equation (3.129) with respect to r gives the radial strain rate

$$\dot{\varepsilon}_{rr}(r) = \frac{C}{a^2}.$$

(3.130)

Inverting the flow law (3.105), the effective stress is

$$\sigma_{rr} - \sigma_{\theta\theta} = \frac{2}{A^{1/n}}\dot{\varepsilon}_{rr}^{1/n} =$$

$$= \frac{2C^{1/n}}{A^{1/n}}\frac{1}{r^{2/n}}.$$

(3.131)

Substituting this expression in the balance equation (3.111) gives

$$\frac{\partial \sigma_{rr}}{\partial r} + \frac{2C^{1/n}}{A^{1/n}}\frac{1}{r^{2/n+1}} + \rho g \cos\theta = 0.$$

(3.132)

Integrating with respect to r, and using the condition that the radial stress must be equal to the undisturbed solution (3.96) for large values of r to determine the integration constant, yields the stress solution

$$\sigma_{rr}(r) = -P_i + \rho g r \cos\theta + \frac{nC^{1/n}}{A^{1/n}}\frac{1}{r^{2/n}}.$$

(3.133)

The constant, C, can be determined from the condition that on the wall of the tunnel the radial normal stress must balance the water pressure, giving

$$C = \frac{A}{n^n}(P_e - \rho g a \cos\theta)^n a^n.$$

(3.134)

The rate of closure, $S = u_r(a)/a$, now becomes

$$S = \frac{A}{n^n}(P_e - \rho g a \cos\theta)^n. \tag{3.135}$$

Comparison with equation (3.126) shows that both methods yield the same result, except for the constants. This difference can be attributed to neglecting the tangential component of velocity in Nye's analysis. Given the uncertainty in the rate factor, the difference is probably not very important.

4 Modeling Glacier Flow

4.1 INTRODUCTION

The force that makes ice flow in the direction of decreasing surface elevation is the driving stress as defined in equation (3.21). This action is opposed by reactions, or resistive forces. Resistance to flow may originate at the glacier bed and at the lateral margins, or resistance may be associated with gradients in longitudinal stress (c.f. Section 3.1). Generally, because of the nonlinearity of the flow law, a velocity solution cannot be derived analytically, except in simplified cases where flow resistance is offered by only one of these potential sources. While this may appear overly restrictive, these limiting cases do apply to certain flow regimes found in Nature. Flow in the interior of ice sheets is mostly controlled by a balance between driving stress and drag at the glacier bed, and the corresponding lamellar flow model provides a good approximation of ice flow. On mountain glaciers, lateral drag arising from friction between the ice and valley walls may provide resistance to flow in addition to basal drag. In first approximation, this effect can be incorporated by introducing a shape factor to reduce basal drag in accordance with the role of lateral drag, but otherwise adopting the lamellar flow model. On Whillans Ice Stream in West Antarctica, basal friction is vanishingly small due to the presence of a water-saturated weak layer of sediment (for example, Whillans and van der Veen, 1997) and the driving stress is balanced almost entirely by lateral drag originating at the margins where the ice stream moves past nearly stagnant ice (van der Veen et al., 2007). This model of flow controlled by lateral drag also applies to floating ice shelves in comparatively narrow fjords. Finally, on free-floating ice shelves, the only resistance to flow is associated with gradients in longitudinal stress. In this chapter, these "end member" solutions are discussed.

Before continuing with the discussion of the various analytical velocity solutions, it is convenient to introduce a dimensionless vertical coordinate to account for changes in ice thickness, H, in the direction of flow:

$$s = \frac{h - z}{H}. \tag{4.1}$$

At the surface, z = h and s = 0, whereas at the bed, z = h − H and s = 1. By using this vertical coordinate, the model domain becomes a rectangular two-dimensional grid.

To find a relationship between horizontal gradients in the (x, z) coordinate system and those in the (x, s) coordinate system, consider Figure 4.1. Let F be an arbitrary

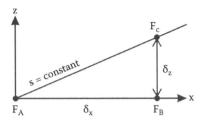

FIGURE 4.1   Transformation of gradients in the (x, z) coordinate system to those in the (x, s) coordinate system.

function of both coordinates x and z. The gradient in F evaluated along the surface s = constant is related to that evaluated at z = constant, as

$$\frac{F_C - F_A}{\delta x} = \frac{F_C - F_B}{\delta z}\frac{\delta z}{\delta x} + \frac{F_B - F_A}{\delta x}. \tag{4.2}$$

In the limit of δx and δz approaching zero, this expression becomes

$$\left(\frac{\partial F}{\partial x}\right)_s = \left(\frac{\partial F}{\partial z}\right)_x\left(\frac{\partial z}{\partial x}\right)_s + \left(\frac{\partial F}{\partial x}\right)_z, \tag{4.3}$$

where the subscripts refer to the coordinate kept constant in calculating the gradients. From the definition of s it follows that $z = h - Hs$, and

$$\left(\frac{\partial z}{\partial x}\right)_s = \frac{\partial h}{\partial x} - s\frac{\partial H}{\partial x} =$$

$$= \Delta_s. \tag{4.4}$$

Also,

$$\left(\frac{\partial F}{\partial z}\right)_x = \left(\frac{\partial F}{\partial s}\right)_x\left(\frac{\partial s}{\partial z}\right)_x =$$

$$= -\frac{1}{H}\left(\frac{\partial F}{\partial s}\right)_x. \tag{4.5}$$

Substitution of these relations in equation (4.3) gives

$$\left(\frac{\partial F}{\partial x}\right)_z = \left(\frac{\partial F}{\partial x}\right)_s + \frac{\Delta_s}{H}\left(\frac{\partial F}{\partial s}\right)_x. \tag{4.6}$$

In the following, the subscripts on the derivatives are omitted where no confusion is likely.

In the (x, s) coordinate system, equation (3.8) expressing force balance in the x-direction in terms of full stresses, becomes (omitting the second term involving the gradient in the cross-flow direction, because plane flow is considered here)

$$\frac{\partial \sigma_{xx}}{\partial x} + \frac{\Delta_s}{H} \frac{\partial \sigma_{xx}}{\partial s} - \frac{1}{H} \frac{\partial \sigma_{xz}}{\partial s} = 0. \tag{4.7}$$

Substituting the partitioning of full stresses into lithostatic and resistive components, and multiplying by H, gives

$$H \frac{\partial R_{xx}}{\partial x} + \Delta_s \frac{\partial R_{xx}}{\partial s} - \rho g H \frac{\partial h}{\partial x} - \frac{\partial R_{xz}}{\partial s} = 0. \tag{4.8}$$

Integrating this expression with respect to s, from a general depth, s, in the ice, to the surface, s = 0, yields the following expression for the shear stress at that depth:

$$R_{xz}(s) = s\tau_{dx} - \frac{\partial}{\partial x}\left(\int_s^0 H R_{xx}\, d\overline{s} \right) + \Delta_s R_{xx}(s), \tag{4.9}$$

where the boundary condition (3.19) for a stress-free surface has been used. Equation (4.9) describes the balance of forces acting on an ice column extending from the surface to the depth s.

The next step is to express strain rates in terms of velocity gradients in the (x, s) coordinate system. Applying the transformation formula (4.6), along-flow stretching is given by

$$\dot{\varepsilon}_{xx} = \left(\frac{\partial u}{\partial x}\right)_z =$$

$$= \left(\frac{\partial u}{\partial x}\right)_s + \frac{\Delta_s}{H}\left(\frac{\partial u}{\partial s}\right)_x. \tag{4.10}$$

This expression cannot be evaluated directly because the last term on the right-hand side is related to the shear strain rate, which is unknown until the balance equation is solved.

To find an expression for the velocity gradient $(\partial u/\partial s)_x$ consider the definition of the shear strain rate

$$\dot{\varepsilon}_{xz} = \frac{1}{2}\left[\left(\frac{\partial u}{\partial z}\right)_x + \left(\frac{\partial w}{\partial x}\right)_z\right]. \tag{4.11}$$

In terms of gradients in the (x, s) coordinate system, the two terms on the right-hand side are

$$\left(\frac{\partial u}{\partial z}\right)_x = -\frac{1}{H}\left(\frac{\partial u}{\partial s}\right)_x, \tag{4.12}$$

and

$$\left(\frac{\partial w}{\partial x}\right)_z = \left(\frac{\partial w}{\partial x}\right)_s + \frac{\Delta_s}{H}\left(\frac{\partial w}{\partial s}\right)_x. \tag{4.13}$$

The last term on the right-hand side of this equation is linked to the vertical strain rate:

$$\left(\frac{\partial w}{\partial s}\right)_x = -H\left(\frac{\partial w}{\partial z}\right)_x =$$

$$= -H\dot{\varepsilon}_{zz} = \tag{4.14}$$

$$= H\dot{\varepsilon}_{xx},$$

where incompressibility of glacier ice has been invoked to link the vertical strain rate to the horizontal strain rate. Equation (4.11) can now be rewritten as

$$\left(\frac{\partial u}{\partial s}\right)_x = -H\left[2\dot{\varepsilon}_{xz} - \left(\frac{\partial w}{\partial x}\right)_s - \Delta_s\dot{\varepsilon}_{xx}\right]. \tag{4.15}$$

For most modeling applications, along-flow gradients in the vertical velocity are small and may be neglected. Similarly, even where the stretching rate may be significant, when multiplied by Δ_s the result will be small compared to the vertical shear strain rate, and equation (4.15) may be approximated by

$$\left(\frac{\partial u}{\partial s}\right)_x = -2H\dot{\varepsilon}_{xz}. \tag{4.16}$$

With these preliminary manipulations out of the way, analytical solutions for simplified flow regimes can be derived.

4.2   LAMELLAR FLOW

In the simplest model for glacier flow, the driving stress is taken to be opposed entirely by basal drag; longitudinal stresses and lateral shear are neglected. This model is referred to as the lamellar flow model or the shallow ice approximation (for example, Greve and Blatter, 2009, Section 5.4).

If drag at the glacier base provides the sole resistance to flow, the force-balance equation (3.22) reduces to

$$\tau_{dx} = \tau_{bx}. \tag{4.17}$$

The shear stress at depth follows from consideration of the horizontal balance equation (4.9), with the last two terms on the right-hand side neglected:

$$R_{xz}(s) = s\tau_{dx}. \tag{4.18}$$

In other words, in the lamellar flow model, the shear stress, R_{xz}, always increases linearly from zero at the glacier surface to the maximum value, equal to the driving stress, at the glacier bed.

By definition of lamellar flow, the only nonzero strain rate is the vertical shearing rate, $\dot{\varepsilon}_{xz}$. Using the constitutive relation (3.47) to link the shear stress to shear strain rate, the balance equation becomes

$$B\dot{\varepsilon}_{xz}^{1/n} = s\tau_{dx}, \tag{4.19}$$

where B represents the viscosity factor and n the exponent in Glen's flow law (2.10). Except in the vicinity of ice divides, the vertical velocity, w, is usually small, and along-flow gradients in this quantity may be neglected. The vertical gradient in along-flow velocity is then related to the shear strain rate as (equation (4.16))

$$\left(\frac{\partial u}{\partial s}\right)_x = -2H\dot{\varepsilon}_{xz}. \tag{4.20}$$

Combining equations (4.19) and (4.20) yields the following expression for the vertical gradient in the horizontal velocity, u:

$$\frac{\partial u}{\partial s} = 2AH\tau_{dx}^n s^n, \tag{4.21}$$

with $A = B^{-n}$ (c.f. Section 2.2).

The velocity at depth can be found by integrating equation (4.21) with respect to s from the bed (s = 1) to some level s in the ice:

$$u(s) = \frac{2AH}{n+1}(1-s^{n+1})\tau_{dx}^n + U_b, \tag{4.22}$$

where U_b represents the basal sliding velocity, that is, the velocity at the glacier bed. The first term on the right-hand side corresponds to the component of velocity associated with internal deformation.

The velocity at the glacier surface (s = 0) is

$$u(0) = \frac{2AH}{n+1}\tau_{dx}^n + U_b. \tag{4.23}$$

Setting the sliding velocity equal to zero, the deformational component of velocity at depth can also be written as

$$u(s) = u(0)(1 - s^{n+1}). \tag{4.24}$$

With n = 3, this expression shows that the velocity from internal deformation increases with the fourth power of depth below the surface. Most shearing is concentrated near the glacier base, as shown in Figure 4.2.

In deriving the horizontal velocity profile, the assumption is made that the along-flow gradient in the vertical velocity, w, is negligible. This does not necessarily imply that the vertical velocity itself is negligible, and an expression for w can be derived

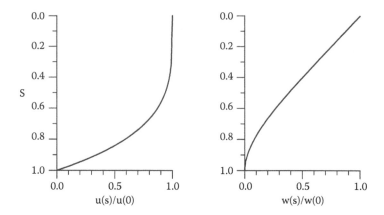

FIGURE 4.2   Horizontal (left panel) and vertical (right panel) component of velocity scaled to the surface velocity as a function of dimensionless depth, for the lamellar-flow model (using n = 3 for the exponent in the constitutive relation).

from the incompressibility criterion, which in the (x, s) coordinate system reads as (equations (4.14) and (4.10))

$$\left(\frac{\partial w}{\partial s}\right)_x = \left(\frac{\partial u}{\partial x}\right)_s + \frac{\Delta_s}{H}\left(\frac{\partial u}{\partial s}\right)_x, \qquad (4.25)$$

where subscripts refer to the coordinate kept constant in evaluating the derivatives, and

$$\Delta_s = \frac{\partial h}{\partial x} - s\frac{\partial H}{\partial x}. \qquad (4.26)$$

For a typical glacier, $\Delta_s/H \approx 10^{-5} - 10^{-6}$ and the second term in equation (4.25) may be neglected. Substituting the solution for the horizontal velocity gives the variation with depth of the vertical velocity

$$\left(\frac{\partial w}{\partial s}\right)_x = \frac{\partial u(0)}{\partial x}[1 - s^{n+1}]. \qquad (4.27)$$

Setting the vertical velocity at the bed equal to zero, this expression can be integrated from the bed (s = 1) to some level s in the ice to yield

$$w(s) = -\frac{n+1}{n+2}\frac{\partial u(0)}{\partial x}\left[1 - s\left(\frac{n+2}{n+1} - \frac{s^{n+1}}{n+1}\right)\right]. \qquad (4.28)$$

The vertical velocity at the surface is

$$w(0) = -\frac{n+1}{n+2}\frac{\partial u(0)}{\partial x}, \qquad (4.29)$$

and the profile of the vertical velocity can be written as

$$w(s) = w(0)\left[1 - s\left(\frac{n+2}{n+1} - \frac{s^{n+1}}{n+1}\right)\right]. \tag{4.30}$$

This profile is shown in the right panel of Figure 4.2 and indicates that the vertical velocity decreases approximately in a linear fashion with depth below the surface.

An important quantity needed in depth-averaged models is the vertical-mean velocity, U. Averaging equation (4.22) over the depth of the glacier gives

$$U = \frac{2AH}{n+2}\tau_{dx}^n + U_b. \tag{4.31}$$

The ice flux through a vertical section of unit width and extending from the base to the surface is given by HU and can be calculated from the glacier geometry (first term on the right-hand side) and a sliding relation for the sliding velocity, U_b.

The flow model described above is often referred to as *lamellar* or *laminar flow* and was first derived by Nye (1952a). The term *lamellar* is preferred to avoid any possible suggestion that glacier flow can become turbulent, as laminar fluid flow in a channel can become turbulent if the Reynolds number surpasses a critical value. To recapitulate, the most important assumptions of the lamellar model are that (1) basal drag is the only resistance to flow, (2) $\dot{\varepsilon}_{xz}$ is the only nonzero strain rate, and (3) the ice is isothermal so that the rate factor, A, may be taken constant with depth. Under these constraints, which imply that the glacier geometry must be simple, with the upper and lower surfaces varying slowly in the direction of flow (that is, the glacier must resemble a plane slab), successive depth layers glide over one another similar to a deck of cards.

In the modeling community, the lamellar flow solution has become known as the Shallow Ice Approximation (SIA), in recognition of the fact that horizontal scales in glaciers and ice sheets are one or more orders of magnitude greater than the vertical scale. This allows introduction of typical quantities that are used as scaling parameters for physical variables and render the governing equations dimensionless. In particular, the horizontal coordinate, x, is nondimensionalized by dividing by the length scale (glacier length or half-width), L, while the dimensionless vertical coordinate is z/H. This introduces the small aspect ratio, H/L, as multiplier for some terms in the governing equations, allowing identification of higher-order terms that are likely to be small and that may be neglected in first approximation. A formal and rigorous treatment of scaling the force balance equations is presented in Hutter (1983) and Greve and Blatter (2009) and shows that, while the formal approach is decidedly more complex and lengthier than the intuitive discussion presented in this section, the resulting solution for the velocity distribution in glaciers is essentially the same.

Real glaciers are much more complex than assumed in the lamellar flow model. However, full calculations of stresses at depth for actual glaciers indicate that the shear stress, R_{xz}, increases almost linearly with depth, even where the driving stress is not entirely balanced by drag at the glacier base (for example, Section 4.3). This means that the lamellar-flow solution is a useful approximation for the large-scale flow, provided that the driving stress in equations (4.22) and (4.31) is replaced by basal drag, τ_{bx}, if other resistive stresses are important. On this zero-order solution, small-scale perturbations may be super-imposed.

The lamellar flow model predicts an upper limit to the surface velocity associated with internal deformation. Equation (4.23) allows the surface velocity to be calculated from the glacier geometry or, better, from basal drag. An associated quantity that allows for comparison of different glaciers is the surface velocity from internal deformation divided by the ice thickness:

$$\frac{u(0)}{H} = \frac{A}{2}\,\tau_{bx}^3, \tag{4.32}$$

using n = 3 for the exponent in the flow law. For a given basal drag, the upper limit of this quantity can be readily calculated by using the value for a rate factor A that corresponds to ice temperatures close to the melting point. In Figure 4.3 this upper

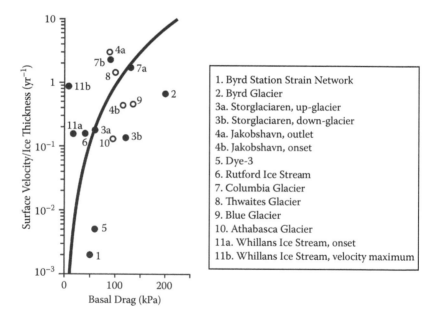

FIGURE 4.3 Relation between surface velocity divided by ice thickness, and basal drag, for various glaciers. Open circles indicate that basal drag is estimated from the driving stress, and filled circles refer to glaciers for which basal drag was estimated from balance of forces. The curve represents the upper limit to the deformational velocity, calculated from equation (4.24) using a rate factor corresponding to ice near the melting temperature. (Compiled from the following data sources: (1) Van der Veen and Whillans, 1989b; (2) Whillans et al., 1989; (3) Hooke et al., 1989; (4) Echelmeyer and Harrison, 1990 and Bindschadler, 1984; (5) Van der Veen and Whillans, 1990; (6) Frolich and Doake, 1988 and Frolich et al., 1987; (7) Van der Veen and Whillans, 1993; (8) McIntyre, 1985; (9) Engelhardt et al., 1978; (10) Raymond, 1969 quoted in Kamb, 1970; (11) Van der Veen, 1999b, Section 5.8.)

limit is shown as the solid curve. Also plotted in this figure is the ratio of the measured surface velocity and the ice thickness as a function of basal drag estimated for various glaciers. The curve divides the glaciers in two categories, namely, those that are definitely sliding over their bed (represented by the points on the left of the curve, such as 11a and 11b), and those that are flowing by internal deformation only (points to the right of the curve; for example, 2 and 5). Points that lie close to the theoretical curve represent glaciers where both basal sliding and internal deformation contribute to the flow.

4.3   INCLUDING LATERAL DRAG

Mountain glaciers generally flow through valleys that may partially impede the flow of the glacier. The friction generated between the moving ice and the rock walls provides resistance to flow, and the driving stress is supported by lateral drag (or sidewall friction) as well as drag at the glacier base. The effect of lateral drag on the flow of a valley glacier can be clearly seen in Figure 4.4, which shows contours of surface velocity obtained from repeat aerial photogrammetry of Columbia Glacier in Alaska. The glacier reaches its maximum speed near the center, and the velocity decreases toward both lateral margins. Transects of velocity are very similar to the depth profile shown in Figure 4.2, with the centerline corresponding to the glacier surface.

Where lateral drag is important, both shear stresses, R_{xz} and R_{xy}, must be included in the force-balance equations. Because of the nonlinear constitutive relation, a solution can be found only for a few exceptional cases, namely, a glacier in a circular basin (or nearly circular basin; Nye, 1965c), a very deep basin (in which case basal

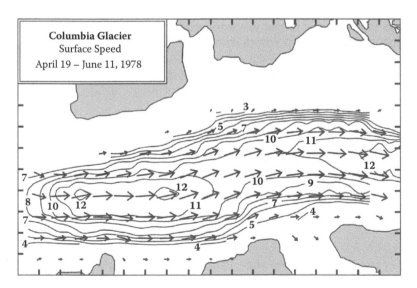

FIGURE 4.4   Surface speed on Columbia Glacier, Alaska, derived from repeat aerial photogrammetry. Speeds are in 100 m/yr and the contour interval is 100 m/yr. Shading represents bare rock.

drag and the vertical shear stress, R_{xz}, may be neglected), or a very wide glacier (in which case lateral drag may be neglected, except close to the margins, and the solution for the inner portion of the glacier reduces to the lamellar-flow solution discussed in Section 4.2). For more realistic geometries, the flow in a cross-section of a valley glacier must be found numerically as done by Nye (1965c) in what is perhaps the first numerical modeling study in glaciology.

The model used here to find the stress distribution in a cross-section is similar to the model described in Van der Veen (1989), except that the glacier geometry varies in the transverse direction rather than in the direction of flow. That is, the ice thickness, H, is assumed to be a function of the cross-flow coordinate, y, but constant in the flow direction, x. The surface elevation, h, is a function of the flow direction only, such that the surface slope, $\partial h/\partial x$, is constant in the direction of flow, and also taken constant across the width of the glacier. For the present discussion of results, the numerical details of this solution scheme are not important (the interested reader can find these details in Van der Veen, 1999b, Section 5.4). Similar results are obtained using off-the-shelf finite element solvers. Zero velocity is prescribed at the lower boundary.

Results of calculations for rectangular and parabolic channels are shown in Figures 4.5 and 4.6. The parabolic cross-sectional profile is chosen such that the thickness at the margins (y = ±W) is one-tenth of the ice thickness at the center, H_o, and

$$H(y) = H_o\left(1 - 0.9\left(\frac{y}{W}\right)^2\right). \tag{4.33}$$

This profile differs somewhat from that of Nye (1965c), who considered parabolic cross-sections with zero ice thickness at the margins. The parabolic profile is probably more representative of real valley basins. The examples shown are for basins with a half-width to centerline depth ratio of 4 (Figure 4.5) and 10 (Figure 4.6).

The upper panels in Figures 4.5 and 4.6 show the normalized driving stress, T_d, normalized basal resistance, T_b, and normalized resistance from lateral drag, T_s. Following Nye (1965c), stresses are normalized with the driving stress at the centerline

$$\tau_{dx}(0) = -\rho g H_o \frac{\partial h}{\partial x}. \tag{4.34}$$

Also shown in the upper panels is the velocity at the glacier surface, normalized with the surface velocity at the centerline as predicted by the lamellar flow theory with n = 3 and no basal sliding:

$$U_d(0) = \frac{1}{2} A H_o \tau_{dx}^3(0). \tag{4.35}$$

The other panels in the figures show the normalized component of velocity in the direction of flow, the normalized vertical shear stress, and the normalized lateral shear stress.

The effect of the width-to-depth ratio is perhaps best seen in the panels showing the normalized velocity. For the glacier in a relatively narrow basin (Figure 4.5),

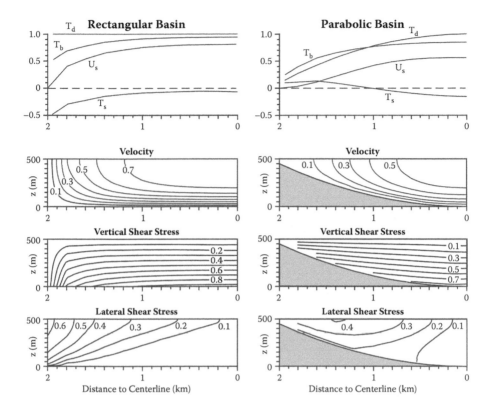

FIGURE 4.5   Effect of lateral drag on a shallow glacier ($W/H_o = 4$) in a rectangular basin (panels on the left) and parabolic basin (panels on the right; shading indicates bedrock). The upper panels show driving stress (T_d), lateral drag (T_s), and basal drag (T_b) nondimensionalized with the driving stress at the center, as well as the surface velocity (U_s) nondimensionalized with the surface velocity at the centerline as predicted by the lamellar-flow theory (equation (4.24)). The other panels show cross-sectional views of the nondimensionalized velocity, vertical shear stress, and lateral shear stress.

calculated velocities are smaller than for the glacier in the wide valley (Figure 4.6). The reason for this is that lateral drag provides more resistance to flow when the glacier basin is narrow. Consequently, drag at the glacier base is smaller than the driving stress. Near the glacier center, where the velocity may be calculated from the lamellar flow theory with the driving stress replaced by basal drag, the surface velocity is therefore less than it would be in the absence of lateral drag.

There is a marked difference between the flow through a rectangular cross-section and the flow through a parabolic basin (Nye, 1965c). This difference is most clearly shown in Figure 4.5, representing relatively narrow glaciers. For the rectangular basin, the curve of the surface velocity is always convex, implying that lateral drag (being proportional to the second cross-flow derivative of the velocity) is always negative, that is, opposing the flow of the glacier. For the parabolic basin, the curve of the surface velocity has an inflection point (in the case of Figure 4.5

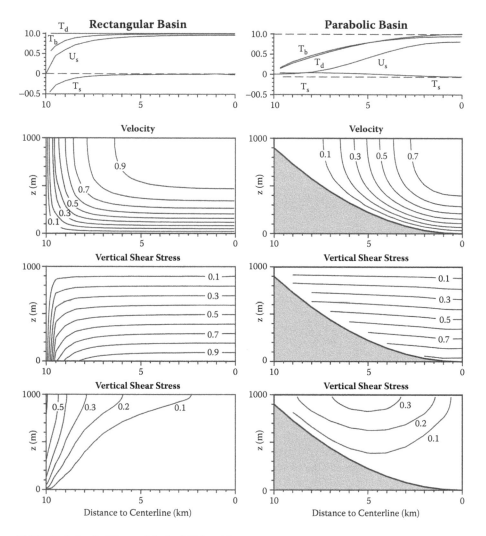

FIGURE 4.6 As Figure 4.5, for $W/H_o = 10$.

at $y = 1$ km), caused by the decreasing driving stress toward the margins (although the surface slope is taken constant across the width of the glacier, the ice thickness, and hence the driving stress, decreases away from the centerline). The concave curvature of the velocity curve indicates positive lateral drag near the margins; in the center, where this curvature is convex, lateral drag is negative. This means that the center ice is dragging the margin ice along, and a region exists near the margins where basal drag exceeds the local driving stress.

All examples show that, at the centerline, the vertical shear stress, R_{xz}, increases approximately linearly with depth, and that lamellar flow theory may be applied to the central parts to calculate the surface velocity. How well this linear dependency applies to the main body of the glacier depends on the width-to-height ratio, W/H_o. To account quantitatively for the effect of the valley walls on the glacier flow, Nye

TABLE 4.1
Shape Factors Corresponding to the Examples in Figures 4.5 and 4.6

	Shape Factor from Centerline Surface Velocity	Shape Factor from Width-Averaged Force Balance
Rectangular basin, $W/H_o = 4$	0.93	0.87
Rectangular basin, $W/H_o = 10$	0.99	0.96
Parabolic basin, $W/H_o = 10$	0.83	0.98
Parabolic basin, $W/H_o = 10$	0.93	0.99

(1965c) introduced the *shape factor*. This factor, commonly denoted by f, represents the fraction of driving stress that is supported by basal drag, the remainder of resistance being due to lateral drag. That is

$$\tau_{bx} = f\,\tau_{dx}. \qquad (4.36)$$

The shape factor can be calculated as that needed to produce the correct surface velocity at the centerline if lamellar flow is assumed, or it can be calculated from force balance for the entire width of the glacier. Both methods yield results that differ by about 10% to 20%, depending on the shape of the glacier basin. The reason for this is that using the centerline velocity to calculate the shape factor yields a local value applicable to the centerline only. If, on the other hand, the large-scale balance of forces is invoked to determine the shape factor, a width-averaged value of f is found. Because in general, local basal drag does not support the same fraction of the local driving stress across the width of the glacier (as shown in the upper panels of Figures 4.5 and 4.6), a difference between both values may be expected. Shape factors calculated using both methods for the examples shown in Figures 4.5 and 4.6 are given in Table 4.1. Shape factors for some other geometries were determined by Nye (1965c) and are given in Table 4.2.

In the model of Nye (1965c), as well as in the model described above, the ice is taken to flow by internal deformation only, with no basal slip. However, observations conducted on Athabasca Glacier in Alberta, Canada, indicate that basal sliding accounts for about 80% of the surface velocity near the center of that glacier. Toward the margins, the sliding velocity decreases (Raymond, 1971). This result prompted Reynaud (1973) and Harbor (1992) to incorporate basal sliding into their numerical models for flow in a valley glacier cross-section. To obtain reasonable agreement between model predictions and the transverse velocity profile derived from the measurements given in Raymond (1971), in particular the observed small sliding velocities near the margin, Harbor (1992) had to make the assumption that the friction increases toward the margin. However, as pointed out by Harbor (1992), there are not enough constraining measurements to validate this assumption.

TABLE 4.2

Shape Factor from Centerline Surface Velocity for Various Cross-Sectional Basins, Calculated by Nye (1965c)

W/H_o	Rectangular Basin	Ellipsoidal Basin	Parabolic Basin
0	0	0	0
1/4	—	0.134	—
1/3	0.204	0.185	—
1/2	0.313	0.281	—
1	0.558	0.500	0.445
2	0.789	0.709	0.646
3	0.884	0.799	0.746
4	—	0.849	0.806
∞	1.000	1.000	1.000

4.4 GLACIER FLOW CONTROLLED BY LATERAL DRAG

Lateral drag originates at the margins, where glacier ice moves past rock walls or, as in the case of ice streams, past nearly stagnant ice. Because most glaciers are comparatively shallow with respect to their width, with ice thicknesses less than about 2 km, the area over which lateral drag acts (the vertical sides of the glacier) is limited. This means that there is a practical upper limit to how much driving stress can be supported by lateral drag. In contrast, basal drag acts over the full width of the glacier and can therefore support a larger driving stress. This difference between both sources of flow resistance can be quantified by considering the force budget averaged over the width of the glacier.

Making the assumption that the lateral shear stress, R_{xy}, is constant with depth, and neglecting gradients in longitudinal stress, the force-balance equation (3.22) reduces to

$$\tau_{dx} = \tau_{bx} - \frac{\partial}{\partial y}(HR_{xy}).\tag{4.37}$$

The lateral resistance averaged over the half width, W, is

$$\overline{-\frac{\partial}{\partial y}(HR_{xy})} = -\frac{1}{W}\int_{-W}^{0}\frac{\partial}{\partial y}(HR_{xy})\,dy =$$

$$= -\frac{1}{W}(HR_{xy}(0) - HR_{xy}(-W)).\tag{4.38}$$

Flow through the cross-section is symmetric around the centerline (y = 0) so that the shear stress, R_{xy}, is zero there. Denoting the lateral shear stress at the margin (y = −W) by τ_s, and the thickness at the margin by H_w, the lateral resistance on a section of glacier of unit width is

$$F_s = \frac{H_w \tau_s}{W}, \tag{4.39}$$

and the width-averaged force-balance equation becomes

$$\overline{\tau}_{dx} = \overline{\tau}_{bx} + \frac{H_w \tau_s}{W}, \tag{4.40}$$

with an overbar denoting a width-averaged value.

To estimate the potential importance of lateral drag, a value for the shear stress at the margins, τ_s, is needed. Ice is often considered a plastic material with a yield stress of about 100–200 kPa (Section 5.1). This observation is based on the rather narrow range of values of basal drag determined for various glaciers (as shown in Figure 4.3). It seems reasonable therefore to use a similar upper limit to the lateral shear stress at the margins. For a typical mountain glacier such as Columbia Glacier, with a half width of about 3 km and a marginal thickness of about 200 m, a shear stress at the margins of 200 kPa can support only 13 kPa of driving stress. Basal drag comparable to the yield stress, on the other hand, can support 200 kPa of driving stress. With driving stresses typically on the order of 50–200 kPa, the practical upper limit to resistance offered by lateral drag limits the role of sidewall friction to opposing about 25% or less of the driving stress (as also indicated by the shape factors given in Table 4.2). A notable exception to this may be the West Antarctic ice streams, where high speeds (up to 800 m/yr on Whillans Ice Stream; Whillans and Van der Veen, 1993b) are achieved in spite of unusually small driving stresses (about 15 kPa; Alley and Whillans, 1991). This combination of high speed, and associated lateral shear at the margins, with small surface slope allows for the possibility that the flow of these ice streams is controlled entirely by lateral drag. Whillans and Van der Veen (1997) present several arguments in favor of ice stream flow controlled by lateral drag. Another instance where lateral drag opposes a significant part of the driving stress is on the lower reach of Jakobshavn Isbræ, on the west coast of Greenland. The fast-moving part of this outlet glacier is very narrow (ca. 6 km), and the ice thickness at the shear margin is about 1500 m. Because of this geometry, lateral drag supports in excess of 200 kPa of driving stress just upglacier of the grounding line (Van der Veen et al., 2011).

With lateral drag providing the sole resistance to flow, the force-balance equation (4.37) reduces to

$$\tau_{dx} = -\frac{\partial}{\partial x}(HR_{xy}). \tag{4.41}$$

Any (small) residual basal drag may be accounted for by defining an "effective" driving stress

$$\hat{\tau}_{dx} = \tau_{dx} - \tau_{bx}. \tag{4.42}$$

Taking the driving stress (or effective driving stress) and ice thickness constant across the width of the glacier, it follows from the balance equation (4.41) that the shear stress, R_{xy}, varies linearly across the glacier and

$$R_{xy}(y) = a y + b. \tag{4.43}$$

At the centerline ($y = 0$) this shear stress is zero, so that $b = 0$. As above, lateral drag at the margins is denoted by $+ \tau_s$ ($y = -W$) and $- \tau_s$ ($y = +W$). From the balance equation (4.40) it follows that

$$\tau_s = \frac{W}{H} \hat{\tau}_{dx}, \tag{4.44}$$

giving the shear stress across the glacier

$$R_{xy}(y) = -\frac{y}{H} \hat{\tau}_{dx}. \tag{4.45}$$

The transverse profile of velocity is found by invoking the constitutive relation linking stresses to strain rates. In the present model the only nonzero strain rate is the rate of lateral shearing

$$\dot{\varepsilon}_{xy} = \frac{1}{2} \frac{\partial U}{\partial y}. \tag{4.46}$$

The flow law then reduces to

$$R_{xy} = B \left(\frac{1}{2} \frac{\partial U}{\partial x} \right)^{1/n}, \tag{4.47}$$

or

$$\frac{\partial U}{\partial y} = 2 \left(\frac{R_{xy}}{B} \right)^{n}. \tag{4.48}$$

Substituting expression (4.47) for the shear stress, and integrating with respect to y, gives

$$U(y) = -\frac{2}{n + 1} \left(\frac{\hat{\tau}_{dx}}{BH} \right)^{n} + C. \tag{4.49}$$

The integration constant, C, follows from the no-slip boundary condition at both margins:

$$C = \frac{2}{n + 1} \left(\frac{\hat{\tau}_{dx}}{BH} \right)^{n} W^{n+1}. \tag{4.50}$$

The transverse velocity profile can now be written as

$$U(y) = U(0)\left(1 - \left(\frac{y}{W}\right)^{n+1}\right), \tag{4.51}$$

with the centerline velocity given by

$$U(0) = \frac{2}{n+1}\left(\frac{\hat{\tau}_{dx}}{BH}\right)^n W^{n+1}. \tag{4.52}$$

These expressions for the transverse profile of velocity are very similar to equations (4.23) and (4.24) derived for lamellar flow. This is because both models are essentially the same with the controlling shear stress, R_{xz} or R_{xy}, respectively, decreasing linearly from the site of friction (the glacier bed, or the glacier margin) to zero at either the glacier surface or its centerline. However, because glaciers are mostly shallow, with relatively large width compared with the depth, the lateral-control model may lead to centerline velocities that are orders of magnitudes larger than the surface velocity predicted by the lamellar-flow theory.

4.5   ICE-SHELF SPREADING

In the models discussed so far in this chapter, gradients in longitudinal stress have been neglected as a source of flow resistance, and the driving stress was assumed to be balanced by drag at the glacier bed and/or lateral drag. However, on floating ice shelves, basal drag is obviously zero, and lateral drag may be small if the shelf has formed in a wide embayment or is floating freely. In that case, all resistance to flow is associated with longitudinal stresses. The resulting flow solution was first derived by Weertman (1957b). Note that in the Weertman model, an ice shelf of constant thickness is considered. This is an unrealistic situation as it would imply a zero surface slope and thus no driving stress. The following derivation does not make any assumption about the geometry of the ice shelf, other than that the ice is in hydrostatic equilibrium with the sea water, and arrives at the same result as did Weertman (1957b).

For a free-floating ice shelf, the force-balance equation (3.22) reduces to a balance between driving stress and gradients in longitudinal stress, such that

$$\tau_{dx} = -\frac{\partial}{\partial x}\int_{h-H}^{h} R_{xx}\ dz, \tag{4.53}$$

where h represents the surface elevation and H the thickness of the ice shelf. Neglecting the depth variation in the longitudinal resistive stress, R_{xx} (as seems reasonable because vertical shear in an ice shelf is negligible; Sanderson and

Doake, 1979), and using equation (3.21) for the definition of the driving stress, the balance equation simplifies to

$$\rho g H \frac{\partial h}{\partial x} = \frac{\partial}{\partial x}(H R_{xx}). \tag{4.54}$$

This equation can be solved for the resistive stress, R_{xx}, by eliminating the surface elevation, h. Because ice shelves are floating in sea water, the height of the ice surface above sea level follows from hydrostatic equilibrium

$$h = \left(1 - \frac{\rho}{\rho_w}\right)H, \tag{4.55}$$

where ρ_w represents the density of sea water and ρ the density of ice. Substituting this expression for h in the balance equation (4.54) gives

$$\frac{1}{2}\rho g\left(1 - \frac{\rho}{\rho_w}\right)\frac{\partial H^2}{\partial x} = \frac{\partial}{\partial x}(H R_{xx}). \tag{4.56}$$

Integrating with respect to x gives the longitudinal resistive stress

$$R_{xx} = \frac{1}{2}\rho g\left(1 - \frac{\rho}{\rho_w}\right)H + C. \tag{4.57}$$

For an (imaginary) ice shelf of zero thickness, the resistive stress, R_{xx}, must be zero so the integration constant, C, must be zero and

$$R_{xx} = \frac{1}{2}\rho g\left(1 - \frac{\rho}{\rho_w}\right)H. \tag{4.58}$$

The next step is to derive an expression for the horizontal strain rates in the ice shelf. The rate of longitudinal spreading, or creep rate, can be found by invoking the constitutive relation. Bridging effects (discussed in Section 3.4) are zero because the weight of the shelf ice is locally supported by the sea water. Following Thomas (1973), the rate of transverse spreading and lateral shearing are linked to the along-flow creep rate as

$$\dot{\varepsilon}_{yy} = \alpha \dot{\varepsilon}_{xx}, \tag{4.59}$$

$$\dot{\varepsilon}_{xy} = \beta \dot{\varepsilon}_{xx}, \tag{4.60}$$

where α and β are functions of horizontal position (depending on shelf geometry) but constant over the ice thickness. The effective strain rate is then

$$\dot{\varepsilon}_e^2 = (1 + \alpha + \alpha^2 + \beta^2)\,\dot{\varepsilon}_{xx}^2, \tag{4.61}$$

and inverting the flow law gives

$$\dot\varepsilon_{xx} = \theta \left(\frac{R_{xx}}{B} \right)^n,$$

(4.62)

with

$$\theta = \frac{(1 + \alpha + \alpha^2 + \beta^2)^{(n-1)/2}}{(2 + \alpha)^n}.$$

(4.63)

For an ice shelf that is spreading only in the direction of flow (that is, an ice shelf in a parallel-sided bay, with zero or small sidewall friction), $\alpha = \beta = 0$. Substituting expression (4.58) for the longitudinal resistive stress into equation (4.62) gives the expression for the creep rate as derived by Weertman (1957b):

$$\dot\varepsilon_{xx} = \left[\frac{\rho g}{4 B} \left(1 - \frac{\rho}{\rho_w} \right) \right]^n H^n.$$

(4.64)

If the ice shelf is floating freely, spreading may be expected to be uniform in both horizontal directions ($\alpha = 1$) with creep rates

$$\dot\varepsilon_{xx} = \dot\varepsilon_{yy} = 3^{-(n+1)/2} \left[\frac{\rho g}{4 B} \left(1 - \frac{\rho}{\rho_w} \right) \right]^n H^n.$$

(4.65)

Few of the Antarctic ice shelves are spreading freely, and most have formed in embayments, with lateral drag providing additional resistance to flow. Thomas (1973) includes lateral drag in his model for ice-shelf spreading by making the assumption that the lateral shear stress, R_{xy}, is constant throughout the entire ice thickness and varies linearly in the transverse direction:

$$R_{xy}(y) = -\frac{y}{W} \tau_s,$$

(4.66)

with W representing the half-width of the ice shelf and τ_s the shear stress at the margins (as in Section 4.4). By adopting this transverse profile for R_{xy}, the assumption is made that resistance from lateral drag is equally important across the entire width of the ice shelf. Adding lateral drag to the balance equation (4.53) gives

$$\tau_{dx} = -\frac{\partial}{\partial x}(H R_{xx}) + \frac{H \tau_s}{W},$$

(4.67)

or

$$\frac{\partial}{\partial x}(H R_{xx}) = \frac{1}{2} \rho g \left(1 - \frac{\rho}{\rho_w} \right) \frac{\partial H^2}{\partial x} - \frac{H \tau_s}{W}.$$

(4.68)

Integrating this expression from the ice-shelf front (x = L) to some distance L − x upglacier, yields

$$
HR_{xx}(L) - HR_{xx}(L-x) =
$$

$$
= \frac{1}{2} \rho g \left(1 - \frac{\rho}{\rho_w} \right) [H^2(L) - H^2(L-x)] - \int_L^{L-x} \frac{H\tau_s}{W} \, d\bar{x}. \tag{4.69}
$$

At the shelf front, R_{xx} may be calculated from expression (4.58), and the longitudinal resistive stress on the ice shelf is

$$
R_{xx}(x) = \frac{1}{2} \rho g \left(1 - \frac{\rho}{\rho_w} \right) H - \frac{1}{H} \int_x^L \frac{H\tau_s}{W} \, d\bar{x}, \tag{4.70}
$$

with the x-axis directed toward the terminus (so that x < L on the ice shelf).

Again, the spreading rate is found by inverting the constitutive relation. For brevity, the creep rate is written as

$$
\dot{\varepsilon}_{xx} = \theta \left(\frac{R_{xx}^{(0)} - \sigma_b}{B} \right)^n, \tag{4.71}
$$

in which θ is defined as above (equation (4.63)). In this expression, $R_{xx}^{(0)}$ represents the stretching stress for a free-floating ice shelf given by equation (4.58), while σ_b represents a "back pressure" arising from lateral shear (Thomas and Bentley, 1978; MacAyeal, 1987). At any point on the ice shelf, this back pressure is defined as the downglacier integrated resistance associated with lateral drag:

$$
\sigma_b(x) = \frac{1}{H} \int_x^L \frac{H\tau_s}{W} \, d\bar{x}. \tag{4.72}
$$

Thus, the more important lateral drag, the larger the back pressure. The result of this additional resistance from lateral drag is a reduced spreading rate when compared with that of a free-floating ice shelf.

The physical significance of the back pressure as introduced above may be somewhat difficult to understand because the above definition involves comparing a real ice shelf to an imaginary "Weertman-type" ice shelf that is spreading freely. Recall, however, that the Weertman solution is derived from the assumption that gradients in longitudinal stress balance the driving stress and integrating the resulting force-balance equation in the direction of flow. Therefore, at any point on the ice shelf, $R_{xx}^{(0)}$ represents the total driving force per unit width (that is, the integrated driving stress) acting on the section of shelf extending beyond a point x to the shelf front at x = L. Thus, comparing the back pressure with the resistive stress that one would find on a free-floating ice shelf is equivalent to comparing the net resistance to flow associated with lateral drag along this stretch of the ice shelf with the net driving force. If both are the same, all flow resistance is due to lateral drag. On the other hand, if the back pressure is zero, the resistance to flow arises entirely from gradients in longitudinal stress (as for a free-floating ice shelf). In other words, the backstress at any point on an ice shelf represents the fraction of the total driving force acting on the section of ice shelf extending from that point to the calving front, that is, supported by lateral drag and/or basal drag acting on that section.

To illustrate the procedure for estimating back stress, consider a flowline extending from the grounding line of Whillans Ice Stream to the front of the Ross Ice Shelf. At the survey stations on the ice shelf shown in Figure 4.7, the three surface strain rates were determined from the deformation of small strain rosettes (tabulated in Thomas et al., 1984; reproduced here in Table 4.3). Although these strain rates represent local values that may not be entirely representative of larger areas, the errors are sufficiently small to favor using these data rather than large-scale averages

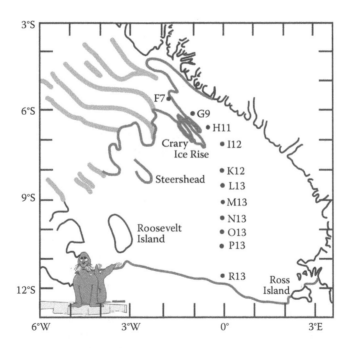

FIGURE 4.7   Location of the stations on the flowline used to calculate back stress on the Ross Ice Shelf. Marginal numbers are grid coordinates, with the Greenwich meridian as grid longitude 0°, and the South Pole at grid latitude 0° (Bentley et al., 1979); 1° in either direction equals 111 km. Light shading indicates lateral shear margins of the ice streams; dark shading represent the grounding line and the ice-shelf front. (From Whillans, I. M., and C. J. van der Veen, *J. Glaciol.,* 39, 483–490, 1993. Reprinted from the *Journal of Glaciology* with permission of the International Glaciological Society and the authors.)

TABLE 4.3

Data Used to Estimate the Role of Gradients in Longitudinal Stress on the Ross Ice Shelf

Station	x (km)	H (m)	U (m/yr)	$\dot{\varepsilon}_{xx}$ (10^{-5} yr^{-1})	$\dot{\varepsilon}_{yy}$ (10^{-5} yr^{-1})	$\dot{\varepsilon}_{xy}$ (10^{-5} yr^{-1})
F17	300.2	760	530	-11.2 ± 15.1	154.2 ± 6.7	43.6 ± 4.1
G9	405.0	620	450	-69.0 ± 2.5	48.0 ± 2.5	-81.7 ± 1.7
H11	481.2	605	432	-1.8 ± 4.9	76.8 ± 4.9	-9.8 ± 0.9
I12	562.0	523	501	49.8 ± 2.9	72.2 ± 5.3	20.0 ± 3.3
K12	664.4	400	547	56.1 ± 1.0	14.9 ± 3.5	18.2 ± 1.4
L13	721.5	365	582	72.3 ± 1.0	1.7 ± 1.0	-16.6 ± 0.3
M13	781.3	360	616	111.4 ± 12.6	-35.4 ± 8.7	26.4 ± 2.7
N13	836.2	360	705	147.5 ± 1.9	-72.5 ± 1.0	-38.3 ± 0.4
O13	893.4	345	815	167.9 ± 5.5	-128.9 ± 3.7	-106.2 ± 2.1
P13	942.2	320	893	134.4 ± 3.7	-50.4 ± 1.9	-63.3 ± 1.3
R13	1057.7	295	998	74.9 ± 1.0	56.2 ± 3.6	-6.7 ± 1.2

Source: Based on data from Thomas, R. H., in *The Ross Ice Shelf: Glaciology and Geophysics,* American Geophysical Union, Antarctic Research Series, Washington, DC, 42, 21–53, 1984.

Note: Strain Rates are defined with respect to a (local) flow-following coordinate system, with x in the direction of flow, and y perpendicular to it.

calculated from velocity gradients. At each station, $R_{xx}^{(0)}$ is estimated from the ice thickness using equation (4.58). Rewriting equation (4.71) as

$$\sigma_b = R_{xx}^{(0)} - B\left(\frac{\dot{\varepsilon}_{xx}}{\theta}\right)^{1/n}, \tag{4.73}$$

the back pressure can then be calculated from measured strain rates.

The calculated back pressure is shown in the upper panel of Figure 4.8. As expected, σ_b increases steadily toward the grounding line (because σ_b represents lateral drag integrated from the ice front toward the interior, and lateral drag cannot be negative). This means that lateral drag is important along the entire ice shelf; if at some site lateral drag would be zero, the curve of σ_b would show a horizontal plateau. Comparison with the curve of $R_{xx}^{(0)}$ in the lower panel of Figure 4.8 shows that both are very similar, indicating that lateral drag is the most important source of flow resistance. $R_{xx}^{(0)}$ as given by equation (4.58) represents the driving stress integrated from the ice-shelf front in the upstream direction. If gradients in longitudinal stress were to be dominant, the longitudinal resistive stress, R_{xx}, calculated from the strain rates (also shown in the lower panel of Figure 4.8), would be (almost) the same as $R_{xx}^{(0)}$. The fact that both curves in the lower panel of Figure 4.8 are so different indicates that gradients in longitudinal stress are unimportant along this flowline.

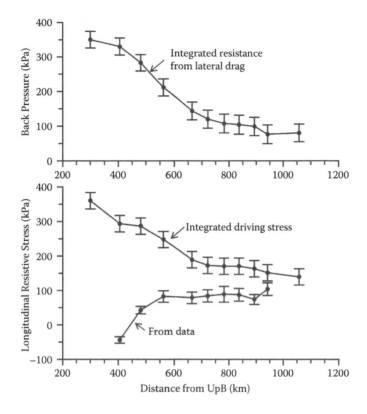

FIGURE 4.8 Calculated back pressure, σ_b, at the stations on the Ross Ice Shelf (upper panel). The lower panel shows the longitudinal resistive stress, R_{xx}, as calculated from equation (4.58) (the value predicted for a free-spreading ice shelf, corresponding to the integrated driving stress), as well as the values calculated from measured strain rates.

4.6 ALONG-FLOW VARIATIONS IN GLACIER FLOW

An important problem in glaciology is the transfer of flow or topographic irregularities at the glacier base to the surface. This topic is of importance because processes at the glacier base may be controlling the mechanics and motion of the glacier and lead to surface undulations or small-scale variations in surface velocity. Direct observation of basal processes is often impossible or very expensive and labor intensive and limited to individual boreholes reaching the glacier bed. The glacier surface, on the other hand, is more readily accessible for measurements, and surface observations are often used to infer conditions at the glacier bed (for example, Thorsteinsson et al., 2003). The difficulty with such procedures is that basal perturbations propagate upward but do so with decreasing amplitude so that, generally, the surface expression is much smaller than the causal bed variations. Inverting this glacier filter to infer basal conditions from surface measurements could lead to calculated patterns of basal drag being dominated by small-scale surface features. Thus, what is needed is an analytical framework that describes the surface response to small amplitude variations in bed topography or basal sliding. This has been discussed many times before (for example, Balise and Raymond, 1985; Balise, 1988; Budd, 1970a, b; Gudmundsson, 1997, 2003, 2008; Hutter, 1981, 1983, Chapter 4; Hutter et al., 1981; Kamb, 1970; Nye, 1969a; Whillans and Johnsen, 1983). Generally, simplifying assumptions are introduced such as linear viscosity, small-amplitude basal varia-tions that allow for a perturbation analysis, or simplified form of the sliding relation linking basal drag to sliding speed. Nevertheless, the analytical treatment becomes quite complex and involved.

The main obstacle to finding analytical solutions that describe glacier flow is the nonlinearity of the constitutive relation (Section 2.2). For this reason, many earlier authors adopted a linear ice rheology in which a particular component of stress depends only on the corresponding component of strain rate (rather than on all other strain rates through the effective strain rate, as in Glen's flow law). The effective viscosity is either taken constant throughout the ice thickness or varies in a prescribed way with depth. The main advantage of a linear rheology is that the stress and velocity solutions may be split in an average and a perturbation component. Doing so leads to a biharmonic solution for the perturbation velocity field, which describes how basal variations are transferred to the glacier surface. Below, this procedure is discussed, based mainly on the analyses of Whillans and Johnsen (1983) and Balise and Raymond (1985). Because the mathematics involved tend to become rather lengthy, the solution method is emphasized more than the analytical procedures.

The geometry is shown in Figure 4.9. Plane flow in the (x, z) plane is considered, with the x-axis along the mean surface of the glacier and the z-axis perpendicular to the x-axis and directed upward. The mean bed is parallel to the glacier surface, as in the plane-slab model.

Adopting a linear rheology allows partitioning of stress and velocity in an aver-age and a perturbation component (denoted by an overbar and a tilde, respectively). Note that the perturbations need not be small compared with the averages. For full stresses, this division gives

$$\sigma_{ij} = \bar{\sigma}_{ij} + \tilde{\sigma}_{ij}, \tag{4.74}$$

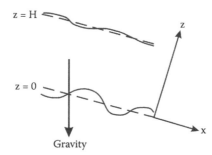

FIGURE 4.9 Geometry of the model for describing along-flow variations in glacier flow. Undulations in the basal and surface topography are exaggerated for clarity.

with the average stress defined as

$$\bar{\sigma}_{ij} = \frac{1}{L} \int_{x-L/2}^{x+L/2} \sigma_{ij} \, dx. \tag{4.75}$$

Here, L is the averaging length, which must be much larger than the ice thickness.

For the chosen system, the component of gravity in the x-direction is $g_x = -\rho g \sin\alpha$, and that in the z-direction is $g_z = -\rho g \cos\alpha$, where ρ represents the density of the ice, g the gravitational acceleration, and α the mean surface (and basal) slope. The equivalent forms of the balance equations (3.8) and (3.10) are then

$$\frac{\partial \sigma_{xx}}{\partial x} + \frac{\partial \sigma_{xz}}{\partial z} = \rho g \sin\alpha, \tag{4.76}$$

$$\frac{\partial \sigma_{xz}}{\partial x} + \frac{\partial \sigma_{zz}}{\partial z} = \rho g \cos\alpha. \tag{4.77}$$

Averaging these equations over the distance L gives

$$\frac{\partial \bar{\sigma}_{xx}}{\partial x} + \frac{\partial \bar{\sigma}_{xz}}{\partial z} = \rho g \sin\alpha, \tag{4.78}$$

$$\frac{\partial \bar{\sigma}_{xz}}{\partial x} + \frac{\partial \bar{\sigma}_{zz}}{\partial z} = \rho g \cos\alpha. \tag{4.79}$$

Subtracting the averaged equations (4.78) and (4.79) from the full equations (4.76) and (4.77), respectively, gives the balance equations for the stress perturbations

$$\frac{\partial \tilde{\sigma}_{xx}}{\partial x} + \frac{\partial \tilde{\sigma}_{xz}}{\partial z} = 0, \tag{4.80}$$

$$\frac{\partial \tilde{\sigma}_{xz}}{\partial x} + \frac{\partial \tilde{\sigma}_{zz}}{\partial z} = 0. \tag{4.81}$$

These equations show that the gravitational body forces are associated with the average stresses. The perturbation solution is independent of the gravitational forces, except where these forces enter into the boundary conditions (Whillans and Johnsen, 1983).

The next step is to derive the biharmonic perturbation equation from which the perturbation velocities can be calculated. First, differentiating equation (4.80) with respect to z and equation (4.81) with respect to x, and subtracting the two resulting equations, gives

$$\frac{\partial^2 \tilde{\sigma}_{xz}}{\partial z^2} - \frac{\partial^2 \tilde{\sigma}_{xz}}{\partial x^2} + \frac{\partial^2}{\partial x \, \partial z}(\tilde{\sigma}_{xx} - \tilde{\sigma}_{zz}) = 0. \tag{4.82}$$

Because strain rates are linked to stress deviators, this equation, containing perturbations of the full stress components, needs to be rewritten in terms of perturbation stress deviators. For plane flow, the spherical stress equals $(\sigma_{xx} + \sigma_{zz})/2$, and the perturbation component of the normal stress deviator is

$$
\begin{aligned}
\tilde{\tau}_{xx} &= \tilde{\sigma}_{xx} - \frac{1}{2}(\tilde{\sigma}_{xx} + \tilde{\sigma}_{zz}) = \\
&= \frac{1}{2}(\tilde{\sigma}_{xx} - \tilde{\sigma}_{zz}).
\end{aligned}
\tag{4.83}
$$

With the deviatoric shear stress equal to the full shear stress, equation (4.82) can also be written as

$$
\frac{\partial^2 \tilde{\tau}_{xz}}{\partial z^2} - \frac{\partial^2 \tilde{\tau}_{xz}}{\partial x^2} + 2\frac{\partial^2 \tilde{\tau}_{xx}}{\partial x \, \partial z} = 0.
\tag{4.84}
$$

The ice is assumed to obey a linear constitutive relation and

$$
\tau_{ij} = 2\eta \dot{\varepsilon}_{ij},
\tag{4.85}
$$

where η represents the effective viscosity of the ice, taken constant here. Using this relation between stress and strain rate, and the definition of strain rates (Section 1.2), allows equation (4.84) to be written in terms of the perturbation velocities:

$$
2\eta \frac{\partial^2}{\partial z^2}\left[\frac{1}{2}\left(\frac{\partial \tilde{u}}{\partial z} + \frac{\partial \tilde{w}}{\partial x}\right)\right] - 2\eta \frac{\partial^2}{\partial x^2}\left[\frac{1}{2}\left(\frac{\partial \tilde{u}}{\partial z} + \frac{\partial \tilde{w}}{\partial x}\right)\right] + 4\eta \frac{\partial^2}{\partial x \, \partial z}\left[\frac{\partial \tilde{u}}{\partial x}\right] = 0.
\tag{4.86}
$$

It is now convenient to introduce a stream function, Ψ, as

$$
\tilde{u} = \frac{\partial \Psi}{\partial z},
$$

$$
\tilde{w} = -\frac{\partial \Psi}{\partial x}.
\tag{4.87}
$$

By defining the stream function this way, the incompressibility condition (1.39) is always satisfied. Replacing velocities in equation (4.86) by derivatives of the stream function gives

$$
\frac{\partial^4 \Psi}{\partial x^4} + \frac{\partial^4 \Psi}{\partial z^4} + 2\frac{\partial^4 \Psi}{\partial x^2 \, \partial z^2} = 0.
\tag{4.88}
$$

This equation does not contain the effective viscosity, η, and the perturbation velocities are independent of the stiffness of the glacier ice. The perturbation stresses are, of course, dependent on the effective viscosity through the flow law (4.85).

The perturbation solution can now be found by solving the biharmonic equation (4.88) for the stream function, Ψ. For example, Whillans and Johnsen (1983) adopt a solution that is periodic in x and varies exponentially with depth

$$\Psi = \left[(a_1 + a_2 z)e^{\omega z} + (a_3 + a_4 z)e^{-\omega z} \right] \sin \omega x +$$

$$+ \left[(a_5 + a_6 z)e^{\omega z} + (a_7 + a_8 z)e^{-\omega z} \right] \cos \omega x. \tag{4.89}$$

The eight constants a_i are to be determined from the boundary conditions, including those associated with variations in the surface and basal topography. To avoid lengthy arithmetic, the example of transfer of basal sliding variations to the surface (Balise and Raymond, 1985; Balise, 1988) is considered next, neglecting variations in the surface and basal topography. This simplifies the general form of the stream function and reduces the number of required boundary conditions to four. The more elaborate procedure followed by Whillans and Johnsen (1983) is essentially the same as that discussed below.

Balise and Raymond (1985) and Balise (1988) investigate the transfer of basal velocity anomalies to the surface of a glacier that is approximated as a planar parallel-sided slab of constant thickness, H. Using the Fourier transform of the biharmonic equation (4.88), surface velocities can be found for any perturbation in basal velocity. To illustrate the solution method and the basic properties of the transfer function, the example of a harmonic surface–parallel basal velocity anomaly is discussed here, following Balise (1988). The formulas given in Balise (1988) and below differ from those in Balise and Raymond (1985) because the latter authors use a coordinate system in which the z-axis is directed downward. Here, as in Balise (1988), the z-axis is positive upward.

The perturbation velocity parallel to the glacier bed is taken to be periodic in the flow direction as

$$\tilde{u}_b(x) = U_b \sin \omega x, \tag{4.90}$$

where U_b represents the amplitude of the basal perturbation. The perturbation velocity perpendicular to the bed is set to zero:

$$\tilde{w}_b(x) = 0. \tag{4.91}$$

The basal boundary conditions compatible with these basal velocity perturbations follow immediately from definition (4.87) of the stream function:

$$\frac{\partial \Psi}{\partial z} = U_b \sin \omega x, \qquad (z = 0) \tag{4.92}$$

and

$$\frac{\partial \Psi}{\partial x} = 0. \qquad (z = 0). \tag{4.93}$$

Variations in the surface and bed topography are neglected and the general solution adopted by Whillans and Johnsen (1983) may be simplified somewhat by omitting the cosine-term on the right-hand side of equation (4.89). That is

$$\Psi(x,z) = U_b\left[(a_1 + a_2 z)e^{\omega z} + (a_3 + a_4 z)e^{-\omega z}\right]\sin \omega x. \qquad (4.94)$$

This expression contains four constants to be determined. Thus, in addition to the two boundary conditions (4.92) and (4.93) applying to the glacier base, two more boundary conditions are needed. These follow from the condition that the glacier surface must be stress free.

Because the x-axis is chosen parallel to the glacier surface, the shear stress at the surface must be zero and

$$\tilde{\tau}_{xz}(x,H) = 0. \qquad (4.95)$$

Using the linear constitutive relation (4.85), this condition is equivalent to

$$\frac{\partial \tilde{u}}{\partial z} + \frac{\partial \tilde{w}}{\partial x} = 0, \qquad (z = H) \qquad (4.96)$$

or in terms of the stream function,

$$\frac{\partial^2 \Psi}{\partial z^2} - \frac{\partial^2 \Psi}{\partial x^2} = 0. \qquad (z = H) \qquad (4.97)$$

The full stress perturbation perpendicular to the surface must also be zero. This means that at the glacier surface, the deviatoric stress perturbation is (from equation (4.83))

$$\tilde{\tau}_{xx}(x,H) = \frac{1}{2}\tilde{\sigma}_{xx}(x,H). \qquad (4.98)$$

At the surface, the perturbation balance equation (4.80) then reduces to

$$2\frac{\partial \tilde{\tau}_{xx}}{\partial x} + \frac{\partial \tilde{\tau}_{xz}}{\partial z} = 0. \qquad (z = H) \qquad (4.99)$$

Again, invoking the linear flow law and the definitions of the stream function gives the following boundary condition:

$$3\frac{\partial^3 \Psi}{\partial x^2 \partial z} + \frac{\partial^3 \Psi}{\partial z^3} = 0. \qquad (z = H) \qquad (4.100)$$

The four constants a_i appearing in expression (4.94) for the stream function can now be determined from the four boundary conditions (4.92), (4.93), (4.97), and (4.100). After some tedious arithmetic, the following result is found (Balise, 1988, p.10):

$$a_1 = \frac{-2\omega H^2}{2 + 4\omega^2 H^2 + e^{2\omega H} + e^{-2\omega H}}, \tag{4.101}$$

$$a_2 = \frac{1 - 2\omega H + e^{-2\omega H}}{2 + 4\omega^2 H^2 + e^{2\omega H} + e^{-2\omega H}}, \tag{4.102}$$

$$a_3 = \frac{2\omega H^2}{2 + 4\omega^2 H^2 + e^{2\omega H} + e^{-2\omega H}}, \tag{4.103}$$

$$a_4 = \frac{1 + 2\omega H + e^{-2\omega H}}{2 + 4\omega^2 H^2 + e^{2\omega H} + e^{-2\omega H}}. \tag{4.104}$$

With the stream function known, the two components of velocity can be determined using equation (4.87). Symbolically, the velocities can be written as

$$\tilde{u}(x,z) = F_u(\omega,z)\, U_b \sin \omega x, \tag{4.105}$$

$$\tilde{w}(x,z) = F_w(\omega,z)\, U_b \cos \omega x. \tag{4.106}$$

In these expressions, F_u and F_w represent the transfer functions, defined as (Balise, 1988, p.11)

$$F_u(\omega,z) = \omega a_1 e^{\omega z} - \omega a_3 e^{-\omega z} + (a_2 + a_2 \omega z)e^{\omega z} + (a_4 - a_4 \omega z)e^{-\omega z}, \tag{4.107}$$

and

$$F_w(\omega,z) = \omega a_1 e^{\omega z} + \omega a_3 e^{-\omega z} + a_2 \omega z e^{\omega z} + a_4 \omega z e^{-\omega z}. \tag{4.108}$$

These transfer functions describe how the basal anomaly is attenuated upward to the glacier surface. Because the amplitude of the perturbation decreases going upward toward the glacier surface, the transfer functions are also referred to as filter functions. The transfer functions are independent of the along-flow coordinate, and the amount of filtering (or relative decrease in amplitude) is independent of the amplitude of the basal perturbation. The only important parameter in the transfer functions is the wavelength of the basal perturbation, $L = 2\pi/\omega$. This is illustrated in Figure 4.10, which shows F_u and F_w as a function of the dimensionless wavelength, L/H, and dimensionless depth, z/H.

For wavelengths longer than about 50 ice thicknesses, the basal perturbation attenuates almost undisturbed to the glacier surface. For very short wavelengths, less than about one ice thickness, the perturbation is damped very rapidly going upward in the ice, affecting only the lower 10% of the glacier. For intermediate wavelengths,

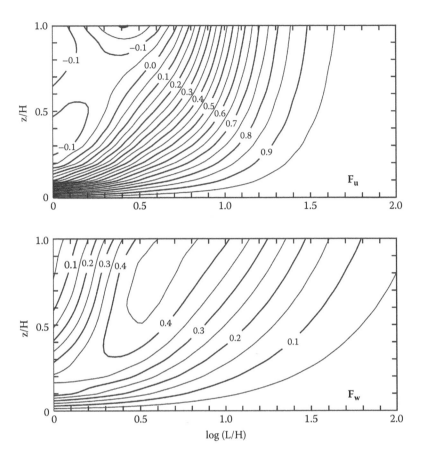

FIGURE 4.10 Transfer (or filter) functions at depth and as a function of the dimensionless wavelength of the harmonic perturbation in the velocity parallel to the bed. F_u represents the transfer function for the horizontal component of velocity and F_w represents the transfer function for the vertical velocity component, both for the case of a linear-viscous glacier.

the transfer of basal perturbations is most complex. This is better illustrated by considering the perturbation velocities at the glacier surface.

The transfer functions evaluated at the surface, $z = H$, are

$$F_u(\omega, H) = \frac{(2 - 2\omega H)e^{\omega H} + (2 + 2\omega H)e^{-\omega H}}{2 + 4\omega^2 H^2 + e^{2\omega H} + e^{-2\omega H}}, \tag{4.109}$$

$$F_w(\omega, H) = \frac{2\omega H e^{\omega H} + 2\omega H e^{-\omega H}}{2 + 4\omega^2 H^2 + e^{2\omega H} + e^{-2\omega H}}. \tag{4.110}$$

These transfer functions are shown in Figure 4.11 and represent the relative amplitude of the surface perturbation velocity, if the amplitude of the basal perturbation is taken equal to unity (= 1). Figure 4.11 shows that three different regimes can be

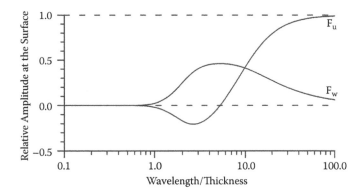

FIGURE 4.11   Surface value of the transfer functions shown in Figure 4.10.

identified, depending on the wavelength of the basal perturbation. For very short wavelengths (L ≤ 0.7 H), there is no surface expression of the basal perturbation. If the wavelength is comparable to the thickness of the glacier (0.7 H ≤ L ≤ 5 H), the perturbation velocity parallel to the surface has a direction opposite that of the basal perturbation. This effect is largest for L/H = 2.75, with the surface velocity about 20% of the basal velocity. The result is a circular motion of the perturbation flow, as shown in Figure 4.12. The velocity perpendicular to the surface reaches a maximum for L/H ≈ 5, while for this wavelength the surface-parallel component of velocity is zero. For increasing wavelengths larger than five ice thicknesses, the surface expression of the basal perturbation becomes increasingly similar to the basal perturbation, and the velocity anomaly does not recirculate (Balise and Raymond, 1985).

The model used to arrive at expressions for the transfer functions may be criticized, most notably for using a linear viscous rheology. Balise (1988) uses a finite-difference model to investigate how a power-law rheology affects the transfer functions. The main difference between the linear and power-law solutions is that the magnitudes of the nonlinear transfer functions, F_u and F_w, are considerably less than the magnitudes of the corresponding linear transfer functions (4.107) and (4.108). Thus, for a given harmonic variation in the bed-parallel velocity, the nonlinear rheology results in less effect at the glacier surface than with the linear rheology. Interestingly, for a harmonic variation in the component of velocity perpendicular to the glacier bed (w_b), both rheologies produce similar results.

The nonlinear transfer functions are determined by Balise (1988) by prescribing a harmonic basal perturbation of certain wavelength and calculating numerically the surface perturbations. The transfer functions so determined have more limited applicability than their linear counterparts. For a linear rheology, an arbitrary basal variation may be considered the sum of (many) harmonic perturbations. Applying the corresponding linear transfer function to each harmonic component allows the various harmonic surface perturbations to be calculated. Adding these components gives the total perturbation velocity at the surface. This Fourier-transform method is possible because the harmonic perturbations do not interact with each other in the linear model. In the more realistic case of nonlinear ice rheology, the effective

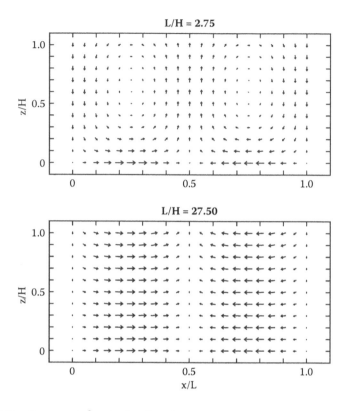

FIGURE 4.12   Pattern of perturbation flow for a short-wavelength (upper panel) and long-wavelength (lower panel) harmonic perturbation in the velocity parallel to the glacier bed.

strain rate depends on all other strain rates and each harmonic component influences all others. In addition, the large-scale flow also affects the perturbation flow. This means that each situation needs to be studied separately, using numerical techniques to solve the nonlinear equations for the complete velocity and stress distributions (Balise, 1988).

A further objection against the model of Balise and Raymond (1985) may be its simple geometry that does not allow for variations in the upper and lower glacier surfaces. As shown by Whillans and Johnsen (1983), perturbations in geometry can be included, by retaining the full form (4.89) of the stream function and imposing additional boundary conditions. Their analysis indicates that sliding variations on a flat bed may lead to important surface relief. For sliding variations with a wavelength of about three ice thicknesses, Whillans and Johnsen (1983) find flow variations similar to those shown in Figure 4.12 (L/H = 2.75), with the surface undulation out of phase with the basal variation. That is, where the velocity perpendicular to the surface is directed outward, a surface high occurs, while a surface low is found where this velocity is directed into the glacier.

The main reason for discussing the earlier analyses of Whillans and Johnsen (1983), Balise and Raymond (1985), and Balise (1988) in some detail is their relative

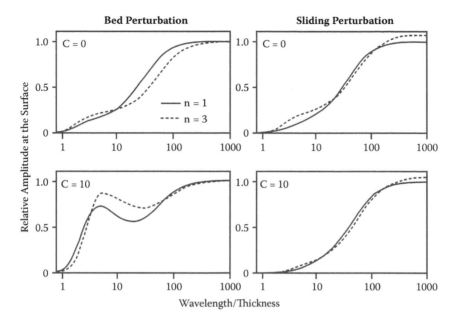

FIGURE 4.13 Relative amplitude of surface topography for a perturbation in basal topography (left panels) and sliding perturbation (right panels), for the case of no or little sliding (upper panels) and for large basal sliding (lower panels), and for linear (n = 1) and nonlinear (n = 3) ice rheology. (From Raymond, M. J., and G. H. Gudmundsson, *J. Geoph. Res.*, 110, B08411, 2005. Copyright by the American Geophysical Union.)

simplicity. More recent treatments are perhaps more rigorous but are also of greater mathematical complexity, which may obscure the underlying physics. For example, Gudmundsson (2008) presents analytical solutions describing the effects of perturbations in basal boundary conditions for ice streams moving over soft sediment. Linear viscosity is assumed for the ice. Scaling is applied to the governing momentum equations to arrive at a system of equations that is solved using Fourier and Laplace methods. While in detail, the conclusions from that analysis and similar recent ones differ from earlier studies, from the perspective of using surface measurements to infer conditions at the glacier bed, the conclusions are similar, namely, that short-scale variations at the glacier bed are difficult to detect at the glacier surface, if not completely absent.

Figures 4.13 and 4.14 show transfer functions obtained by Raymond and Gudmundsson (2005) using finite-element solutions to the full momentum equations for both linear (n = 1) and nonlinear (n = 3) ice rheology. Perturbations in bed topography (panels on the left) or basal sliding (panels on the right) were prescribed, and their effect on surface topography (Figure 4.13) and surface speed (Figure 4.14) was investigated in the absence of significant basal motion (C = 0 or 1; upper panels) and for the case of large sliding (C = 10; lower panels). For sliding variations the transfer functions for both the surface topography and surface velocity increase steadily with wavelength of the perturbation. For transfer of basal topographic variations to the surface, three spatial scales should be considered as also suggested by the transfer

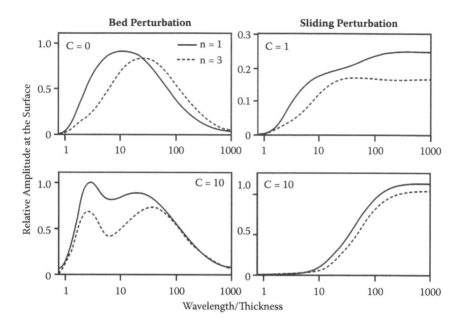

FIGURE 4.14   Relative amplitude of surface velocity for a perturbation in basal topography (left panels) and sliding perturbation (right panels), for the case of no or little sliding (upper panels) and for large basal sliding (lower panels), and for linear (n = 1) and nonlinear (n = 3) ice rheology. (From Raymond, M. J., and G. H. Gudmundsson, *J. Geoph. Res.*, 110, B08411, 2005. Copyright by the American Geophysical Union.)

functions shown in Figure 4.11. On short scales (wavelength less than about three ice thicknesses), surface expressions of basal perturbations are small, more so when nonlinear rheology is assumed. When basal sliding is zero, the transfer function for surface topography (Figure 4.13, upper left panel) increases rapidly for intermediate wavelengths and reaches unity for long wavelengths. Where sliding is important, this transfer function reaches a maximum at about seven ice thicknesses, then decreases, and increases again for longer wavelengths (Figure 4.13, lower left panel). The associated perturbation in surface velocity reaches a maximum for intermediate wavelengths and decreases to zero for long wavelengths (Figure 4.14, left panels). This behavior likely reflects that for very short perturbations, flow anomalies are concentrated in the basal layers, while for very long perturbations, the ice can simply follow the bed topography without adjusting its flow. At intermediate scales, longitudinal stresses and other flow interactions lead to the greatest surface expression.

The implication is that basal perturbations with spatial scales of less than about one ice thickness cannot be detected using surface measurements. Also, small errors in measured surface velocities may lead to unrealistically large variations in sliding velocity. Consider, for example, the situation in which surface measurements are spaced one ice thickness apart. If the uncertainty in surface velocity is, say, 1 cm/yr, the corresponding error in basal velocity is about 90 cm/yr. Increasing the measurement interval to two ice thicknesses would reduce the error in basal velocity to about

6 cm/yr. Achieving a similar accuracy by making more precise surface measurements at the original spacing of one ice thickness would require a reduction in the measurement error to less than 1 mm/yr. Such an accuracy cannot be accomplished in many cases.

While the discussion in this section focuses on the transfer of basal variations to the glacier surface, these theories and models can also be used to explain observed variations in internal layers detected by radar. For example, Whillans and Johnsen (1983) apply their model to the Byrd Station Strain Network, West Antarctica, and find that the phase relationship between internal layer distortion and surface topography agrees with model predictions, while Whillans and Jezek (1987) successfully apply a similar theoretical model to explain folding of internal layers along the Dye-3 strain grid. Hindmarsh and others (2006) develop a mechanical theory to show that where wavelengths of basal topography are comparable with or less than the ice thickness, internal layers tend to override the basal topography with layers above topographic highs being essentially flat. For longer wavelengths, internal layers are draped over the topography, with the layers essentially following the bed topography. For the most part, however, the use of internal layer structure has not been widely used in glaciology to constrain or validate theoretical and numerical models.

4.7   FLOW NEAR AN ICE DIVIDE

Most deep-ice-drilling programs, aimed at obtaining long paleoclimatic records, are conducted at or near ice divides on ice caps. The rationale is that, provided the divide has not migrated in the past, ice at the divide is not affected by horizontal flow and may be assumed to have originated at the same location. This means that variations with depth of parameters measured in the core, such as annual layer thickness or isotope concentration, reflect temporal changes in the local climate conditions (precipitation, temperature, etc.). In contrast, for a drilling site removed from the divide, there is a need to model ice flow, including past history, along the flowline extending from the divide to the drilling location. The lack of horizontal perturbations to flow at the divide often outweighs the problems associated with establishing an accurate depth-age scale for an ice divide. The reason for this difficulty is that conventional models for glacier flow, in particular the lamellar flow theory (Section 4.2), do not apply to the divide region where the vertical shear stress vanishes.

The ice divide, or flow divide, is defined as the plane extending from the base to the surface, through which there is no transport of mass. The ice crest is the location where surface elevation is highest. Thus, the divide separates ice flowing in opposite directions. On smaller ice caps or at local ice domes that are spreading radially, the divide reduces to a single line from surface to bed. In the absence of basal topography, or where a bed topographic high underlies the ice crest, the flow divide occurs at the highest surface elevation (Figure 4.15, upper panel). However, most divides are not symmetric because there may be a gradient in net accumulation across the ice sheet, or because of an asymmetric basal topography. In that case, the flow divide and ice crest are co-located at the surface only, but the divide plane slopes such that the flow divide at the base does not directly underly that at the surface (Figure 4.15, lower panel).

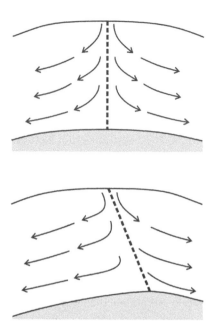

FIGURE 4.15 Illustrating different ice divide geometries. Where the bed topography is symmetric around the vertical line through the ice crest (highest surface elevation), the ice divide (or flow divide) is also vertical (dashed line in upper panel). Where the bed topography is asymmetric, the ice divide is generally sloping and coincides with the ice crest only at the surface (dashed line, lower panel).

The distribution of stresses and velocities near an ice divide has been discussed in a number of studies using numerical methods to solve the full balance equations (Raymond, 1983; Paterson and Waddington, 1984; Reeh and Paterson, 1988; Dahl-Jensen, 1989; Szidarovsky et al., 1989; Hvidberg, 1996). The main conclusion that can be drawn from these studies is that the divide region is relatively small. At a distance of about four ice thicknesses from the divide, flow on the flank is adequately described by the lamellar flow assumption. Following Raymond (1983) an approximate analytical solution can be derived.

For an isothermal divide on a horizontal bed, model results show that at the divide the vertical velocity varies approximately parabolically with depth and

$$w(s) = w(0)(1 - s)^2, \tag{4.111}$$

with $w(0)$ the vertical velocity at the surface ($s = 0$). For steady state, this velocity must balance the accumulation rate and thus the vertical velocity is negative throughout the column, decreasing to zero at the bed (neglecting basal melting). The corresponding vertical strain rate is

$$\dot{\varepsilon}_{zz} = 2\frac{w(0)}{H}s, \tag{4.112}$$

and increases linearly from zero at the surface to the maximum value at the bed. Note that w(0)/H represents the depth-averaged strain rate, that is, the difference between the vertical velocity at the surface and at the bed, divided by the ice thickness.

Incompressibility dictates that the horizontal stretching rate equals minus the vertical strain rate. With the horizontal velocity being zero at the ice divide ($x = 0$), integrating equation (4.112) with respect to x and adding a minus sign gives the vertical profile of the horizontal velocity

$$u(s) = u(0)(1 - s). \tag{4.113}$$

In the integration, the ice thickness is kept constant, and thus small variations in geometry in the vicinity of the divide are neglected. The horizontal velocity at the surface increases with distance from the divide as

$$u(0) = -2\frac{w(0)}{H}x. \tag{4.114}$$

Because the vertical velocity is directed downward ($w(0) < 0$), the horizontal velocity is positive.

A comparison between the divide solution and the lamellar flow solution is shown in Figure 4.16. Because the horizontal velocity varies linearly with depth in the divide region, vertical shear is uniform throughout the column and the normalized velocity is smaller than on the flank. In contrast, the normalized vertical velocity in the divide region is greater than on the flank. While differences in the two profiles of the vertical velocity shown in the right panel of Figure 4.16 are less pronounced than those in the profiles in the horizontal velocity, this difference has interesting consequences for the shape of isochrones (layers of constant age) in the divide region.

The left panel in Figure 4.17 shows the vertical thinning rate, $\dot{\varepsilon}_{zz}$, and shows that thinning of layers occurs more uniformly with depth at the ice divide than on

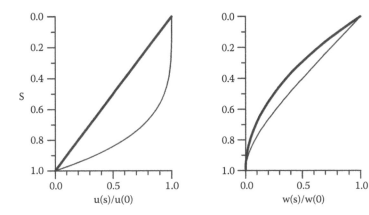

FIGURE 4.16   Horizontal (left panel) and vertical (right panel) component of velocity scaled to the surface velocity, as a function of dimensionless depth, for the approximate divide solution (heavy curves) and the lamellar flow solution (light curves).

the flanks, where thinning is greatest in the upper half of the glaciers, and rapidly decreases near the bottom. As a result, downward flow is impeded at the divide, and ice at a given depth will be of younger age than ice at the same depth on the flank. To quantify this difference, the age of the ice at depth can be estimated from

$$a(s) = \int_0^s \frac{H}{w(\overline{s})} d\overline{s}. \tag{4.115}$$

Integrating this expression gives

$$a(s) = \frac{H}{w(0)} \frac{1}{1-s}. \tag{4.116}$$

For the lamellar flow solution (4.31), the integral is more complex and estimated from numerical integration. The right panel in Figure 4.17 shows both age-depth relations. Note that these results apply only to an ice divide that has been stationary and with constant thickness and accumulation rate, which probably are not very realistic assumptions for real divide regions.

On the flank, ice moves downward more quickly than near the divide, producing local upward arches in isochrones under the divide, called Raymond bumps. Such bumps have been observed in radar-sounding records (that record internal stratigraphic layers) in several locations including Fletcher Promontory, a large ice rise on the Ronne Ice Shelf, West Antarctica (Vaughan et al., 1999); on Siple Dome, a large ice dome adjacent to the Ross Ice Shelf and separating Kamb and Echelmeyer ice streams (Nereson et al., 1998); and on Roosevelt Island on the Ross Ice Shelf (Conway et al., 1999). Radar transects across the divide in Greenland do not show such upwarping arches (Jacobel and Hodge, 1995), which

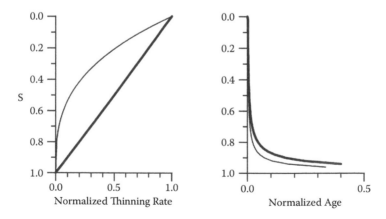

FIGURE 4.17   Normalized thinning rate (left panel) and normalized age (right panel) as a function of dimensionless depth, for the approximate divide solution (heavy curves) and the lamellar flow solution (light curves).

may be because the divide migrated during the last glacial cycle (Marshall and Cuffey, 2000).

The Raymond (1983) model applies to isothermal flow in the vicinity of an ice divide. Since that study, several investigations have been carried out to investigate the importance of ice temperature, ice rheology, and divide migration on isochrone geometry. Reduced downward advection of cold surface ice may keep the basal ice under the divide warmer than away from the divide, thereby softening the ice. The model results of Hvidberg (1996) suggest that this thermal softening lowers the amplitude of the Raymond bumps.

Further refinements to modeling divide flow have been made. Pettit and Waddington (2003) argue that at low deviatoric stresses found near divides, a linear flow law is more appropriate than the usual Glen's law with exponent $n = 3$. Their model results indicate that adopting a linear flow law yields velocity profiles that are similar at the divide and at the flank so that internal layers or isochrones do not show the Raymond bump. Pettit et al. (2007) further investigated the role of crystal fabric, recognizing that near an ice divide, a preferred crystal fabric with c-axes clustered vertically is often found. Such a fabric tends to increase the amplitude of the Raymond bump. The model of Martin et al. (2009) applies to isothermal conditions but includes calculation of fabric development induced by ice deformation, and a nonlinear anisotropic flow law. Their model produces a number of features observed in radar echograms, including concave surface undulations on both sides of the divide, synclines in isochrones on both sides of the Raymond bumps, and double-peaked Raymond bumps close to the bed.

5 Equilibrium Profiles of Glaciers

5.1 PERFECT PLASTICITY

The rate of deformation of glacier ice under its own weight is described by the constitutive relation discussed in Chapter 2. According to the commonly used flow law (2.10), the rate of deformation increases with the third power of the applied stress. In Figure 5.1 this relation is shown schematically. Also shown are two other common forms of rheologic behavior, namely, Newtonian (linear) viscosity and perfect plasticity.

For a Newtonian material, the rate of deformation is linearly dependent on the applied stress, with the constant of proportionality termed the viscosity. In many models for glacier flow, ice is approximated as a Newtonian material to find an analytical solution to the problem under study (see, for example, the model for along-flow variations in ice flow described in Section 4.6). This approximation is valid only where the effective stress is constant over the entire thickness of the glacier. Where this assumption is not valid, effects of the depth-variation in effective stress, and perhaps temperature also, can be included in the linear-viscosity model by prescribing a depth dependency of the viscosity (for example, Whillans and Johnsen, 1983).

A material exhibits perfect plasticity if deformation is negligible when the applied stress is below some critical value, the *yield stress*. For stresses larger than the yield stress, the material deforms "instantly" to relieve the applied stress. As a result, the stress in the material never exceeds the yield stress. Comparison of the curves shown in Figure 5.1 suggests that, at least for larger stresses, glacier flow may be treated as a problem in plasticity (Orowan, 1949; Nye, 1951; Reeh, 1982). In effect, this is equivalent to taking the limit $n \to \infty$ for the exponent in the flow law. Making the perfect plasticity approximation allows the geometry of a glacier to be determined with a minimum of information. While this may not be realistic, it is often the best one can do, especially when reconstructing the large ice sheets that covered the American and European continents during the last glacial period.

Neglecting the effects of gradients in longitudinal stress and lateral drag, the driving stress is balanced by drag at the glacier bed, and the shear stress, τ_{xz}, increases linearly with depth from zero at the surface to the value of the basal drag at the bed (c.f. Section 4.2). Assuming perfect plasticity, the stress in the ice cannot exceed the

113

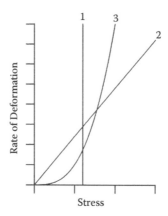

FIGURE 5.1   Rate of deformation as a function of applied stress for three commonly used rheological models. Curve 1 corresponds to a perfectly plastic material, curve 2 represents a linear-viscous material (Newtonian viscosity), and curve 3 corresponds to glacier ice deforming according to Glen's law with exponent n = 3.

yield stress, τ_o, and basal drag must be equal to τ_o. Equating basal drag to the driving stress, it follows immediately that the driving stress is constant, so that

$$-\rho g H \frac{\partial h}{\partial x} = \tau_o, \tag{5.1}$$

where H represents the ice thickness and h the surface elevation. It may be noted that the assumptions made above are perhaps unnecessarily strict. A more general derivation is presented by Nye (1951). His analysis, although more complex because normal stress components are also included, gives essentially the same result for the shear stress, and also leads to equation (5.1) linking the driving stress to the yield stress.

Making the assumption that the glacier rests on a horizontal bed ($\partial h/\partial x = \partial H/\partial x$), equation (5.1) can be integrated to give

$$H^2 = C - \frac{2\tau_o}{\rho g} x. \tag{5.2}$$

The integration constant, C, can be determined from the condition that the ice thickness must be zero at the edge of the glacier. Denoting the half-width of the glacier by L, this constant is found to be

$$C = \frac{2\tau_o}{\rho g} L, \tag{5.3}$$

and

$$H^2 = \frac{2\tau_o}{\rho g}(L - x). \tag{5.4}$$

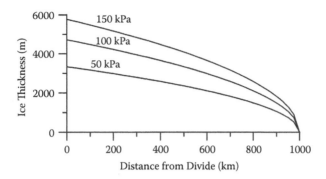

FIGURE 5.2   Flowline profiles of a perfectly plastic ice sheet resting on a flat bed. Labels give the value for the yield stress used in each calculation.

The thickness at the divide $(x = 0)$ is given by

$$H_o = \left(\frac{2\tau_o}{\rho g} L \right)^{1/2}, \tag{5.5}$$

and the profile of the plastic ice sheet can be written as

$$H = H_o \left(1 - \frac{x}{L} \right)^{1/2}. \tag{5.6}$$

The elevation decreases parabolically toward the glacier edge; the surface slope at the margin becomes infinite. Examples calculated for different values of the yield stress are shown in Figure 5.2.

In deriving the parabolic profile, continuity of mass is not considered. Of course, integrated over the entire glacier, snowfall must be larger than ablation to prevent the glacier from wasting away. However, any increase in snowfall has no effect on the ice-sheet profile but is compensated by an increase in ice discharge toward the glacier edge. This admittedly unrealistic behavior is the result of the decoupling between stresses and strain rates in the plastic approach.

For steady-state conditions, the discharge velocity, U, can be derived from conservation of mass. The mass flux through a vertical cross-section of unit width is equal to the discharge velocity (the vertically averaged velocity in the mean direction of flow) times the local ice thickness. To maintain steady state, the mass flux at any location must be equal to the integrated upstream accumulation. That is,

$$H U = \int_0^x M(\bar{x}) d\bar{x}, \tag{5.7}$$

where M represents the net surface accumulation (snowfall minus melting). Taking M constant along the flowline, it follows that the mass flux must increase linearly

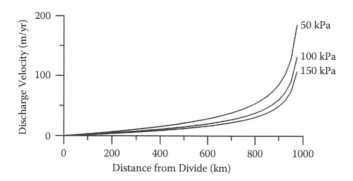

FIGURE 5.3   Discharge velocities required to maintain steady state for the ice-sheet profiles shown in Figure 5.2. The accumulation rate was set to 0.1 m of ice per year.

with distance from the divide. Eliminating the ice thickness using equation (5.6), the discharge velocity is found to be

$$U = \frac{Mx}{H_o}\left(1 - \frac{x}{L}\right)^{1/2}. \tag{5.8}$$

The discharge velocities corresponding to the profiles in Figure 5.2 are shown in Figure 5.3.

At the terminus of the glacier, the model fails. As $x \rightarrow L$, the ice thickness goes to zero and the discharge velocity must go to infinity to maintain the required discharge. This shortcoming is not unique to the plastic model. Most steady-state models in which the ice thickness approaches zero at the glacier terminus predict velocities that become unrealistically large near the snout.

The plasticity approach can be extended to include the second horizontal dimension and to reconstruct the steady-state shape of a three-dimensional ice sheet (Reeh, 1982). On a flat bed, the parabolic profile (5.6) applies to flowlines that are perpendicular to the elevation contours. Because the margin constitutes an elevation contour (namely, H = 0), flowlines are perpendicular to the margin and remain straight when traced toward the interior. This means that where the margin is convex, the upstream region of the ice sheet exhibits divergent flow, and ice divides may develop that run inland from the margins. A concave margin, on the other hand, indicates convergent flow upstream, which may feed into ice streams (Reeh, 1982). Where the bed is not horizontal, deflection of flowlines may occur; an analytical expression for the profile cannot be readily derived, and numerical methods have to be used.

Perhaps the most valuable application of the perfect-plasticity approximation is the reconstruction of former ice sheets. Out of necessity, such reconstructions are often based on sparse geological observations of glacial landforms such as end moraines that indicate the maximum extent of the former glacier and other geological markings that indicate the nature of the former flow regime as well. For example, striated pavements suggest that basal sliding was important, while till sheets may be indicative of streaming flow. Combining various sources of information, Denton and

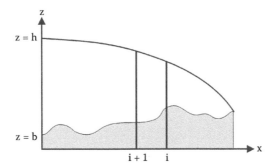

FIGURE 5.4 Geometry used in the numerical solution scheme for reconstructing the profile of a glacier on an arbitrary bed topography, using the perfectly plastic approximation. Note that the gridpoint number, i, increases toward the interior.

Hughes (1981) and Hughes (1985) present possible reconstructions of the Cenozoic ice sheet during the time of maximum extent.

The reconstruction scheme is based on equation (5.1), which is solved numerically on discrete gridpoints, starting at the glacier margin and proceeding upglacier. To do so, the continuous derivative of the surface elevation is discretized as (see Figure 5.4 for the notation used)

$$\frac{\partial h}{\partial x} = \frac{h_i - h_{i+1}}{\Delta x},$$
(5.9)

noting that the x-axis is directed from the ice-sheet center toward the margin. The subscript i refers to gridpoints, with i = 1 denoting the position of the ice margin (where calculations start), while Δx is the horizontal spacing between adjacent gridpoints. Equation (5.1) can thus be rewritten to give (Schilling and Hollin, 1981; Hughes, 1985)

$$h_{i+1} = h_i + \left(\frac{\tau_0}{H}\right)\frac{\Delta x}{\rho g}.$$
(5.10)

To be fully correct, the yield stress and ice thickness, appearing on the right-hand side of this expression, should be evaluated at i + 1/2. Multiplying equation (5.10) by the ice thickness gives

$$(h_{i+1} - h_i)H_{i+1/2} - \frac{2\,\Delta x\,\bar{\tau}_0}{\rho g} = 0.$$
(5.11)

Denoting the bed elevation by b_i (positive when above sea level), the ice thickness is $H_i = h_i - b_i$. Evaluating the thickness over the grid interval from the average of the two values at the surrounding gridpoints, the following equation is found:

$$h_{i+1}^2 - h_{i+1}(b_i + b_{i+1}) + h_i(b_{i+1} - H_i) - \frac{2\Delta x\,\bar{\tau}_0}{\rho g} = 0,$$
(5.12)

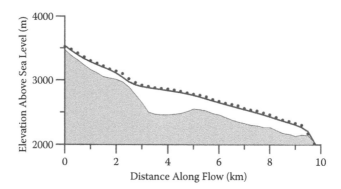

FIGURE 5.5 Reconstructed and measured profile of the Rhône Glacier, France. Measured surface (dots) and basal topography are from Stroeven, A., R. van de Wal, and J. Oerlemans (1989). The reconstructed profile was calculated using the perfectly plastic approximation with a yield stress of 200 kPa.

where the overbar indicates that the average value of the yield stress over the grid interval is to be used.

The advantage of using equation (5.12) instead of the more simple version (5.10) is that the first equation can also be applied at the margin to calculate the surface elevation at the next upglacial gridpoint (i.e., h_2). This is not the case when equation (5.10) is used, because the ice thickness is zero at the margin. Instead, h_2 must be estimated from, for example, the parabolic solution (equations (5.5) and (5.6)) if the bed near the margin is flat, or from the solutions that apply to a sloping bed derived in Nye (1952b) (c.f. Schilling and Hollin, 1981).

A comparison between a "reconstructed" and an actual glacier profile is shown in Figure 5.5. The glacier considered is the Rhône Glacier in the French Alps, for which the surface and bed elevations are tabulated in Stroeven et al. (1989). The reconstructed surface profile was obtained using a yield stress of 200 kPa. This may seem like a large value for a relatively small valley glacier. However, the width of the glacier is about 1 to 2 km (Stroeven et al., 1989), and lateral drag may be expected to be important. Because the driving stress is equated to the yield stress, τ_o, basal drag is fτ_o, where f represents the shape factor (c.f. Section 4.3). An appropriate value for the shape factor is 0.7 to 0.8, giving a basal drag of 140 to 160 kPa.

In the examples discussed so far, the yield stress is taken constant for the entire glacier. However, where basal sliding occurs, the drag at the glacier bed may be expected to be smaller than would be the case for a frozen bed. To include this in the perfectly plastic model, a smaller yield stress can be prescribed in areas where sliding is believed to have occurred (as inferred, for example, from glacial striae). The effect of lowering the yield stress is to lower the surface slope. To illustrate this, Figure 5.6 shows two reconstructed profiles, one calculated with a yield stress decreasing toward the ice-sheet edge, and one with a constant yield stress (chosen such that for both ice sheets, the average yield stress is the same).

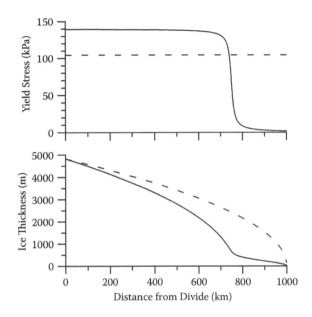

FIGURE 5.6 Comparison of ice-sheet profiles calculated using a constant yield stress (dashed curve) and using a yield stress decreasing near the margin. The average yield stress is the same for both glaciers.

As noted earlier, the models discussed above do not invoke mass continuity. One way to satisfy conservation of mass is to use either the lamellar flow theory or a sliding relation to link the ice velocity to basal drag and hence to the yield stress. The ice velocity is calculated from the continuity equation for mass. Thus, an expression for the yield stress along the glacier can be derived and substituted into equation (5.12) to allow the glacier profile to be calculated (for example, Hughes, 1985). Although the solution scheme becomes more complicated, the essential idea is similar to the models described above.

5.2 CONTINUITY EQUATION

An important requirement for (numerical) models of glaciers is that no ice may be created or lost: thickness changes at any particular point must be entirely due to the ice flow and local snowfall or melting. Integrated over the entire glacier, the average rate of thickness change must equal the total amount of ice added at the surface through snowfall minus losses from melting and calving at the glacier terminus. This conservation of mass is expressed by the continuity equation. Because the density of ice is usually taken constant (thus neglecting densification in the upper firn layers), mass conservation corresponds to conservation of ice volume.

To derive the continuity equation for an ice column extending from the bed to the surface, consider Figure 5.7. For clarity, flow in one direction only (along the x-axis) is considered; extension to include the second horizontal direction is straightforward.

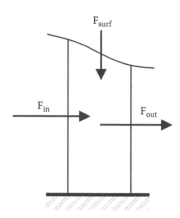

FIGURE 5.7   Mass fluxes in and out of a vertical column extending from the bed to the surface.

By definition, the ice flux through any vertical section is equal to HU, being the amount of ice flowing through the section per unit time and unit width in the cross-flow direction. The ice flux into the column of unit width is thus

$$F_{in} = HU(x),$$ (5.13)

and the flux out of the column is

$$F_{out} = HU(x+\Delta x).$$ (5.14)

The accumulation rate, M, is the amount of snowfall or ablation expressed in meters of ice depth per unit time and unit area (if basal melting occurs, this can also be included in M; usually the rate of basal melting is negligible compared to the surface accumulation or ablation). Hence, the ice input at the surface is (per unit width):

$$F_{surf} = M\Delta x.$$ (5.15)

When these three fluxes are not in balance, the column must become either thicker or thinner. The rate of thickness change is $\partial H/\partial t$, and the corresponding change in volume of the column is (per unit time and unit width)

$$\frac{\partial H}{\partial t} \Delta x.$$ (5.16)

Conservation of mass (or volume) is now expressed by

$$\frac{\partial H}{\partial t} \Delta x = F_{in} - F_{out} + F_{surf} =$$

$$= HU(x) - HU(x+\Delta x) + M\Delta x.$$ (5.17)

Dividing both sides by Δx and taking the limit $\Delta x \to 0$ gives the continuity equation

$$\frac{\partial H}{\partial t} = -\frac{\partial(HU)}{\partial x} + M. \tag{5.18}$$

This equation states that the thickness may change as a result of flow divergence (first term on the right-hand side) and due to local accumulation or ablation (second term on the right-hand side). To account for flow in the second horizontal direction, a term describing divergence of flow in that direction can be added to the right-hand side.

A more formal derivation of the continuity equation is based on the incompressibility condition (1.39)

$$\frac{\partial u}{\partial x} + \frac{\partial v}{\partial y} + \frac{\partial w}{\partial z} = 0, \tag{5.19}$$

where u, v, and w represent the three components of velocity in the three orthogonal directions x, y, and z. The z-axis is chosen vertical and positive upward, while x and y represent the two horizontal directions. Because the density of ice is assumed constant, this equation can be considered the local continuity equation (c.f. Lliboutry, 1987c, p. 50).

Integrating equation (5.19) with respect to z from the base ($z = b$) to the surface ($z = h$) gives

$$w(h) - w(b) = -\int_b^h \frac{\partial u}{\partial x}\, dz - \int_b^h \frac{\partial v}{\partial y}\, dz. \tag{5.20}$$

Using Leibnitz's rule, the order of integration and differentiation can be changed, and

$$w(h) - w(b) = -\frac{\partial}{\partial x}\left(\int_b^h u\, dz\right) + u(h)\frac{\partial h}{\partial x} - u(b)\frac{\partial b}{\partial x} +$$

$$-\frac{\partial}{\partial y}\left(\int_b^h v\, dz\right) + v(h)\frac{\partial h}{\partial y} - v(b)\frac{\partial b}{\partial y}. \tag{5.21}$$

The vertical velocity at the surface is, by definition, the rate of change in surface elevation that is not due to local accumulation (that is, if the glacier is locally thickening because snow is added at the surface, the vertical velocity is zero). Thus

$$w(h) \equiv \frac{dh}{dt} - M =$$

$$= \frac{\partial h}{\partial t} + u(h)\frac{\partial h}{\partial x} + v(h)\frac{\partial h}{\partial y} - M. \tag{5.22}$$

Similarly, the vertical velocity at the glacier base is

$$w(b) \equiv \frac{db}{dt} + M_b =$$

$$= \frac{\partial b}{\partial t} + u(b)\frac{\partial b}{\partial x} + v(b)\frac{\partial b}{\partial y} + M_b. \qquad (5.23)$$

Here, M_b represents the basal melting rate, taken positive when melting occurs (note that basal melting causes the lower ice surface to move upward).

Substituting expressions (5.22) and (5.23) for the vertical velocity at the surface and base, into the depth-integrated continuity equation (5.21) gives

$$\frac{\partial h}{\partial t} - M - \frac{\partial b}{\partial t} + M_b = -\frac{\partial(HU)}{\partial x} - \frac{\partial(HV)}{\partial y}, \qquad (5.24)$$

with the depth-averaged components of velocity, U and V, defined as

$$U = \frac{1}{H} \int_b^h u\,dz, \qquad (5.25)$$

and

$$V = \frac{1}{H} \int_b^h v\,dz. \qquad (5.26)$$

The ice thickness is given by $H = h - b$, so that equation (5.24) can be rewritten in its familiar form

$$\frac{\partial H}{\partial t} = -\frac{\partial(HU)}{\partial x} - \frac{\partial(HV)}{\partial y} + M - M_b. \qquad (5.27)$$

The vertically integrated continuity equation (5.27) expresses conservation of mass for an ice column extending over the entire ice thickness. The left-hand side is the time derivative of the local thickness, making the continuity equation a prognostic equation that can be used to determine ice thickness at a later time when the terms on the right-hand side are known. Divergence of ice flux and surface mass balance can be estimated from diagnostic equations that do not contain a time derivative and that specify a balance of quantities (for example, forces) at any moment in time. In the case of numerical ice-flow models, velocities and surface mass balance are calculated from the current geometry, and the continuity equation then allows the change in glacier geometry to be estimated. Therefore, equation (5.27) is the central equation that is solved in time-marching numerical models (c.f. Chapter 9).

Where surface velocities and ice thickness has been measured, equation (5.24) can be used to estimate the local rate of thickness change. This procedure was applied by Reeh and Gundestrup (1985) to flow along a strain grid to estimate the mass balance of the Greenland Ice Sheet at Dye-3. Taking the x-axis in the direction of flow, the continuity equation becomes

$$\frac{\partial H}{\partial t} = M - \frac{1}{f}\left[H\left(\frac{\partial U_o}{\partial x} + \frac{\partial V_o}{\partial y} \right) + U_o \frac{\partial H}{\partial x} \right],$$
(5.28)

where the subscript o indicates measured surface velocities and f is a correction factor accounting for the depth variation in horizontal ice velocity. Perhaps the greatest uncertainty in this calculation derives from the unknown shape of the velocity profiles. Reeh and Gundestrup (1985) argue that f ranges between 1.13 and 1.10. For isothermal lamellar flow, $f = (n + 2)/(n + 1) = 1.25$ for $n = 3$.

5.3   STEADY-STATE PROFILES ALONG A FLOWLINE

Using the continuity equation derived in the previous section, the steady-state profile of an ice sheet can be determined if the discharge velocity can be calculated from the geometry. An analytical solution for the ice thickness along a flowline can be found for only a few very simple cases, including an ice sheet on a flat bed and constant surface accumulation (Vialov, 1958), or for an ice sheet on a flat bed with uniform accumulation in the interior region and uniform ablation in the marginal region (Weertman, 1961c; Paterson, 1972). For more realistic basal topography and surface accumulation or ablation, the continuity equation needs to be integrated numerically to find the steady-state profile.

For lamellar flow, the depth-averaged discharge velocity is given by equation (4.31), with basal sliding set to zero

$$U = -\frac{2AH}{n+2}(\rho g H)^n \left| \frac{\partial h}{\partial x} \right|^{n-1} \frac{\partial h}{\partial x}.$$
(5.29)

On a flat bed, $\partial h/\partial x = \partial H/\partial x$, and

$$U = -A_o H^{n+1} \left| \frac{\partial H}{\partial x} \right|^{n-1} \frac{\partial H}{\partial x},$$
(5.30)

with A_0 containing all constants

$$A_o = \frac{2A}{n+2}(\rho g)^n.$$
(5.31)

Using equation (5.30) to eliminate the velocity from the continuity equation yields one equation with one unknown (the ice thickness) that can be solved.

For steady state and flow in the x-direction only, the continuity equation (5.27) reduces to

$$\frac{\partial (HU)}{\partial x} = M.$$

(5.32)

For surface accumulation, M, constant along the flowline, this may be integrated to give

$$HU = Mx,$$

(5.33)

where the boundary condition of no ice flow across the divide (at x = 0) has been invoked.

Substituting expression (5.30) for the vertical mean ice velocity into the integrated continuity equation (5.33) gives

$$-A_o H^{n+2} \left|\frac{\partial H}{\partial x}\right|^{n-1} \frac{\partial H}{\partial x} = Mx.$$

(5.34)

Rearranging

$$\left(H^{1+2/n} \frac{\partial H}{\partial x}\right)^n = -\frac{M}{A_o}x,$$

(5.35)

or

$$\frac{\partial}{\partial x}(H^{2+2/n}) = -\frac{2n+2}{n}\left(\frac{M}{A_o}\right)^{1/n} x^{1/n}.$$

(5.36)

Integrating this expression, and using the boundary condition that the thickness is zero at the ice-sheet margin (at x = L), the profile is found to be

$$H^{2+2/n} = H_o^{2+2/n}\left[1 - \left(\frac{x}{L}\right)^{1+1/n}\right],$$

(5.37)

with

$$H_o^{2+2/n} = \frac{2n+2}{n+1}\left(\frac{M}{A_o}\right)^{1/n} L^{1+1/n},$$

(5.38)

the ice thickness at the divide.

The steady-state solution (5.37) is usually written as

$$\left(\frac{H}{H_o}\right)^{2+2/n} + \left(\frac{x}{L}\right)^{1+1/n} = 1.$$

(5.39)

This profile was first derived by Vialov (1958) and is commonly referred to as the Vialov profile. It applies to a glacier that flows by internal deformation only, over a flat bed and with constant rate of surface accumulation.

The steady-state profile of a glacier that is sliding over a (horizontal) bed can be found if Weertman-type sliding is adopted. Basal sliding is discussed in Chapter 7, but for now it suffices to know that this relation links the sliding speed to basal drag raised to the power m. Equating basal drag with driving stress, and assuming a horizontal bed, gives

$$U = -A_{sl} H^m \left| \frac{\partial H}{\partial x} \right|^{m-1} \frac{\partial H}{\partial x}, \tag{5.40}$$

in which A_{sl} is a sliding parameter that depends on the bed roughness; the sliding exponent m equals $(1 + n)/2$. Note that, apart from the different prefactors, the only difference between this sliding relation and the velocity from internal deformation is a factor H.

Again, the steady-state profile is found by substituting expression (5.40) for the sliding velocity into the continuity equation (5.32) and integrating twice along the flowline. The result is

$$\left(\frac{H}{H_o} \right)^{2+1/m} + \left(\frac{x}{L} \right)^{1+1/m} = 1, \tag{5.41}$$

with

$$H_o^{2+1/m} = \frac{2m + 2}{m + 1} \left(\frac{M}{A_{sl}} \right)^{1/m} L^{1+1/m}. \tag{5.42}$$

Note that in the limit $n \to \infty$, both the profile (5.41) and the profile given by (5.39) reduce to the parabolic profile derived under the perfectly plastic approximation.

The thickness at the ice divide can be calculated from equation (5.38) for the Vialov profile, and from (5.42) for the Weertman-sliding profile. For both profiles, this thickness depends only weakly on the mass balance, M. A doubling of the surface accumulation leads to an increase in H_o of only 9% for the Vialov profile and about 15% for the Weertman-sliding profile (if the length of the ice sheet is kept fixed, and using $n = 3$ or $m = 2$). Apparently, the increased accumulation is compensated for by increased discharge, without affecting the geometry of the glacier very much, and the steady-state geometry is rather insensitive to changes in M. In the perfectly plastic approach, the profile is completely independent from M.

A generalization to equilibrium profiles for nonuniform surface mass balance is given by Bueler and others (2005), based on parameterization of the mass flux. On a horizontal bed the mass flux, Q, is given by the left-hand side of equation (5.34). Considering the region $0 \le x \le L$ extending from the ice divide to the margin, the thickness gradient is negative and the mass flux positive, and this expression can be rewritten as

$$H^{n+2}\left(\frac{\partial H}{\partial x}\right)^n = -\frac{Q}{A_o}. \tag{5.43}$$

Raising both sides to the power 1/n gives

$$H^{(n+2)/n}\frac{\partial H}{\partial x} = -\left(\frac{Q}{A_o}\right)^{1/n}, \tag{5.44}$$

or

$$\frac{n}{2n+2}\frac{\partial}{\partial x}[H^{(2n+2)/n}] = -\left(\frac{Q}{A_o}\right)^{1/n}. \tag{5.45}$$

Integrating from the ice divide (x = 0) to some point x downglacier gives the ice thickness

$$H(x)^{(2n+2)/n} = H_o^{(2n+2)/n} - \frac{2n+2}{n}\frac{1}{A_o^{1/n}}\int_0^x Q(\bar{x})^{1/n}\,d\bar{x}. \tag{5.46}$$

Many functional relations can be adopted for the flux, including the one corresponding to a constant mass balance (which would give the Vialov profile (5.37)). Bueler and others (2005) adopt the following function:

$$Q(x) = C\left[\left(\frac{x}{L}\right)^{1/n} + \left(1-\frac{x}{L}\right)^{1/n} - 1\right]^n, \tag{5.47}$$

where C is a constant to be determined from the boundary condition that the thickness at the margin is zero. The integral of the ice flux is then

$$\int_0^x Q(\bar{x})^{1/n}\,d\bar{x} = C^{1/n}L\frac{n}{n+1}\left[\left(\frac{x}{L}\right)^{(n+1)/n} - \left(1-\frac{x}{L}\right)^{(n+1)/n} + 1 - \frac{n+1}{n}\frac{x}{L}\right]. \tag{5.48}$$

Substituting in equation (5.46) and invoking H(L) = 0 gives

$$C = H_o^{2n+2}A_o\left[2L\left(1-\frac{1}{n}\right)\right]^{-n}, \tag{5.49}$$

and the steady-state profile is given by

$$H(x) = \frac{H_o}{(n-1)^{n/(2n+2)}}\left[(n+1)\frac{x}{L} - 1 + n\left(1-\frac{x}{L}\right)^{1+1/n} - n\left(\frac{x}{L}\right)^{1+1/n}\right]^{n/(2n+2)}. \tag{5.50}$$

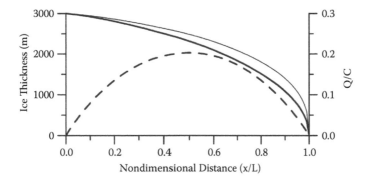

FIGURE 5.8 Steady-state profiles on a horizontal bed, corresponding to the Vialov solution (5.37) for constant mass balance (light curve; equation (5.37)), and to the Bueler profile (5.50); the dashed line (scale on the right) shows the normalized ice flux corresponding to the Bueler profile.

The mass balance corresponding to this steady-state profile is found by differentiating the flux, Q:

$$M(x) = \frac{C}{L}\left[\left(\frac{x}{L}\right)^{1/n} + \left(1 - \frac{x}{L}\right)^{1/n} - 1\right]^{n-1}\left[\left(\frac{x}{L}\right)^{(1-n)/n} - \left(1 - \frac{x}{L}\right)^{(1-n)/n}\right]. \quad (5.51)$$

A comparison between the Vialov profile (5.37) and the Bueler profile (5.50) is shown in Figure 5.8; the Weertman profile (5.41) is almost indistinguishable from the Vialov profile and not shown in this figure. The dashed line in this figure represents the normalized ice flux, Q/C, adopted by Bueler and others (2005) and reaches a maximum at L/2. Because the surface mass balance equals the divergence of the ice flux, this means that M is negative (ablation) over the lower half of the model ice sheet. For the Vialov profile, M is constant and the ice flux increases linearly from zero at the divide to the value ML at the ice edge. A consequence of this difference between the two profiles is that for the Vialov profile the ice velocity must go to infinity at the margin, whereas in the Bueler model, the velocity at the margin is bounded. Or, equivalently, for the Bueler solution the driving stress at the margin remains finite but is unbounded in the Vialov solution (Greve and Blatter, 2009, p. 89).

5.4 STEADY-STATE PROFILE OF AN AXISYMMETRIC ICE SHEET

In most ice-sheet models discussed in this book, an orthogonal Cartesian coordinate system is used with the three axes perpendicular; the z-axis is chosen vertical (positive upward), while the x-axis and y-axis lie in the horizontal plane. Their orientations may be chosen arbitrarily (but perpendicular to each other), although it is customary to chose the x-axis along the mean direction of flow. In some instances, however, it is more convenient to use a different coordinate system. One example is the axisymmetric ice sheet: such an ice sheet can be rotated around the vertical z-axis without altering the geometry or flow in a fixed location. Another example

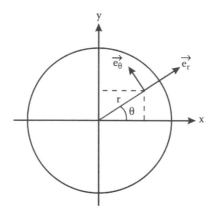

FIGURE 5.9   Relation between Cartesian coordinates, x and y, and radial coordinates, r and θ.

is the modeling of the Northern Hemisphere glacial ice sheets. Rather than using a Cartesian coordinate system, a system based on geographical coordinates (latitude and longitude) is sometimes used (for example, Birchfield et al., 1981; Peltier and Hyde, 1987). A third example, creep closure of englacial tunnels, is discussed in Section 3.6.

The coordinate system used to model an axisymmetric ice sheet is the cylindrical system, in which the coordinates of a point are given by the radius, r, the azimuth angle, θ, and elevation, z (see also Section 3.6). The relation between these coordinates and the Cartesian coordinates follows directly from geometrical considerations (Figure 5.9), and is

$$r = \sqrt{x^2 + y^2},$$

$$\tan\theta = \frac{y}{x},$$

(5.52)

while the vertical z-coordinate is the same in both systems. To find the model equations in the new coordinate system, one could use the expressions (5.52) to write x and y in terms of r and θ and substitute the result in the Cartesian form of the equations. A simpler and faster way is to write the equations first in a form that is independent of the coordinate system used, and next to use standard mathematical formulas to arrive at the form appropriate for the new coordinate system.

Consider first the depth-integrated ice velocity. For deformational flow the x-component of velocity is proportional to the slope of the surface elevation in the x-direction. Similarly, the y-component of velocity is proportional to the surface slope in the y-direction. Then

$$U = -f(H)\left|\frac{\partial h}{\partial x}\right|^{n-1}\frac{\partial h}{\partial x},$$

(5.53)

and

$$V = -f(H) \left| \frac{\partial h}{\partial y} \right|^{n-1} \frac{\partial h}{\partial y}, \tag{5.54}$$

with

$$f(H) = \frac{2AH}{n+2} (\rho g H)^n. \tag{5.55}$$

These two equations can be combined into one vector equation

$$\vec{U} = -f(H) |\nabla h|^{n-1} \nabla h. \tag{5.56}$$

Here, $\vec{U} = (U, V)$ represents the horizontal velocity vector with components U and V in the (x, y) coordinate system. The symbol ∇ is called the nabla operator (Section 1.1). When applied to a scalar quantity such as the surface elevation, ∇h represents the gradient of h, or the surface slope vector with components $\partial h / \partial x$ and $\partial h / \partial y$ in the (x, y) coordinate system (actually, ∇h is a three-dimensional vector; however, the surface elevation is independent of the vertical z-direction, and the z-component of the vector ∇h is zero and omitted here). Because equation (5.56) is written in vector form, it is independent of the coordinate system used, provided, of course, that the velocity and gradient vectors are referenced to the same coordinate system.

Most mathematics textbooks provide expressions for ∇h applicable to the three most-often used coordinate systems (the Cartesian system, cylindrical coordinates, and polar coordinates). For the cylindrical (r, θ) system, the gradient of the surface is given by (omitting the z-component)

$$\nabla h = \frac{\partial h}{\partial r} \vec{e}_r + \frac{1}{r} \frac{\partial h}{\partial \theta} \vec{e}_\theta, \tag{5.57}$$

where $\vec{e}_r$ and $\vec{e}_\theta$ represent the unit vectors in the r- and θ-directions, respectively (c.f. Figure 5.9). Using this expression in equation (5.56), the two components of ice velocity in the r- and θ-directions can be determined.

The second equation to be considered is the continuity equation (5.27). The first two terms on the right-hand side represent the divergence of ice flux. This is a scalar function, derived from the vector representing the ice flux. In mathematical notation, $\mathrm{div}(H\vec{U}) = \nabla \cdot (H\vec{U})$. Although the nabla symbol is used again, it here represents a vector operator that, when applied to a vector, gives a scalar quantity (the dot signifies the distinction with the earlier use). The continuity equation can now be rewritten as

$$\frac{\partial H}{\partial t} = -\nabla \cdot (H\vec{U}) + M. \tag{5.58}$$

Again, this form of the continuity equation is independent of the coordinate system, provided the nabla operator is referenced to the same coordinate system as the flux vector. In cylindrical coordinates, the divergence of flux becomes

$$\nabla \cdot (H\vec{U}) = \frac{1}{r}\frac{\partial}{\partial r}(rHU_r) + \frac{1}{r}\frac{\partial U_\theta}{\partial \theta}. \tag{5.59}$$

The two components of velocity, U_r and U_θ, can be found by substituting equation (5.57) for the gradient of the surface slope into equation (5.56) for the velocity vector. The radial component of velocity, U_r, is in the direction of $\vec{e}_r$, while the tangential component of velocity, U_θ, is in the direction of $\vec{e}_\theta$.

For an axisymmetric ice sheet, the equations can be simplified because derivatives with respect to θ are zero. Thus, the velocity component in the θ-direction, U_θ, is zero; that is, the ice flows radially outward from the ice divide with a velocity given by (assuming a flat bed so that the surface elevation may be replaced with the ice thickness):

$$U_r = -A_o H^{n+1} \left|\frac{\partial H}{\partial r}\right|^{n-1} \frac{\partial H}{\partial r}. \tag{5.60}$$

For steady state, the continuity equation (5.58) now becomes

$$-\frac{1}{r}\frac{\partial}{\partial r}\left(r A_o H^{n+2} \left|\frac{\partial H}{\partial r}\right|^{n-1} \frac{\partial H}{\partial r}\right) = M. \tag{5.61}$$

As in the previous section, this equation can be integrated twice with respect to r, to yield the thickness profile

$$\left(\frac{H}{H_o}\right)^{2+2/n} + \left(\frac{r}{R}\right)^{1+1/n} = 1, \tag{5.62}$$

where $r = R$ represents the (circular) edge of the ice sheet with $H(R) = 0$. The ice thickness at the divide is given by

$$H_o^{2+2/n} = \frac{2n+2}{n+1}\left(\frac{M}{2A_o}\right)^{1/n} R^{1+1/n}. \tag{5.63}$$

Comparing the axisymmetric solution (5.62) with the flowline solution (5.39) shows that both profiles are similar (Figure 5.10). The only difference is the thickness at the ice divide. For the same accumulation rate, M, flow parameter, A_o, and horizontal extent ($R = L$), the axisymmetric ice sheet is thinner by a factor of $(1/2)^{1/n} = 0.79$ (for $n = 3$). This difference is due to the fact that the flowline model does not account for divergence of flow (that is, the flowband is of constant width and, locally, there is no ice lost due to flow in the transverse direction). The axisymmetric solution includes flow divergence associated with the radially outward flow.

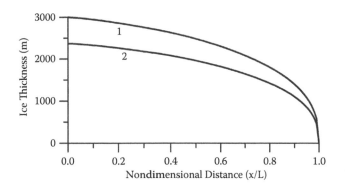

FIGURE 5.10   Steady-state profiles on a horizontal bed, corresponding to the lamellar-flow solution for an ice sheet of constant width (curve 1; equation (5.37)), and for an axisymmetric ice sheet (curve 2; equation (5.62)). For profile 1 the thickness at the center was set to 3000 m; for profile 2 this thickness was reduced by a factor of 0.79. The solution (5.42) for Weertman-type sliding is indistinguishable from curve 1.

5.5   STEADY-STATE PROFILE OF A FREE-FLOATING ICE SHELF

Sanderson (1979) describes how the steady-state profile of an ice shelf can be calculated by numerical integration of the continuity equation. As shown by Van der Veen (1983), an analytical expression for the ice-shelf profile can be derived for the simple case of a free-floating ice shelf spreading in one direction only. The assumption of plane flow implies that the ice thickness, velocity, and strain rate vary only in the direction of flow, and resistance to flow from lateral drag is neglected. Furthermore, vertical shear is neglected (Sanderson and Doake, 1979), and the ice velocity is taken constant with depth.

Under these simplifying assumptions, the along-flow rate of stretching is a function of the ice thickness only, as given by equation (4.64)

$$\dot{\varepsilon}_{xx} = CH^n, \tag{5.64}$$

with the constant, C, defined as

$$C = \left(\frac{\rho g (\rho_w - \rho)}{4 B \rho_w} \right)^n, \tag{5.65}$$

in which ρ and ρ_w represent the density of ice and sea water, respectively, and B denotes the viscosity parameter in Glen's flow law (2.12).

The next step is to consider the continuity equation (5.32). Using the definition for the stretching rate as the along-flow gradient in the along-flow component of velocity, U,

$$\dot{\varepsilon}_{xx} = \frac{\partial U}{\partial x}, \tag{5.66}$$

the continuity equation for steady state (i.e., $\partial H/\partial t = 0$) can also be written as

$$H\dot{\varepsilon}_{xx} + U\frac{\partial H}{\partial x} = M. \tag{5.67}$$

For constant accumulation rate, M, the ice flux increases linearly with distance along the flowline (equation (5.33)) and

$$HU = Mx + H_o U_o, \tag{5.68}$$

in which $H_o U_o$ represents the ice flux across the grounding line (at $x = 0$) into the ice shelf. Using this expression to eliminate the velocity from equation (5.67) and substituting (5.64) for the stretching rate gives the following equation:

$$(Mx + H_o U_o)\frac{\partial H}{\partial x} = MH - CH^{n+2}. \tag{5.69}$$

The only unknown in this equation is the ice thickness, H. To solve this equation analytically, three different cases must be considered, namely, $M = 0$ (surface accumulation equals basal melting), $M > 0$ (surface accumulation is larger than basal melting), and $M < 0$ (basal melting exceeds surface accumulation).

If the rate of basal melting equals the rate of surface accumulation, $M = 0$, and the continuity equation (5.69) reduces to

$$\frac{\partial H}{\partial x} = -\frac{C}{H_o U_o}H^{n+2}. \tag{5.70}$$

Integration yields

$$H = \left(\frac{(n+1)C}{H_o U_o}x + H_o^{-(n+1)}\right)^{-1/(n+1)}, \tag{5.71}$$

where the boundary condition $H = H_o$ at the grounding line ($x = 0$) has been used. For $n = 3$, the thickness is proportional to $x^{-1/4}$, and the strongest decrease in ice thickness is found in the vicinity of the grounding line. Farther away from the grounding line, the thickness gradient becomes very small and the thickness tends to reach a constant value (as illustrated by profile a shown in Figure 5.11). Equation (5.70) restricts neither the ice thickness nor the length of the ice shelf, and some criterion must be used to prevent the shelf from becoming infinitely long. In calculating the profile shown in Figure 5.11, the maximum length of the ice shelf was set to 500 km.

When surface accumulation is larger than basal melting (or when basal freezing occurs), M is positive, and equation (5.69) can be written as

$$\frac{\partial H}{\partial x} = \frac{MH - CH^{n+2}}{Mx + H_o U_o}. \tag{5.72}$$

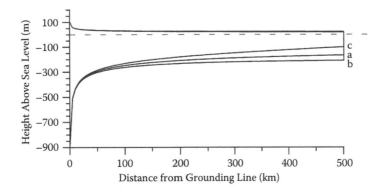

FIGURE 5.11 Equilibrium profile of ice shelves spreading in the along-flow direction only. The thickness at the grounding line equals 1000 m and the speed across the grounding line equals 250 m/yr for all three profiles. The accumulation rate is M = 0 (profile a), M = +0.25 m/yr (profile b), and M = −0.25 m/yr (profile c).

Collecting all terms containing the unknown ice thickness on the left-hand side, this equation is rewritten as

$$\left(\frac{1}{MH} + \frac{\dfrac{C}{M^2}H^n}{1 - \dfrac{C}{M}H^{n+1}} \right) \partial H = \frac{1}{Mx + H_o U_o}\, \partial x. \tag{5.73}$$

Recalling that $\partial(\ln H) = 1/H\, \partial H$, both sides can be integrated to

$$\frac{1}{M}\ln H - \frac{1}{M(n+1)}\ln\left(1 - \frac{C}{M}H^{n+1}\right) =$$

$$= \frac{1}{M}\ln\left(Mx + H_o U_o\right) + \frac{1}{M(n+1)}\ln D. \tag{5.74}$$

Note how the integration constant (the last term on the right-hand side) is written. Using some elementary rules for manipulating natural logarithms, the following relation is found:

$$H^{n+1}\left(1 - \frac{C}{M}H^{n+1}\right)^{-1} = D\left(Mx + H_o U_o\right)^{n+1}. \tag{5.75}$$

The integration constant, D, follows from the boundary condition that the thickness at the grounding line (at x = 0) must be equal to H_o and

$$D = \frac{1}{U_o^{n+1}\left(1 - \dfrac{C}{M}H_o^{n+1}\right)}. \tag{5.76}$$

After some more algebraic manipulation, the following expression for the ice thickness is found

$$H = \left(\frac{C}{M} - \frac{U_o^{n+1}\left(\frac{C}{M}H_o^{n+1}-1\right)}{(Mx + H_o U_o)^{n+1}} \right)^{-1/(n+1)}. \tag{5.77}$$

Although this expression looks much more complicated than the profile obtained for $M = 0$ (equation (5.71)), the essential characteristics of the profile are retained, as can be seen in Figure 5.11 (profile b). Near the grounding line, the ice thickness decreases rapidly and then reaches a more or less constant value. For a free-floating ice shelf, the thickness must decrease with distance from the grounding line (otherwise the ice would float back toward the grounding line), or $\partial H/\partial x < 0$. From equation (5.72) it follows that this requires that the ice thickness is greater than a critical value, H_{cr}, given by

$$H_{cr} = \left(\frac{M}{C} \right)^{1/(n+1)}. \tag{5.78}$$

However, as can be seen in Figure 5.11, this critical thickness is reached asymptotically (for the profile shown in Figure 5.11, the critical thickness is 307 m). Therefore, to prevent the shelf from becoming unrealistically large, the maximum length for the profile shown in this figure was set to 500 km, the same length used for profile a ($M = 0$ m/yr).

The third possibility to be considered is an ice shelf subject to basal and/or surface melting, such that M is negative. Defining $\hat{M} = -M$, equation (5.69) becomes

$$\frac{\partial H}{\partial x} = -\frac{\hat{M}H + CH^{n+2}}{H_o U_o - \hat{M}x}. \tag{5.79}$$

Following a similar procedure as above, the ice-shelf profile is found to be

$$H = \left(\frac{U_o^{n+1}\left(1 + \frac{C}{\hat{M}}H_o^{n+1}\right)}{(H_o U_o - \hat{M}x)^{n+1}} - \frac{C}{\hat{M}} \right)^{-1/(n+1)}. \tag{5.80}$$

Note the similarity between this profile and equation (5.77). Again, the thickness must decrease away from the grounding line. Because all quantities in equation (5.79) are positive, the condition $\partial H/\partial x < 0$ requires that the length of the shelf must be smaller than the critical length, L_{cr}, given by

$$L_{cr} = \frac{H_o U_o}{\hat{M}}. \tag{5.81}$$

For profile c shown in Figure 5.11, this critical length is 533 km. However, to allow for a comparison with the other two profiles, the maximum length was set to 500 km.

The profiles derived above pertain to the very special situation of an ice shelf spreading in one direction, with resistance to flow arising from gradients in longitudinal stress only. This assumption of plane flow is not realistic for a free-floating ice shelf because, if not restricted laterally, the shelf will spread in both horizontal directions. Thus, the plane-flow model applies only to an ice shelf in a parallel-sided bay where lateral drag is negligible. While the two-dimensional profile of an ice shelf spreading in both horizontal directions cannot be easily determined, the effect of lateral spreading on the centerline profile of the shelf can be estimated as discussed below. A more complete treatment of radial ice-shelf flow can be found in Morland (1987) and Morland and Zainuddin (1987).

For uniform spreading in both horizontal directions, the creep rates are given by equation (4.65)

$$\dot{\varepsilon}_{xx} = \dot{\varepsilon}_{yy} = C_2 H^n, \tag{5.82}$$

with

$$C_2 = 3^{-(n+1)/2} \left(\frac{\rho g (\rho_w - \rho)}{2 B \rho_w} \right)^n. \tag{5.83}$$

Assuming steady state, the continuity equation is

$$H \frac{\partial U}{\partial x} + U \frac{\partial H}{\partial x} + H \frac{\partial V}{\partial y} + V \frac{\partial H}{\partial y} = M. \tag{5.84}$$

Along the centerline of the shelf, it follows from symmetry arguments that the transverse velocity component, V, is zero and the last term on the left-hand side is zero. Substituting (5.82) for the strain rates gives

$$2 C_2 H^{n+1} + U \frac{\partial H}{\partial x} = M. \tag{5.85}$$

Because the flow is two-dimensional, the continuity equation cannot be integrated along the flowline to find an expression for the ice velocity, U (as was done to arrive at equation (5.33) for the case of plane flow). This means that an analytical solution cannot be derived for the centerline profile of an ice shelf spreading in both horizontal directions. However, this profile can be found by numerically integrating the continuity equation (5.85) along the flowline analogous to the method of Sanderson (1979). Rewriting (5.85) as

$$\frac{\partial H}{\partial x} = \frac{1}{U} (M - 2 C_2 H^{n+1}), \tag{5.86}$$

the thickness gradient at the grounding line can be estimated from the prescribed thickness, H_o, and velocity, U_o, at the grounding line. This allows the thickness at some distance Δx away from the grounding line to be calculated. From this new thickness, the creep rate $\dot{\varepsilon}_{xx} = \partial U / \partial x$ is calculated using equation (5.82). This allows the

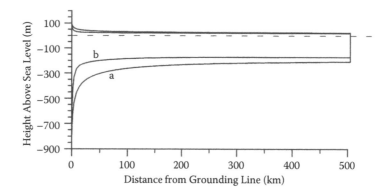

FIGURE 5.12 Equilibrium profile of an ice shelf spreading in the along-flow direction only (profile a) and the profile of an ice shelf spreading in both horizontal directions (profile b). Constants used are $H_o = 1000$ m, $U_o = 250$ m/yr, and M = 0.25 m/yr.

velocity to be extrapolated to the new point. With all quantities on the right-hand side of equation (5.86) known, the thickness gradient can be estimated, and the thickness at the next gridpoint determined. This procedure is repeated until some criterion for the maximum length or minimum thickness is reached. Because the thickness decreases very rapidly near the grounding line, the horizontal step, Δx, must be very small (a few meters) in this region to minimize numerical errors. Farther away, the value of Δx may be gradually increased to, say, 5 km. Figure 5.12 shows a centerline profile calculated using this numerical scheme. Also shown is the plane-flow profile given by equation (5.77). Lateral spreading significantly reduces the thickness of the shelf, as a result of mass being lost in the cross-flow direction. The essential characteristics of the profile are similar to those of the plane-flow solutions, however.

5.6   FLOW CONTROLLED BY LATERAL DRAG

The equations for ice-shelf spreading derived in Section 4.5 include the effect of lateral drag, and it was shown that the stretching rate at some point on the shelf depends on the lateral drag integrated over the distance from that point to the ice-shelf terminus. This means that to solve the continuity equation numerically to find the ice-shelf profile, the integration must start at the shelf front (with prescribed thickness and velocity) and proceed upglacier to the grounding line, instead of starting at the grounding line as in the model discussed in the previous section (Sanderson, 1979). An analytical expression for the profile cannot be derived except for the special case of lateral drag supporting all (or most) of the driving stress. This model may apply to a floating ice shelf in a parallel-sided embayment, or to an ice stream decoupled completely from its bed.

If the driving stress is fully supported by drag at the margins, the transverse velocity profile is given by equation (4.51)

$$U(y) = U_c \left(1 - \left(\frac{y}{W} \right)^{n+1} \right),\tag{5.87}$$

where y represents the transverse distance with y = ±W at the lateral margins (the half-width of the flowband, W, is taken constant here), and U_c the centerline velocity, given by equation (4.52)

$$U_c = \frac{2}{n+1}\left(\frac{\tau_{dx}}{BH}\right)^n W^{n+1}. \tag{5.88}$$

Integrating this expression over the full width of the flowband, the width-averaged discharge velocity is found to be

$$U = \frac{n+1}{n+2}U_c. \tag{5.89}$$

Combining equations (5.89) and (5.88), the width-averaged velocity can be written as

$$U = -A_o\left|\frac{\partial h}{\partial x}\right|^{n-1}\frac{\partial h}{\partial x}, \tag{5.90}$$

with

$$A_o = \frac{2}{n+2}\frac{W^{n+1}}{B^n}(\rho g)^n. \tag{5.91}$$

Substituting expression (5.90) for the width-averaged velocity into the continuity equation (5.68) allows an analytical expression for the profile to be determined for two cases in which the flow is controlled entirely by lateral drag, namely, an ice stream on a flat bed and with constant half-width, and a floating ice shelf in a parallel-sided embayment.

Consider first an ice stream moving over a soft bed unable to offer any significant resistance to flow. As argued by Whillans and Van der Veen (1997), this may be the case for Whillans Ice Stream in West Antarctica, where lateral drag appears to provide all resistance to flow. Taking the bed to be horizontal, the average discharge velocity is

$$U = -A_o\left|\frac{\partial H}{\partial x}\right|^{n-1}\frac{\partial H}{\partial x}. \tag{5.92}$$

As in the previous section, this expression is substituted into the continuity equation (5.68) and integrated in the direction of flow to yield the following profile

$$\left(\frac{H}{H_o}\right)^{1+1/n} = 1 - \frac{U_o^{1+1/n}}{MA_o^{1/n}}\left(\left(\frac{Mx}{H_o U_o}+1\right)^{1+1/n}-1\right). \tag{5.93}$$

Here $H_o U_o$ represents the ice flux at the head of the ice stream (at x = 0). In the upper panel of Figure 5.13 this ice-stream profile is shown. One prominent feature is the convex curvature of the surface, with the surface slope increasing toward the mouth of the ice stream. This increase in slope is needed to allow the velocity to increase with distance along the flowline, as required for steady state (equation (5.33)).

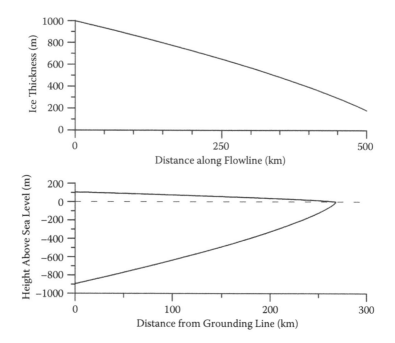

FIGURE 5.13 Equilibrium profile on an ice stream on a horizontal bed (upper panel) and an ice shelf (lower panel), both controlled entirely by lateral drag. Values of constant used are H_o = 1000 m, U_o = 250 m/yr, W = 15 km, and M = 0.15 m/yr.

For an ice shelf, the surface elevation is related to the ice thickness through the flotation criterion

$$h = \left(1 - \frac{\rho}{\rho_w}\right)H, \tag{5.94}$$

and the velocity becomes

$$U = -A_1 \left|\frac{\partial H}{\partial x}\right|^{n-1} \frac{\partial H}{\partial x}. \tag{5.95}$$

The constant, A_1, is related to A_o as

$$A_1 = \left(1 - \frac{\rho}{\rho_w}\right)A_o. \tag{5.96}$$

Thus, the shelf thickness is also given by equation (5.93) but with A_o replaced by A_1. Again, the surface slope increases toward the ice front to maintain mass continuity. Consequently, the resulting profile, shown in Figure 5.13, is essentially different from that of a free-floating ice shelf (Figure 5.11), which is characterized by a concave-upward surface profile.

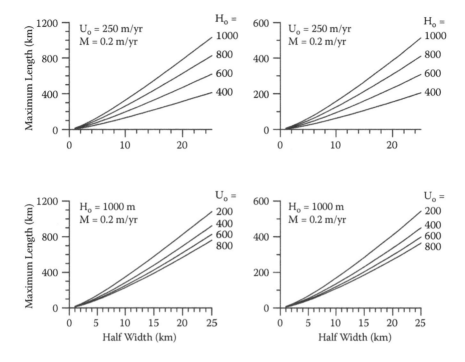

FIGURE 5.14 Maximum length of an ice stream on a horizontal bed (left panels) and of an ice shelf (right panels), both controlled by lateral drag, as a function of the half width, for different values of H_o (in m) (upper panels) and U_o (in m/yr) (lower panels).

The maximum length of the ice stream or ice shelf follows by setting the thickness to zero in equation (5.93). The result is

$$L_{max} = \frac{H_o U_o}{M}\left(\left(\frac{M A_i^{1/n}}{U_o^{1+1/n}} + 1\right)^{n/(1+n)} - 1\right),\tag{5.97}$$

where $A_i = A_o$ for an ice stream on a flat bed, and $A_i = A_1$ for a floating ice shelf. The parameter most critical in determining L_{max} is the half-width, W (Figure 5.14). This is because the ice velocity is proportional to the fourth power of W (if $n = 3$). Thus, a small increase in width leads to a large increase in speed. To maintain the discharge needed for steady state (which is independent of the width), the thickness gradient (or surface slope) must decrease, allowing the flowband to become longer. Similar arguments can be applied to explain the dependency of L_{max} on the thickness and velocity at the head of the flowband. The accumulation rate, M, has only a minor effect on the maximum length. From equation (5.97) it follows that L_{max} is proportional to $M^{1/4}$, and large changes in accumulation rate are needed to significantly alter the profile along the flowband.

6 Glacier Thermodynamics

6.1 CONSERVATION OF ENERGY

The temperature distribution in a glacier is important because the rate of deformation increases rapidly as the ice becomes warmer (Section 2.2). Further, where the basal temperature reaches the pressure-melting temperature, basal melting may lead to the development of an interstitial layer of meltwater and the onset of basal sliding, resulting in ice velocities that can be orders of magnitude greater than those associated with internal deformation (c.f. Chapter 7). The starting point when considering the heat budget of a glacier is the first law of thermodynamics, which states that, in a closed system, no energy can be created or destroyed. Using this law, and some elementary thermodynamics, the thermodynamic or temperature equation can be derived. The following discussion is largely based on the extensive treatment of this topic given in Brown (1991). Readers not interested in the nuts and bolts of thermodynamics can skip this section and take equation (6.17) for granted.

The first law of thermodynamics may be reworded to say that the change in internal energy, dE, of a closed system with a given volume must be equal to the sum of the work done on the system, dW, and heat added to the volume, dQ. Thus, the rate of change of internal energy is given by

$$\frac{dE}{dt} = \frac{dW}{dt} + \frac{dQ}{dt}.$$
(6.1)

In glaciers, work energy is associated with the production of deformational heat, while Q represents heat released into the ice volume by conductive heat fluxes.

The kinetic energy of glaciers may be neglected, and the internal energy represents the thermal energy, which is related to the temperature of the ice. Defining the specific internal energy, e, as the thermal energy per unit mass, the rate of change in E can be written as

$$\frac{dE}{dt} = \frac{d(\rho e)}{dt} =$$

$$= \rho \frac{de}{dt} + e \frac{d\rho}{dt},$$
(6.2)

where ρ represents the density of the ice. For an incompressible fluid, there is no transfer between internal energy and pressure, and

$$e = C_p T,$$
(6.3)

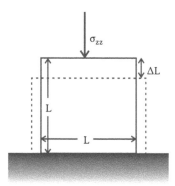

FIGURE 6.1 Deformation of a cube (L × L × L) under uniaxial compression.

with C_p the specific heat capacity and T the temperature. The specific heat capacity is the amount of energy needed to raise the temperature of a unit mass of ice by 1°C (or 1 K), under constant pressure. Further, for an incompressible material, the density may be considered constant, so that the second term on the right-hand side of equation (6.2) can be neglected and the change in internal energy is related to the temperature change as

$$\frac{dE}{dt} = \rho C_p \frac{dT}{dt}. \tag{6.4}$$

Work done on the system is associated with internal deformation of the ice. Assuming that no energy is expended toward mineralogical changes (for example, recrystallization), the work done can be found by considering deformation of a volume of ice (c.f. Means, 1976, pp. 266–267). Referring to Figure 6.1, consider a cube subject to a compressive stress, σ_{zz}. This stress causes the volume to shorten by a small amount ΔL. Recalling the definition of strain as the change in length per unit of original length, the change in length can be written as

$$\Delta L = \varepsilon_{zz} L. \tag{6.5}$$

With work defined as the product of force and displacement, the work done on the cube equals the total force acting on the cube multiplied by the change in length. Thus,

$$W = (\sigma_{zz} L^2)(\varepsilon_{zz} L). \tag{6.6}$$

Dividing by the volume of the cube gives the work per unit volume,

$$W_o = \sigma_{zz} \varepsilon_{zz}. \tag{6.7}$$

Differentiating with respect to time gives the rate of deformational heat production,

$$\frac{dW_o}{dt} = \dot{W}_o = \sigma_{zz} \dot{\varepsilon}_{zz}. \tag{6.8}$$

Similar expressions apply to work done by other stresses, and the total rate of heating is

$$\dot{W}_d = \dot{\varepsilon}_{ij}\,\sigma_{ij}, \tag{6.9}$$

using the summation convention over repeat indices.

In addition, heat may be added by freezing of meltwater. If the amount of meltwater that refreezes per unit time and volume is M_f, release of latent heat equals

$$\dot{W}_L = L_f\,M_f, \tag{6.10}$$

where L_f represents the specific latent heat of fusion.

The conductive heat flux in the ice, $\vec{F}_c$, is linked to the temperature gradient according to Fourier's law for heat conduction

$$\vec{F}_c = -k\,\nabla T, \tag{6.11}$$

where k denotes the thermal conductivity. The amount of heat released by this flux in a volume V enclosed by the closed surface A equals the total heat flux through the enclosing surface, A, and

$$\int_V \frac{dQ}{dt}\,dV = -\iint_A \vec{F}_c \bullet \vec{n}\ dA. \tag{6.12}$$

Here $\vec{n}$ represents the unit vector perpendicular to the enclosing surface, A. Applying Gauss's theorem to rewrite the surface integral on the right-hand side of equation (6.12) as a volume integral gives

$$\int_V \frac{dQ}{dt}\,dV = \int_V \mathrm{div} \bullet \vec{F}_c\ dV =$$

$$= k \int_V \nabla^2 T\ dV. \tag{6.13}$$

Per unit volume, heating from conductive heat fluxes is then

$$\frac{dQ}{dt} = k\,\nabla^2 T. \tag{6.14}$$

The thermodynamic equation (6.1) can now be written as a temperature equation. Expression (6.4) links the internal energy to the temperature, while the heating terms are given by equations (6.9), (6.10), and (6.14). The result is

$$\rho C_p \frac{dT}{dt} = k\,\nabla^2 T + \dot{\varepsilon}_{ij}\,\sigma_{ij} + L_f\,M_f. \tag{6.15}$$

The time derivative on the left-hand side represents the total derivative and can be written as the sum of the local time change in temperature and advective changes associated with a spatially nonuniform temperature distribution. That is,

$$\frac{dT}{dt} \equiv \frac{\partial T}{\partial t} + \vec{u} \cdot \nabla T, \qquad (6.16)$$

in which $\vec{u} = (u, v, w)$ represents the three-dimensional velocity vector. Written out in full, conservation of heat can now be expressed as

$$\frac{\partial T}{\partial t} = - u\frac{\partial T}{\partial x} - v\frac{\partial T}{\partial y} - w\frac{\partial T}{\partial z} +$$

$$+ K\frac{\partial^2 T}{\partial x^2} + K\frac{\partial^2 T}{\partial y^2} + K\frac{\partial^2 T}{\partial z^2} +$$

$$+ \frac{1}{\rho C_p}(\dot{\varepsilon}_{xx}\sigma_{xx} + \dot{\varepsilon}_{yy}\sigma_{yy} + \dot{\varepsilon}_{zz}\sigma_{zz} + 2\dot{\varepsilon}_{xy}\sigma_{xy} + 2\dot{\varepsilon}_{xz}\sigma_{xz} + 2\dot{\varepsilon}_{yz}\sigma_{yz}) + \quad (6.17)$$

$$+ \frac{L_f M_f}{\rho C_p}.$$

In this equation, $K = k/(\rho C_p)$ represents the thermal diffusivity.

To recapitulate the foregoing, for incompressible ice with constant density, ρ, conservation of energy can be expressed in terms of change in temperature of the ice as done in equation (6.17). This temperature may change as a result of advection of heat (first line), diffusion of heat (second line), heat generated by internal deformation (third line), and latent heat released by refreezing meltwater (last line). Except where noted explicitly, the thermal parameters are considered constants, with values given in Table 6.1.

TABLE 6.1
Values of Thermal Parameters of Pure Ice at 0°C

Specific heat capacity	C_p	2097 J kg^{-1}K^{-1}
Specific latent heat of fusion	L_f	333.5 kJ kg^{-1}
Thermal conductivity	k	2.10 Wm^{-1}K^{-1}
Thermal diffusivity	K	1.09 · 10^{-6} m^{2}s^{-1}
		34.4 m^{2}yr^{-1}
Density	ρ	917 kg m^{-3}

Source: Yen, Y.-C., CRREL Report 81–10, Cold Regions Research and Engineering Laboratory, Hanover, NH, 27 pp., 1981.

In general, equation (6.17) cannot be solved analytically. Further, to be fully correct, the thermodynamic equation must be solved simultaneously with the velocity distribution because the two are coupled through the temperature dependence of the rate factor in the constitutive relation. However, by making some simplifying assumptions, the basic characteristics of the temperature distribution in glaciers can be investigated.

6.2   STEADY-STATE TEMPERATURE PROFILES

The first of only a handful of analytical solutions of the temperature equation was derived by Robin (1955). He made the following assumptions:

1. The basal temperature is below the pressure-melting temperature; thus the solution only applies to cold glaciers, frozen to their bed.
2. Horizontal diffusion of heat is much smaller than vertical diffusion; in most parts of a glacier, horizontal temperature gradients are much smaller than vertical gradients, so this assumption is probably valid for most glaciers.
3. Horizontal advection of heat is neglected; this is only true where the horizontal components of velocity are small, that is, near the divide of an ice sheet.
4. Heat generation from internal deformation may be taken into account by increasing the geothermal heat flux at the base of the glacier; because in a glacier frozen to its bed, vertical shear is largest in the lower layers (Figure 4.2), where the shear stress is largest, this assumption is a reasonable first approximation.

Under these assumptions, the temperature equation (6.17) reduces to

$$\frac{\partial T}{\partial t} = K \frac{\partial^2 T}{\partial z^2} - w \frac{\partial T}{\partial z}. \tag{6.18}$$

The first term on the right-hand side represents redistribution of heat through (molecular) diffusion, while the second term describes redistribution of heat by downward ice flow (advection).

To solve equation (6.18) analytically, two more assumptions are needed, namely

5. The ice sheet is in thermal steady state with $\partial T/\partial t = 0$.
6. The ice-sheet profile is also in steady state (otherwise the temperature would not be steady), and the vertical velocity can be simply linked to the accumulation rate; at the surface, the vertical velocity must balance accumulation, while at the base, the vertical velocity must be zero.

The simplest profile that satisfies assumption 6 is

$$w(z) = -\frac{Mz}{H}, \tag{6.19}$$

where M represents the surface mass balance expressed in meters of ice per year ($M > 0$ for accumulation). As shown in Figure 4.2, for lamellar flow the vertical velocity decreases almost linearly with depth in the upper 80% of the ice thickness. Near the base, the vertical velocity decreases more rapidly to zero at the bed. Thus, the profile (6.19) adopted by Robin (1955) may be expected to overestimate downward heat advection in the upper portion of the glacier.

The steady-state temperature can now be found by solving the simplified heat equation

$$K \frac{\partial^2 T}{\partial z^2} + \frac{Mz}{H} \frac{\partial T}{\partial z} = 0. \tag{6.20}$$

To find a solution, two boundary conditions are needed. At the surface, the temperature equals the prescribed surface temperature, T_s. At the base of the glacier, the vertical temperature gradient must match the geothermal heat flux, corrected if necessary to account for heat generated by internal deformation. Denoting the basal heat flux by G, the basal temperature gradient is

$$\left(\frac{\partial T}{\partial z} \right)_b = - \frac{G}{k}. \tag{6.21}$$

Invoking these two boundary conditions, integrating equation (6.20) twice with respect to z gives the temperature profile

$$T(z) - T_s = - \frac{G}{k} \int_0^z \exp(-\bar{z}^2 q^2) \, d\bar{z}, \tag{6.22}$$

with

$$q^2 = \frac{M}{2KH}. \tag{6.23}$$

For surface accumulation M is positive, and this profile can also be written as

$$T(z) - T_s = - \frac{G\sqrt{\pi}}{2kq} [erf(zq) - erf(Hq)]. \tag{6.24}$$

In this expression, erf(z) represents the error function, defined as

$$erf(z) = \frac{2}{\sqrt{\pi}} \int_0^z \exp(-\bar{z}^2) \, d\bar{z}. \tag{6.25}$$

This function is tabulated in mathematical handbooks (for example, Abramowitz and Stegun, 1965) or available as standard routine in mathematical software packages.

In Figure 6.2 temperature profiles are shown for various values of the surface mass balance for a 3000 m thick ice sheet with a basal temperature gradient equal to

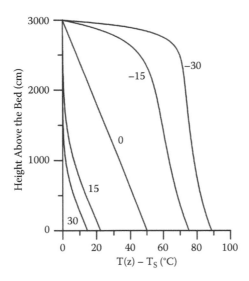

FIGURE 6.2 Steady-state temperature profiles according to the Robin temperature model. Labels indicate the surface accumulation rate in centimeters of ice depth per year. Profiles for positive surface accumulation were calculated using the analytical solution (6.24), while profiles for surface ablation were obtained by numerically integrating the thermodynamic equation (6.18) until a steady state was reached.

−0.0167 K/m. Profiles for negative values of M (that is, surface ablation) were calculated by numerically integrating the temperature equation (6.18) using (6.19) for the vertical velocity at depth, until a steady state was reached.

As shown by the curves in Figure 6.2 surface accumulation (or ablation) has a large effect on the steady-state temperature distribution. For M = 0, the temperature increases linearly with depth, with a gradient equal to the prescribed basal gradient. Surface accumulation results in considerable cooling of the glacier. For M = 0.3 m/yr, the upper half of the glacier is almost isothermal with a temperature equal to the surface temperature. Surface ablation, on the other hand, results in a much warmer glacier. This is a direct consequence of the parameterization of the vertical velocity used in this model; if M < 0, the ice flows upward, bringing warmer deep ice to the surface. Of course, these profiles are not very realistic; if surface ablation occurs, the surface temperature may be expected to be close to the melting temperature; and with temperatures increasing with depth, this would imply melting at all depths, which violates the first assumption on which the temperature solution (6.22) is based.

The temperature difference between the surface and the base is found by setting z = 0 in equation (6.24) and

$$T_b - T_s = \frac{G\sqrt{\pi}}{2kq}\,\mathrm{erf}\left(\frac{MH}{2K}\right)^{1/2}. \tag{6.26}$$

The contour diagram in Figure 6.3 shows how the value of the basal gradient and the glacier thickness affect this temperature difference. Increasing the basal heat flux

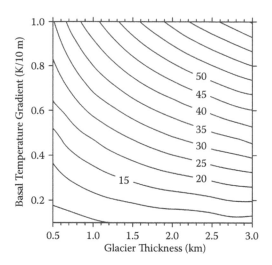

FIGURE 6.3 Temperature difference between the base and surface of a glacier as a function of the glacier thickness and temperature gradient at the base.

(or, equivalently, increasing the basal temperature gradient) increases the basal temperature, and thus the difference with the surface temperature. While the geothermal heat flux is poorly known, an average value is about $5 \cdot 10^{-2}$ W/m², corresponding to a basal temperature gradient of -0.02 K/m. Similarly, increasing the glacier thickness raises the basal temperature. Mathematically this follows immediately from the analytical solution. Physically, this can be understood by recalling that ice is a very good insulator; a thicker layer of ice traps more of the geothermal heat and allows less heat to escape through the upper surface.

The restriction that the basal temperature must be below the melting temperature can be relaxed, as was done by Weertman (1961b). Denoting the melting rate by M_b (positive if melting occurs), the vertical velocity at the base equals $-M_b$, and this term should be added to the right-hand side of equation (6.19) for the vertical velocity at depth. However, basal melt rates are typically of the order of mm/yr and this correction to (6.19) may be neglected. This means that the temperature profile derived above can still be applied if the basal temperature gradient is modified to include latent heat loss associated with melting. That is,

$$\left(\frac{\partial T}{\partial z}\right)_b = -\frac{G}{k} + \frac{L_f M_b}{k}. \tag{6.27}$$

This gradient can be estimated from the temperature difference between the upper surface and base of the glacier. Rearranging equation (6.26) and replacing G/k with the basal temperature gradient, gives

$$\left(\frac{\partial T}{\partial z}\right)_b = -\frac{2\,q(T_b - T_s)}{\sqrt{\pi}}\left[\operatorname{erf}\left(\frac{MH}{2K}\right)^{1/2}\right]^{-1}. \tag{6.28}$$

If melting or freezing occurs, the basal temperature is equal to the pressure melting temperature and the rate of basal melting can be estimated from equations (6.27) and (6.28).

The assumption that strain heating can be accounted for by increasing the geothermal heat flux is necessary to obtain an analytical solution for the temperature at depth. However, when solving the temperature equation numerically, as was done above for the case of surface ablation, this assumption is not necessary. Using the lamellar flow theory, discussed in Section 4.2, strain heating at depth can be calculated explicitly. To do so, the shear stress, τ_{xz}, is taken to increase linearly from zero at the surface to the maximum value, equal to the driving stress, at the base. The associated strain rate can be found by invoking the constitutive relation (2.10). Neglecting all other components of stress and strain rate, heat generated by internal deformation is then

$$W_d(z) = \frac{2A}{\rho C_p}\left(\frac{h-z}{H}\right)^4 \tau_{dx}^4,\tag{6.29}$$

where τ_{dx} represents the driving stress, h the elevation of the surface, and A the flow parameter. As discussed in Section 2.2, the flow parameter depends on the temperature of the ice. In the following calculation, the empirical relation (2.14) obtained by Hooke (1981) is used.

Strain heating affects only the temperature in the lower layers of an ice sheet. This is illustrated in Figure 6.4, which shows temperature profiles obtained by integrating

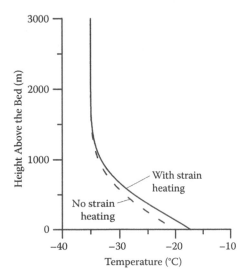

FIGURE 6.4 Steady-state temperature according to the Robin model. The solid line shows the profile calculated with inclusion of strain heating at depth (as estimated from the lamellar flow theory), and the dashed curve represents the profile without strain heating. Surface accumulation is 0.3 m/yr.

the temperature equation numerically until a steady state is reached. The full curve is the profile calculated with inclusion of strain heating at depth, while the dashed curve represents the profile without strain heating. All other parameters were kept the same for both calculations (τ_{dx} = 100 kPa, H = 3000 m, G = 0.042 W/m^2, T_s = −35°C). At the base, the temperature difference between both profiles is about 4°C. In this particular example, the effect of strain heating is small because the ice is relatively cold. The flow parameter, A, increases rapidly as the ice warms. This means that the warmer the ice, the more heat is generated by internal deformation, further raising the temperature, and so forth. It has been suggested that this positive feedback may lead to a runaway increase in temperature and deformation rate, called *creep instability*.

The most important shortcoming of the Robin temperature model is that horizontal advection of heat is neglected. How this affects the temperature distribution is discussed in the next section.

6.3 EFFECT OF HORIZONTAL HEAT ADVECTION

Observations in boreholes away from ice divides often show a reversed temperature gradient in the upper layers. The temperature decreases with depth to a minimum value, below which the ice becomes warmer with depth (c.f. the examples shown in Cuffey and Paterson, 2010, p. 415). In the Robin model, however, the upper layers are nearly isothermal or, where surface accumulation is sufficiently large, the temperature increases with depth from the surface down (Figure 6.2). This deficiency of the model is the direct consequence of neglecting horizontal advection of heat. To understand this, consider streamlines in an ice sheet, as schematically shown in Figure 6.5. These admittedly simplified streamlines represent the paths followed by ice originating at the surface. For the borehole AB, ice at depth C originated at a higher elevation than the surface ice at A. Because the surface temperature decreases with increasing elevation, this means that the ice at depth C is colder than the surface ice. In other words, the temperature decreases with depth until the effect of geothermal heat becomes important.

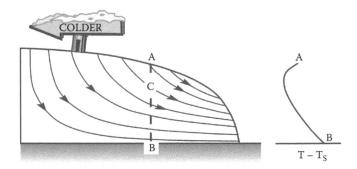

FIGURE 6.5   Simplified representation of streamlines in an ice sheet, illustrating the origin of the reversal of the temperature gradient at depth when horizontal advection becomes important. (Reproduced from Oerlemans, J., and C. J. van der Veen, *Ice Sheets and Climate*, Reidel Publ. Co., Dordrecht, 217 pp., 1984. With permission from Springer Verlag.)

Robin (1955) recognized this problem and derived an expression for the inverse temperature gradient near the surface by assuming that all terms in the heat-balance equation, with the exception of horizontal advection, are negligible. A more extensive treatment was given by Weertman (1968), in an effort to improve agreement between theoretical temperature profiles and that measured in the deep borehole at Camp Century in Greenland. The location of that borehole is far away from the ice divide, and horizontal advection of heat may be important.

Weertman (1968) estimated the horizontal temperature gradient at depth from the Robin profile given by equation (6.24). By choosing the x-axis in the mean direction of flow, only advection in the x-direction needs to be included in the temperature equation. Differentiating equation (6.24) with respect to x gives

$$\frac{\partial T}{\partial x} = \frac{\partial T_s}{\partial x} - \frac{\sqrt{\pi}}{2q^2}\gamma_b \frac{\partial q}{\partial x} \times$$

$$\times \left[\mathrm{erf}(zq) + \mathrm{erf}\left(\frac{z}{H}\right) + \left(\frac{2}{\pi}\right)^{1/2} zq\exp(-z^2 q^2) + \left(\frac{2}{\pi}\right)^{1/2} Hq\exp(-H^2 q^2) \right].$$

$$(6.30)$$

In this expression, γ_b represents the vertical temperature gradient at the base of the glacier. After some arithmetic, equation (6.30) can be rewritten as

$$\frac{\partial T}{\partial x} = \frac{\partial T_s}{\partial x} + \frac{T(z) - T_s}{2}\left(\frac{1}{H}\frac{\partial H}{\partial x} - \frac{1}{M}\frac{\partial M}{\partial x}\right) +$$

$$- \frac{\gamma_b}{2q\sqrt{2}}\left(\frac{1}{H}\frac{\partial H}{\partial x} - \frac{1}{M}\frac{\partial M}{\partial x}\right)[zq\exp(-z^2 q^2) - Hq\exp(-H^2 q^2)].$$

$$(6.31)$$

In his analysis, Weertman neglected the second line of this expression because it is negligibly small.

Also needed to estimate advection of heat is the horizontal velocity at depth. As in the preceding section, this velocity can be calculated from the lamellar flow theory. That is, the shear stress is taken to increase linearly with depth, which allows the shear strain rate to be calculated using the constitutive relation. Integrating this shear strain rate numerically, the horizontal component of velocity can be obtained (c.f. Section 4.2). Because the rate factor is temperature dependent, this procedure must be repeated for each time step when integrating the temperature equation to a steady state.

Inspection of equation (6.31) shows that the horizontal temperature gradient is controlled by three parameters, namely, the gradient in the surface temperature, the ice-thickness gradient, and horizontal gradients in surface accumulation. In most situations, the surface temperature and accumulation increase toward the margin of an ice sheet, while the thickness decreases away from the ice divide. This means that the horizontal temperature gradient at depth is generally smaller than that at the surface and may become negative near the base of the glacier.

TABLE 6.2

Camp Century Data

Ice thickness	H	1387.4 m
Thickness gradient	$\partial H/\partial x$	$-3.6 \cdot 10^{-3}$
Driving stress	τ_{dx}	45 kPa
Surface temperature	T_s	$-24\,°C$
Surface temperature gradient	$\partial T_s/\partial x$	$2.2 \cdot 10^{-5}\,°C/m$
Vertical temperature gradient at the base	γ_b	$-0.0177\,°C/m$
Surface accumulation	M	0.36 m/yr
Gradient in surface accumulation	$\partial M/\partial x$	$2.5 \cdot 10^{-7}\,yr^{-1}$

Source: Weertman, J., *J. Geoph. Res.,* 73, 2691–2700, 1968.

To investigate the importance of horizontal heat advection, the situation near the Camp Century borehole is considered. Data needed for the temperature calculation are taken from Weertman (1968) and given in Table 6.2. Using equation (2.14) to calculate the flow parameter, A, produces a horizontal velocity at the surface that is about a factor 3 less than that observed (3.3 m/yr). Therefore, A is multiplied by a constant factor to give the correct surface velocity. Figure 6.6 shows the results of five numerical integrations of the temperature equation, each differing by the processes included. Because the effects are so small, the difference between calculated temperatures and the Robin profile are shown in this figure. The Robin profile was obtained from equation (6.24) using the basal temperature gradient given in Table 6.2 and not including strain heating. To avoid introducing too many interactions at once, strain heating and the horizontal component of velocity were calculated only once for the calculations labeled 1–4, using the Robin temperature distribution to calculate the flow parameter.

Curve 1 in Figure 6.6 represents the calculation in which horizontal advection of heat is not included; differences between the calculated temperature profile and the Robin profile are entirely due to strain heating. Because the deformation rate and the shear stress are largest near the glacier base, deformational heating affects the lower part of the glacier only, raising the basal temperature by about 0.56°C. The effect of a horizontal gradient in surface temperature is illustrated by curve 2. Because the upglacial ice is colder, the temperature of the entire column decreases and a temperature inversion emerges. The inversion is rather small, however, somewhat more than 0.1°C, as compared with the 0.6°C inversion measured in the Camp Century borehole. The gradients in thickness and surface accumulation result in only minor changes in the temperature, as shown by the curves 3 and 4. Finally, including the interaction between temperature and ice flow by calculating the flow parameter each time step raises the temperature a few tenths of a degree (curve 5).

The modifications to the Robin theory proposed by Weertman (1968) have only a minor effect, and calculated corrections to the Robin temperature profile are less than about 0.5°C. As noted by Weertman (1968), this should not be very surprising because the horizontal velocity is small (3.3 m/yr at the surface) so that the third

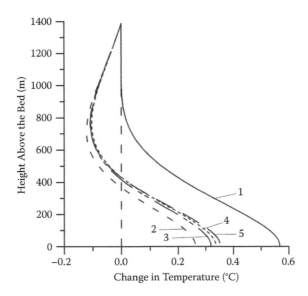

FIGURE 6.6 Effect of horizontal advection and strain heating on the Robin temperature profile (6.24), calculated using the data for Camp Century given in Table 6.2. For curves 1–4, strain heating and the horizontal component of velocity were calculated at the start of the integration only, using the Robin temperature profile to estimate the flow parameter at depth; for curve 5, strain heating and horizontal velocity were calculated each time step, using the result of the temperature calculation to estimate the flow parameter. Curve 1 shows the difference between the temperature calculated with inclusion of strain heating (no advection) and the Robin solution. Curve 2 represents the result of a calculation in which the horizontal temperature gradient at depth is taken equal to the temperature gradient at the surface. Curve 3 also includes the effect of the horizontal thickness gradient on the temperature gradient. Curves 4 and 5 represent calculations that include all terms of equation (6.31) but differ in how strain heating is included.

assumption of the Robin model (horizontal advection of heat may be neglected) is almost satisfied. The profiles in Figure 6.6 show that the effect of heat generated by internal deformation is as important as the effect of horizontal advection of heat.

The aim of Weertman's study was to improve agreement between the predicted and the measured temperature profile at Camp Century. Because the refinements of the temperature model did not substantially improve agreement, Weertman (1968) concluded that the borehole temperature reflects nonsteady conditions. That is, the ice temperature may currently be changing due to changing climatic conditions in the (recent) past. This point is taken up in more detail in the next section.

6.4 THERMAL RESPONSE OF A GLACIER TO CHANGES IN CLIMATE

The temperature distribution in ice sheets is determined by heat fluxes through the surface, the geothermal heat flux at the base, and heat generated by internal deformation. Considering the geothermal flux constant, changes in the ice temperature are mostly initiated at the glacier surface, either by a change in surface temperature or by a change in surface accumulation or ablation. Changes in the thickness of the glacier or in the rate of deformation also affect the temperature distribution.

To study the effect of climate changes, consider as a start a simple harmonic variation in surface temperature:

$$T_s(t) = \Delta T \sin(\omega t), \tag{6.32}$$

where ΔT represents the amplitude of the surface perturbation and t denotes time. Neglecting for the time being downward advection (so vertical heat exchange is by diffusion only), the solution of the temperature equation (6.18) that satisfies this boundary condition, is

$$T(z,t) = \Delta T\, e^{-\bar{z}} \sin(\omega t - \bar{z}), \tag{6.33}$$

with

$$\bar{z} = (h-z)\left(\frac{\omega}{2K}\right)^{1/2}. \tag{6.34}$$

The amplitude of the perturbation decreases exponentially with depth, with the attenuation increasing with frequency. High-frequency variations penetrate only to shallow depths before being diffused away, while perturbations with a long time scale propagate much farther down into the glacier. Also, the time lag between the maximum perturbation at depth and at the surface increases with depth.

The depth at which the magnitude of the perturbation is 5% of the surface variation is

$$D(5\%) = \left(\frac{2K}{\omega}\right)^{1/2} \ln(20). \tag{6.35}$$

The phase shift between the perturbation at depth and that at the surface can be found by setting the time derivative of the temperature given by equation (6.33) to zero and is equal to $\ln(20)/\omega$. For example, the annual cycle has a period of 1 year ($\omega = 6.3$ yr^{-1}), and the 5% depth is 10 m. The maximum temperature perturbation at this depth occurs almost half a year after the temperature maximum at the surface. Because the annual cycle is almost completely diffused away before reaching a depth of 10 m, measured temperatures at this depth are often used as proxy for the annual mean surface temperature.

In the discussion so far, only vertical diffusion of heat was considered, allowing an analytical expression for the seasonal wave propagation to be derived. When advection is important, the temperature equation has to be integrated numerically to determine how the surface perturbation travels downward. Advection of heat greatly increases the depth to which the surface wave propagates. This is illustrated in Figure 6.7, which shows the downward penetration of a perturbation in surface temperature (with a period of 1000 yr), for the case of diffusion only (left panel) and

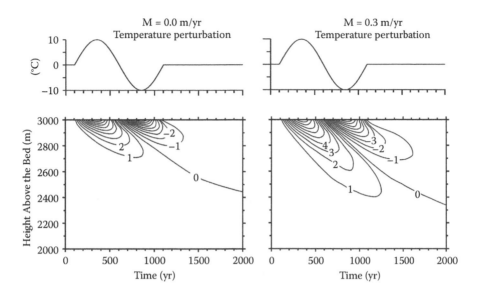

FIGURE 6.7   Downward penetration of a perturbation in the surface temperature by vertical conduction only (left panel) and including downward advection (right panel). Contours show the difference between the calculated temperature and the steady-state temperature at the start of the integration.

when downward advection is included. The downward motion advects the perturbation much deeper before it is diffused away. From this it may be concluded that surface accumulation is very important in determining the thermal response of a glacier.

As can be seen most clearly in the panel on the right of Figure 6.7, the cold wave does not penetrate as deeply as does the preceding warm wave. This is because part of the "excess" cold is used to remove the heat from the warm ice and bring its temperature to the initial Robin temperature. Only after the ice has reached this temperature does further cooling occur. If successive harmonic perturbations were to be prescribed, both the cold and the warm wave would penetrate to the same depth.

The vertical velocity is taken proportional to the rate of surface accumulation (equation (6.19)). Changes in accumulation affect the downward advection of cold surface ice to greater depth and modify the temperature distribution near the base of the glacier. This follows immediately from the steady-state profiles shown in Figure 6.2 and is further illustrated in Figure 6.8. Changes in the upper half of the glacier are negligibly small, while after an increase in surface accumulation, the increased downward transport of colder surface ice results in a lowering of the basal temperature. This calculation is not entirely realistic, however, because changes in accumulation affect the surface temperature also. If the air is colder than the snow surface, precipitation will tend to cool the surface. To account for this, the surface temperature needs to be calculated explicitly as a function of surface heat fluxes.

In the preceding calculations the surface temperature was prescribed as upper boundary condition, with the implicit understanding that this temperature equals

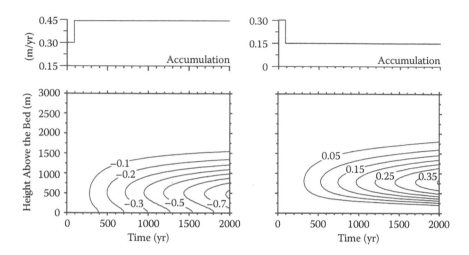

FIGURE 6.8 Effect of a stepwise increase (left panel) and decrease (right panel) on the temperature distribution in a glacier. Contours show the difference between the calculated temperature and the steady-state temperature at the start of the integration.

the air temperature. This is an appropriate approach when considering the temperature distribution in an ice sheet over long periods of time (for example, for modeling glacial cycles), mainly because the temperature at about 10 m below the surface is a good proxy for the annual air temperature if no surface melting occurs. For studying ice-sheet behavior on long time scales, detailed modeling of the thermal regime in the upper firn layers is not required (nor computationally feasible). However, for some applications, such detailed modeling is needed. For example, Van der Veen and Jezek (1993) calculate the temperature profile in the upper 65 m of the firn near Vostok Station in East Antarctica to study the seasonal changes in remotely sensed brightness temperatures. Another application is to investigate how climate conditions, and changes in these, affect surface ablation and englacial temperatures (for example, Greuell and Oerlemans, 1989a, b). In these studies, the surface energy balance replaces the air temperature as the upper boundary condition. The surface temperature of the firn is calculated from the requirement that the warming or cooling of the surface is directly proportional to the net flux of heat through the surface.

A schematic picture of the heat exchange between the atmosphere and the surface of a glacier is shown in Figure 6.9. Heat exchange is due to incoming and outgoing radiation (R_n), the latent heat flux ($L_f E$, with E the moisture flux), heat exchange associated with precipitation (Q_m), and the sensible heat flux (H). At the surface, the total energy flux is

$$Q_s = R_n + L_f E + H + Q_m. \tag{6.36}$$

Fluxes are taken positive when directed toward the glacier, that is, when providing heat to the glacier. It should be noted that Q_s does not represent the net flux through

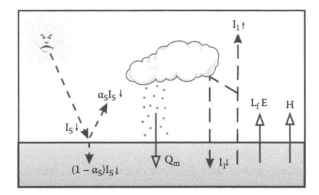

FIGURE 6.9 Schematic representation of the energy exchange between the atmosphere and glacier surface. Arrows with open heads indicate heat fluxes that can be directed either upward or downward, depending on the temperature difference between the glacier surface and the overlying air.

the surface layer, but only the net heat exchange between the atmosphere and glacier. The surface balance also depends on heat exchange with deeper layers of firn (c.f. Section 6.8).

The surface gains heat by incoming solar radiation, infrared long-wave radiation originating in the atmosphere, and loses heat by emitting long-wave radiation. The sensible heat flux acts as a heat source if the overlying air is warmer than the glacier surface, but represents a loss if the surface is warmer than the air above. Similarly, heat associated with precipitation can act as either a source or a sink, depending on the temperature of the precipitation reaching the surface. Where surface melting occurs, the downward percolation of meltwater serves to redistribute energy, warming deeper layers where refreezing occurs and latent heat is released. In addition, Phillips et al. (2010) propose that cryo-hydrologic warming could be a potential mechanism for rapid thermal response of glaciers to climate warming. Water present in englacial conduits (the cryo-hydrologic system) has a temperature that is at or above the pressure melting point. Thus, heat can be conducted from the conduits to the surrounding firn or ice. If some of the meltwater is retained after the melt season, substantial warming of the surrounding ice could occur. The magnitude of this warming depends on the spacing of meltwater conduits. Phillips et al. (2010) apply a numerical model to a location near the equilibrium line in West Greenland and find a substantial warming of ~10°C at 500 m below the surface 5 to 10 years after the onset of surface melting and cryo-hydrologic warming.

By coupling calculated surface fluxes to the thermodynamic equation (6.17), the change in firn temperature throughout the year or under altered climate conditions can be (numerically) calculated. An example of such a calculation is discussed in Section 6.8, but first, parameterizations that allow the surface fluxes to be calculated from observable meteorological variables are needed. In the next two sections, expressions for the radiative and turbulent surface heat fluxes are discussed. A more extensive treatment of this topic can be found in Greuell and Genthon (2003).

6.5   RADIATION BALANCE AT THE SURFACE OF A GLACIER

An extensive discussion of the radiation balance of the earth and atmosphere can be found in Peixoto and Oort (1992). Below, the most important results needed in energy-balance models of the glacier surface are discussed.

The earth receives its energy from the sun in the form of short-wave radiation with wavelengths ranging from ultraviolet, through the visible part of the spectrum, to the infrared. The average amount of solar radiation at the top of the atmosphere is called the *solar constant*, S (= 1395 W/m^2). To be exact, S is defined as the amount of solar radiation incident per unit area and per unit time on a surface normal to the direction of propagation and situated at the earth's mean distance from the sun (Peixoto and Oort, 1992). Of interest for the radiation balance is the amount of radiation through a horizontal surface. At the top of the atmosphere, this horizontal irradiance is

$$I_o = S \cos Z, \qquad (6.37)$$

where Z represents the zenith angle of the sun. To be fully correct, this expression should include a factor that accounts for the varying distance between the earth and the sun. However, this correction factor is small (1.0344 in early January and 0.9646 in early July; Peixoto and Oort, 1992) and is usually neglected. The zenith angle, Z, is a function of geographical latitude, ϕ, solar declination, δ (the angular distance of the sun north [positive] or south [negative] of the equator), and hour angle, h (the angle through which the earth must rotate to bring the local meridian directly under the sun). Using spherical trigonometry, the following relation can be derived (for example, Sellers, 1965, p. 15)

$$\cos Z = \sin \phi \sin \delta + \cos \phi \cos \delta \cos h. \qquad (6.38)$$

Tables of the solar declination are given in any standard ephemeris, while the hour angle follows directly from the local time in Greenwich Mean Time (GMT).

Not all solar radiation reaches the earth's surface. Part is absorbed in the atmosphere, while part is scattered by aerosols and clouds in the atmosphere. Scattering results in redirection of radiation, and part of the scattered radiation may reach the surface as so-called diffuse radiation. Goddard (1975) proposes the following expression for the average intensity of short-wave radiation reaching the surface:

$$I_s \downarrow = (1 - n\alpha_c) S T_r^{\sec Z} \cos Z, \qquad (6.39)$$

where T_r represents an atmospheric turbidity coefficient (= 0.93 for Arctic pack ice simulations), n the cloud cover ($0 \le n \le 1$), and α_c the cloud albedo (ranging from 0.2 for cirrus clouds to around 0.7 for nimbostratus clouds; Houghton, 1985).

A more detailed parameterization for the short-wave solar radiation reaching the surface is used by Konzelmann et al. (1994), namely,

$$I_s \downarrow = f_{mr} \tau_{cs} \tau_{cl} S \cos Z. \qquad (6.40)$$

Here τ_{cs} represents the clear-sky transmission and τ_{cl} the cloud transmission. The factor f_{mr} accounts for multiple clear-sky reflection. The clear-sky transmission is based on Beer's law for absorption of radiation in the atmosphere and is

$$\tau_{cs} = \gamma \exp(-\beta \tau_L m_r), \tag{6.41}$$

where m_r represents the relative optical air mass, τ_L the Linke turbidity factor, and $\gamma = 0.84$ and $\beta = 0.027$. The optical air mass is a measure of the distance that the solar radiation travels through the atmosphere, taking into account the lower density of the upper atmosphere, and is a function of zenith angle and elevation (or pressure). The Linke turbidity factor represents radiative transfer associated with scattering and absorption by air molecules, aerosols, ozone, and water vapor and is a function of the air temperature, vapor pressure, zenith angle, and elevation (as well as the ozone concentration, taken constant by Konzelmann et al., 1994). The multiple reflection term accounts for solar radiation reflected from the surface upward, and reflected downward again by clouds. Konzelmann et al. (1994) argue that high albedos introduce a significant increase in the global radiation compared with areas with low surface albedo, and therefore write the multiple-reflection coefficient as

$$f_{mr} = \frac{1}{1 - \alpha_a \alpha_s}, \tag{6.42}$$

where α_a and α_s represent the albedo of the air and the surface, respectively. Given that clouds reflect much more of the short-wave solar radiation, thus increasing the multiple reflection, than does a clear sky, it is not clear why this parameterization does not include the cloudiness, n. Finally, the cloud transmission coefficient, τ_{cl}, was determined by comparing daily mean global radiation measured at several sites on the Greenland Ice Sheet with calculated global radiation under clear-sky conditions. The measurements show that absorption of solar radiation by clouds is largest at low elevations and decreases with increasing elevation of the surface. In other words, the optical depth of clouds decreases with elevation. For the Greenland data, Konzelmann and others (1994) derive as best fit

$$\tau_{cl} = 1 - 0.78 \, n^2 \, e^{-0.00085h}, \tag{6.43}$$

with $0 \le n \le 1$, and h as the elevation of the surface in meters.

Perhaps the greatest uncertainty in estimating incoming solar radiation is associated with the role of clouds. As noted by Greuell and Genthon (2003), most parameterizations involve cloud cover only (n in equation (6.43)). However, what determines absorption of radiation by clouds is their optical thickness, which is more difficult to estimate. Fitzpatrick and others (2004) define a cloud transmittance, trc, as the ratio of incoming short-wave radiation at the surface when clouds are present to the amount of radiation that would reach the surface in the absence of clouds. They propose the following parameterization:

$$trc = \frac{a(\tau) + b(\tau)\cos Z}{1 + (c - d\alpha)\tau}, \tag{6.44}$$

where $a(\tau)$ and $b(\tau)$ are functions of the cloud optical depth, τ, Z is the solar zenith angle, and α is the surface albedo.

Part of the solar radiation reaching the glacier surface is reflected back to the atmosphere, as described by the albedo. For snow, the albedo is a function of grain size and density and may also be different for direct solar radiation and for diffuse solar radiation. Freshly fallen snow may reflect as much as 95% of the incoming radiation, while bare ice typically reflects about 50% to 60% of the radiation. For wet snow, the albedo is even smaller. Many different parameterizations for the albedo have been proposed, depending on the type of glacier studied. For example, in their study of ablation on alpine glaciers, Greuell and Oerlemans (1986) use an expression that links the albedo to the surface density, the rate of surface melting, cloudiness, and zenith angle. For polar ice caps, Greuell and Oerlemans (1989a) propose a much simpler formula, namely,

$$\alpha_s = \alpha_o + 0.06\,n, \tag{6.45}$$

where $\alpha_o = 0.8$ in the absence of meltwater at the surface and $\alpha_o = 0.54$ if meltwater is present. Greuell and Konzelmann (1994) model the albedo on the Greenland Ice Sheet as a function of the snow density at the surface, ρ_s, and cloud amount, n:

$$\alpha_s = \alpha_i + \frac{\rho_s - \rho_i}{\rho_{fs} - \rho_i}(\alpha_{fs} - \alpha_i) + 0.05(n - 0.5), \tag{6.46}$$

where the subscript fs refers to fresh snow ($\rho_{fs} = 300$ kg/m^3, $\alpha_{fs} = 0.85$) and the subscript i to glacier ice ($\rho_i = 910$ kg/m^3, $\alpha_i = 0.58$). In central East Antarctica, on the other hand, the albedo varies little (Schwerdtfeger, 1970), and a constant value may be used.

To be fully correct, albedo parameterizations should account for the age of the surface snow; as the snow settles after deposition, its structure and reflective properties change. In addition, the onset of melting has a large effect on the albedo. For example, on Haig Glacier in Alberta, Canada, Shea et al. (2005) measured a decrease in albedo from around 0.9 in mid-May to around 0.2 in mid-August. As melting ceased toward the end of the summer, the albedo increased rapidly to around 0.6 in early October and continued to rise throughout the winter. Further, the albedo depends on incidence angle and wavelength of the incoming radiation. For these reasons, then, it is no surprise that many different parameterizations have been proposed. For more extensive discussions, see Brock and others (2000) and Greuell and Genthon (2003). Table 6.3 gives typical albedo values for different types of snow and ice surfaces.

Part of the short-wave radiation penetrates into the upper firn layers. The amount of absorption depends on the wavelength of the radiation as well as on the physical properties of the firn. Holmgren (1971, Part E) describes experiments conducted in the upper 18 cm of firn on Devon Island Ice Cap. These data show that minimum extinction occurs for wavelengths near the middle part of the visible spectrum, while for longer wavelengths, the extinction coefficient increases markedly.

TABLE 6.3

Typical Values for the Albedo of Snow and Ice, Based on a Literature Review by S.J. Marshall

Surface Type	Recommended	Range
Fresh dry snow	0.85	0.75–0.98
Old clean dry snow	0.80	0.70–0.85
Old clean wet snow	0.60	0.46–0.70
Old debris-rich dry snow	0.50	0.30–0.60
Old debris-rich wet snow	0.40	0.30–0.50
Clean firn	0.55	0.50–0.65
Debris-rich firn	0.30	0.15–0.40
Superimposed ice	0.65	0.63–0.66
Blue ice	0.64	0.60–0.65
Clean ice	0.35	0.30–0.46
Debris-rich ice	0.20	0.06–0.30

Source: Reprinted from Cuffey, K. M., and W. S. B. Paterson, *The Physics of Glaciers* (4th ed.), Butterworth-Heinemann, Burlington, MA, 693 pp., 2010. With permission from Elsevier.

For coarse-grained snow, the attenuation of solar radiation is about twice as deep as for fine-grained snow. Greuell and Oerlemans (1989a) make the assumption that all radiation with wavelengths greater than 0.8 μm (about 36% of the solar radiation penetrating the surface) is completely absorbed in the surface layer. The remaining 64% is extinguished at depth according to Beer's law. Thus, solar radiation absorbed by the glacier surface is

$$I_s(h) = 0.36 (1 - \alpha_s) I_s \downarrow, \tag{6.47}$$

while the radiative flux at depth is

$$I_s(h - z) = 0.64 (1 - \alpha_s) e^{-\beta(h-z)} I_s \downarrow. \tag{6.48}$$

The extinction coefficient, β (in m^{-1}), decreases linearly with increasing firn density as

$$\beta = -0.012 \rho_s + 14.706. \tag{6.49}$$

In a later study, Greuell and Konzelmann (1994) obtained values for β that are about twice as large, by optimizing the fit between calculated and measured temperature profiles.

Kuipers Munnike et al. (2009) compare results from an energy-balance model with measurements of the surface energy balance, including subsurface firn temperature, at Summit, Greenland. They find that, on average, 6.3% of the incoming solar radiation is absorbed at depths greater than 0.5 cm below the surface. Penetration of

short-wave radiation into the firn has an important effect on modeled firn temperatures. Not including this term in the energy-balance model results in a decrease in
firn temperature of ~4°C in the upper 1 m.

The second source of energy for the glacier surface is long-wave radiation emitted
by water vapor and other greenhouse gases in the atmosphere. Of these gases, water
vapor accounts for about 80% of the total long-wave radiation (Male, 1980). Hence,
variations in incoming long-wave radiation are mainly due to changes in the water
content of the atmosphere and in its temperature. In most studies, the atmospheric
radiation is related to humidity and temperature at "screen level," usually anywhere
from 1 to 10 m above the surface. This appears to be a reasonable approximation
because most of the long-wave radiation reaching the surface originates from the
lower part of the atmosphere (Geiger, 1966).

For clear-sky conditions, the incoming long-wave radiation is most often estimated using the expression due to Brunt (1939)

$$I_l \downarrow = \sigma T_a^4 (a + b\sqrt{e_a}), \tag{6.50}$$

where $\sigma = 5.6704 \cdot 10^{-8}$ W/m^{2}K^4 represents the Stefan–Boltzmann constant, T_a the
air temperature at about 2 m above the surface, and e_a the water vapor pressure at
that level. Greuell and Oerlemans (1986) use a somewhat different parameterization,

$$I_l \downarrow = \varepsilon_a \sigma T_a^4, \tag{6.51}$$

with the clear-sky emissivity related to water vapor pressure (in Pa) and air temperature (in K) as

$$\varepsilon_a = 0.70 + 5.95 \cdot 10^{-7} e_a e^{1500/T_a}. \tag{6.52}$$

Greuell and Oerlemans (1986) argue that the contribution from clouds to the
downward long-wave radiation may be treated as a separate radiation flux, parameterized as

$$I_{lc} \downarrow = \varepsilon_c n f_w \tau_w \sigma T_{cl}^4. \tag{6.53}$$

In this expression, T_{cl} represents the temperature of the clouds, ε_c the emissivity of
clouds (about 1.0 for dense clouds, and 0.5 for cirrus clouds), f_w the fraction of black-
body radiation emitted in the atmospheric window (8–14 μm) that can be transmitted to the surface, and τ_w the transmittance of the atmosphere for this atmospheric
window. The net incoming long-wave radiation according to Greuell and Oerlemans
(1986) is given by the sum of equations (6.51) and (6.53).

Goddard (1975) uses a simpler expression to estimate incoming long-wave radiation. The effect of clouds is included by introducing an effective sky emissivity, ε_2,
which equals 1 for completely overcast skies, and 1.14 under cloud-free conditions, and

$$I_l \downarrow = \varepsilon_2 C \sigma T_a^6. \tag{6.54}$$

The parameter C is an empirical constant (= $9.35 \cdot 10^{-6}$). This expression does not take into account the effect of water vapor on the atmospheric radiation. A comparable formula is used by Konzelmann et al. (1994). The effective emittance is the weighted mean of the clear-sky emittance, ε_{cs}, and the emittance of a completely overcast sky, ε_{oc}, so that, for completely overcast skies, the parameterization does not include the clear-sky emittance:

$$I_l \downarrow = \left[\varepsilon_{cs}(1 - n^p) + \varepsilon_{oc} n^p \right] \sigma T_a^4. \tag{6.55}$$

The clear-sky emittance is calculated from a somewhat modified form of the expression derived analytically by Brutsaert (1975):

$$\varepsilon_{cs} = 0.23 + b \left(\frac{e_a}{T_a} \right)^{1/m}. \tag{6.56}$$

Using various data collected in Western Greenland, Konzelmann et al. (1994) find an optimal fit between model prediction and measurements for b = 0.443 and m = 8 in equation (6.56), ε_{oc} = 0.952 and p = 4 in (6.55) if instantaneous cloud observations and hourly radiation measurements are used, and ε_{oc} = 0.963 and p = 3 if daily mean observations are used.

The third and final radiative flux that needs to be considered is long-wave radiation emitted by the glacier surface. This radiation, with wavelengths in the range of 5–40 μm, is confined to the snow surface and can be estimated from the Stefan–Boltzmann radiation law

$$I_l \uparrow = \varepsilon_0 \sigma T_s^4. \tag{6.57}$$

The emissivity of snow, ε_0, varies typically between 0.98 and 1.0. Goddard (1975) includes the effect of clouds on the outgoing radiation (part of which is reflected back to the surface) by introducing a long-wave cloud-reduction coefficient, k_c:

$$I_l \uparrow = (1 - n k_c) \varepsilon_0 \sigma T_s^4. \tag{6.58}$$

In simulations for Arctic pack ice, Goddard (1975) uses k_c = 0.75 under clear-sky conditions and k_c = 0.73 for cloudy conditions.

In general, the degree of sophistication used in modeling the radiative fluxes depends on the extent of meteorological observations. If few such observations are available, there is not much point in using very sophisticated models that cannot be tested or calibrated, and a parameterization such as equation (6.39) should be adequate. On the other hand, where extensive measurements have been conducted, including measurements of radiation, more complicated radiation models may be warranted. Of course, one could argue that if actual radiation measurements are available, one might as well use these observations when considering the energy exchange between the atmosphere and glacier surface. However, the implicit assumption behind the more complex models is that these models can also be applied to time periods when fewer meteorological observations are available, as well as to different locations.

The expressions discussed above apply only to situations where the glacier surface is not shielded by surrounding mountains. For small alpine glaciers, the surrounding valley walls may block part of the solar radiation as well as cause significant changes in the long-wave radiative fluxes. By obscuring part of the sky, valley walls reduce the incoming atmospheric radiation. General expressions to describe these effects cannot be given because they are determined by the particular geometry of the glacier and its surroundings.

6.6   TURBULENT HEAT FLUXES

The sensible heat flux is due to the difference in temperature of the surface of the glacier and the air above. The heat exchange is accomplished mainly by turbulent motion in the boundary layer, except perhaps in the first few millimeters above the surface, where molecular diffusion may be more important. For the latent heat flux, the transfer of heat is indirect and associated with evaporation at the surface and subsequent condensation at higher levels in the atmosphere (or vice versa). If the rate of evaporation equals E, the amount of heat needed for evaporation is, per unit time, $L_f E$, where L_f represents the latent heat of evaporation or condensation (that is, the amount of heat needed to evaporate 1 kg of water or condensate the corresponding amount of water vapor). This heat is extracted from the immediate surroundings (for example, the glacier surface) and released where condensation takes place. The ratio of the sensible heat flux to the latent heat flux is called the Bowen ratio.

The motion in the boundary layer is strongly influenced by the friction generated at the surface, and the wind profile exhibits large vertical shear because the speed must be zero at the surface. As a result, and because of the heating or cooling of the surface, turbulent flow develops, in which random velocity perturbations are superimposed on the average flow. This turbulent mixing of air is a very effective process by which properties such as momentum, heat, water vapor, or aerosols are redistributed vertically. To estimate the flux of any quantity (or net vertical transport) at the surface, the assumption is usually made that the flux in the lower few meters of the boundary layer is representative of the flux at the surface. This allows the fluxes to be estimated from standard meteorological observations, such as wind speed, air temperature, and humidity.

To arrive at an expression for the turbulent flux of a quantity F, the continuity equation for that quantity needs to be considered. While molecular diffusion may be important in the thin sublayer very close to the surface, vertical transport is mainly determined by vertical motion in the boundary layer, and if there is no local production or loss, F changes with time as a result of advection only. Denoting the three orthogonal directions by the subscript i, and using the summation convention for repeat indexes, the continuity equation is

$$\frac{\partial F}{\partial t} = - U_i \frac{\partial F}{\partial x_i}, \tag{6.59}$$

where U_i represents the three components of velocity in the boundary layer.

Turbulent flow is characterized by random velocity variations superimposed on the average velocity. Thus, the total velocity in the x_i-direction, U_i, may be written as

$$U_i = \bar{U}_i + \tilde{u}_i, \tag{6.60}$$

where the overbar denotes the average flow, defined as

$$\bar{U}_i = \lim_{T \to \infty} \frac{1}{T} \int_{t_0}^{t_0+T} U_i \, dt. \tag{6.61}$$

The tilde denotes the random fluctuation in velocity. By definition, the time average of this term is zero. In the meteorological literature, this partitioning is referred to as Reynolds decomposition (for example, Tennekes and Lumley, 1972).

A decomposition similar to (6.60) for the velocity can be introduced for the quantity under consideration, F, and substitution in the continuity equation (6.59) yields

$$\frac{\partial}{\partial t}(\bar{F} + \tilde{f}) = -(\bar{U}_i + \tilde{u}_i) \frac{\partial}{\partial x_i}(\bar{F} + \tilde{f}). \tag{6.62}$$

The average flow is assumed to be in steady state, so that taking the time average of this expression gives

$$0 = -\bar{U}_i \frac{\partial \bar{F}}{\partial x_i} - \overline{\tilde{u}_i \frac{\partial \tilde{f}}{\partial x_i}}. \tag{6.63}$$

In the boundary layer, the air may be considered incompressible, that is, having a constant density. By substituting the decomposition of velocity in an average and a fluctuation term into the incompressibility condition and taking the time average, it follows that both the mean flow and the turbulent velocity fluctuations are incompressible (c.f. Tennekes and Lumley, 1972, p. 30). Thus,

$$\frac{\partial \tilde{u}_1}{\partial x_1} + \frac{\partial \tilde{u}_2}{\partial x_2} + \frac{\partial \tilde{u}_3}{\partial x_3} = 0. \tag{6.64}$$

Equation (6.63) can now be rewritten as

$$\bar{U}_i \frac{\partial \bar{F}}{\partial x_i} = -\frac{\partial}{\partial x_i}\left(\overline{\tilde{u}_i \tilde{f}}\right). \tag{6.65}$$

The right-hand side may be considered as the divergence of a flux, with the mean turbulent flux of the quantity F per unit area and unit time given by

$$Q_i = \overline{\tilde{u}_i \tilde{f}}. \tag{6.66}$$

Of interest to the energy budget are the vertical fluxes of sensible and latent heat. The sensible heat flux, H, follows by taking $F = \rho_a C_{pa} T$ and considering the vertical direction only

$$H = \rho_a C_{pa} \overline{\tilde{w}\tilde{t}}. \tag{6.67}$$

In this expression, $\tilde{w}$ represents the fluctuation in the vertical velocity (that is, $\tilde{w} = \tilde{u}_3$ in the notation used above), and $\tilde{t}$ the temperature perturbation. Similarly, the vertical flux of moisture is

$$E = \rho_a \overline{\tilde{w}\tilde{q}}, \tag{6.68}$$

where q represents the specific humidity (in kg of water vapor per kg of air).

The next step is to relate the perturbations to the average field. This is commonly done using the mixing-length theory developed by Prandtl (1932) (c.f. Schlichting, 1968; Peixoto and Oort, 1992). In Prandtl's model, a parcel of air originally at level z moves over a vertical distance ℓ, keeping its original velocity, before being absorbed and assuming the velocity of its surroundings. The difference between the speed of the parcel and its surroundings is a measure for the turbulent fluctuations so that

$$\tilde{u} = \overline{U}(z) - \overline{U}(z + \ell) =$$

$$= -\ell \frac{\partial \overline{U}}{\partial z}. \tag{6.69}$$

It seems reasonable to make the assumption that the mixing length, ℓ, is proportional to the physical space available; that is, ℓ is taken proportional to the distance to the surface as $\ell = kz$, where $\kappa = 0.4$ is the Von Kármán constant. A friction velocity, U_*, is now introduced such that

$$\left| \frac{\partial \overline{U}}{\partial z} \right| = \frac{U_*}{\ell} =$$

$$= \frac{U_*}{\kappa z}. \tag{6.70}$$

Integrating this expression with respect to z gives the often-used logarithmic wind profile

$$\overline{U}(z) = \frac{U_*}{\kappa} \ln \frac{z}{Z_o}. \tag{6.71}$$

Here, Z_o represents the roughness length of the surface, that is, the height above the surface at which the average wind speed is zero. Greuell and Konzelmann (1994) use $Z_o = 0.12$ mm for snow before melting, $Z_o = 1.3$ mm for snow after melting, and $Z_o = 3.2$ mm for ice, based on observations of wind speed over the Greenland Ice Sheet. The friction velocity, U_*, is typically of the order of 0.2 to 0.4 m/s.

If the turbulence is isotropic, the magnitude of velocity perturbations is the same in all directions. In particular

$$|\tilde{w}| = \ell \left|\frac{\partial \bar{U}}{\partial z}\right|. \tag{6.72}$$

For other variables, such as temperature, a similar relation between fluctuation and average vertical gradient is assumed. That is

$$|\tilde{t}| = \ell \left|\frac{\partial \bar{T}}{\partial z}\right|. \tag{6.73}$$

Equation (6.67) for the sensible heat flux can now be written as

$$H = \rho_a C_{pa} \kappa z U_* \frac{\partial \bar{T}}{\partial z}. \tag{6.74}$$

In the following, the overbar is omitted, with the understanding that meteorological quantities refer to time averages (obtained from observations). An expression similar to (6.74) can be given for the vertical flux of moisture.

The final step is to relate the friction velocity and vertical temperature gradient to observations. If U_{ref} represents the measured wind speed at some level Z_{ref}, it follows from the logarithmic wind profile (6.71) that the friction velocity is

$$U_* = \frac{\kappa Z_{ref}}{\ln(Z_{ref}/Z_o)}. \tag{6.75}$$

Analogous to the derivation of the logarithmic wind profile, a logarithmic profile can be derived for the temperature in the boundary layer, with a temperature scale, T_*, given by an expression similar to equation (6.75). For simplicity, often assumptions are made that the roughness length for temperature and wind speed are the same and that both quantities are measured at the same level. This is not necessary but simplifies the expressions for the turbulent heat fluxes.

Using the expressions derived above, the turbulent heat fluxes can now be expressed in terms of measurable meteorological variables. The sensible heat flux becomes

$$H = \rho_a C_{pa} \kappa^2 U_{ref} (T_{ref} - T_s)[\ln(Z_{ref}/Z_o)]^{-2}. \tag{6.76}$$

Note that the surface temperature, T_s, appears in this expression because T_s is not zero (contrary to the wind speed, assumed to be zero at the surface). The corresponding expression for the latent heat flux is

$$L_f E = L_f \rho_a \kappa^2 U_{ref} (q_{ref} - q_s)[\ln(Z_{ref}/Z_o)]^{-2}. \tag{6.77}$$

It may be noted that some authors use the potential temperature, rather than the physical temperature, to calculate the flux of sensible heat. The potential temperature is defined as the temperature that a parcel of air would attain if the pressure is changed adiabatically from the actual pressure, P, to the reference pressure, P_0 (usually taken equal to 1000 mbar). Near the surface, the actual pressure is close to the reference pressure and both temperatures are almost the same and within measurement uncertainty.

The approximation (6.70) implies that the vertical flux of momentum is constant with height (Peixoto and Oort, 1992). This is true where the surface is smooth and the boundary layer is in neutral equilibrium. That is, when the vertical temperature gradient equals minus the dry-adiabatic lapse rate, $\gamma_d = g/C_{pa} \approx 9.8$ K/km. Where the actual temperature gradient is less than $-\gamma_d$, the thermal stratification is unstable, while the stratification is stable where $\partial T/\partial z > -\gamma_d$. In a nonneutral atmosphere, the turbulence acts against the buoyancy forces, which results in an additional upward heat flux in an unstable atmosphere and a downward flux of heat in a stable boundary layer (Busch, 1973). While analytical expressions for the deviations from logarithmic have not been obtained, the Monin–Obukhov similarity theory has been used to establish semi-empirical profiles (for example, Businger, 1973). Because no theoretical solution for the profiles exists, there is still considerable debate as to the form of the correction to the logarithmic profile. Grainger and Lister (1966) conclude that the logarithmic profile applies best to a neutral and extremely stable atmosphere, while a power law may be more appropriate for a moderately stable boundary layer. Holmgren (1971, Part B) argues that observations on Devon Island Ice Cap do not support the Monin–Obukhov theory under stable conditions. Goddard (1975) parameterizes the sensible and latent heat fluxes by adopting a log plus linear profile plus a universal function dependent on the Richardson number. Stearns and Weidner (1993) use an iterative procedure to determine the heat fluxes from measured wind speed and temperature difference. In particular where katabatic winds develop, with a maximum in wind speed a few meters above the surface, important differences between predictions based on the Monin–Obukhov theory and observations have been noted (Denby, 1999; Greuell and Genthon, 2003).

A major uncertainty in determining turbulent fluxes is the roughness length of the surface, Z_0 (Munro, 1989; Braithwaite, 1995). An order of magnitude increase in the value of Z_0 results in a doubling of the calculated turbulent heat fluxes (Brock et al., 2000). By definition, the roughness length is the height above the surface at which the average wind vanishes and, in principle, can be determined by extrapolating the wind speed profile (Munro, 1989). The difficulty with this extrapolation is that the surface of glaciers and ice sheets is irregular with small-scale features such as windblown sastrugi disturbing the flow of air up to a height of about twice the height of the main surface roughness elements (Smeets et al., 1999). Above this roughness-induced sublayer, wind profiles behave as expected for an idealized homogeneous surface. Alternatively, Z_0 can be estimated from the small-scale roughness of the snow surface (Van der Veen et al., 1998, 2009) although it is not immediately evident which spatial scale of roughness elements determines Z_0 (Munro, 1989). A review of estimated roughness values is given by Brock and others (2006) and shows a range from 0.004 mm for snow on the Antarctic Plateau, to up to 80 mm for very rough glacier ice. Typical ranges are summarized in Table 6.4.

TABLE 6.4
Typical Range of Values for the Roughness
Length for Various Snow and Ice Surfaces

Surface Type	Range (mm)
Smooth ice	0.01–0.1
New snow and polar snow	0.05–1
Snow on low-latitude glaciers	1–5
Ice in ablation zone	1–5
Coarse snow with sastrugi	11
Rough glacier ice	20–80

Source: Based on Brock, B. W., I. C. Willis, and M. J. Sharp,
J. Glaciol., 52, 281–297, 2006.

The flux parameterizations (6.76) and (6.77) contain the same roughness length, Z_o. However, close to the surface the vertical exchange of momentum is by molecular diffusion and by pressure forces against obstacles, while transfer of heat and water vapor results from pressure forces only. This difference leads to two different roughness scales. Andreas (1987) derived expressions relating both roughness scales to Z_o. Observations appear to be somewhat ambiguous regarding the applicability of these parameterizations (c.f. Greuell and Genthon, 2003).

6.7   PHYSICAL PROPERTIES OF FIRN

The values of thermal parameters given in Table 6.1 apply to solid glacier ice and may be used in most ice-sheet models. However, for freshly fallen snow and in the upper firn layers of a glacier, the thermal properties are significantly different from those of solid ice. For example, the density at the glacier surface is about 350 kg/m³, compared with 917 kg/m³ for solid ice. These differences are important when considering the evolution of near-surface temperatures and therefore should be included in the calculation of these temperatures. A complete review of the subject and all relevant data can be found in Yen (1981). Below, only the most important relations are summarized.

The heat capacity, C_{pa}, is defined as the amount of heat needed to increase the temperature of a unit mass by 1 degree, keeping the pressure constant. Data from different experimental studies are consistent and show a linear relation between the heat capacity and temperature, T. For the temperature range most relevant to glacier studies (T > 150 K), Yen (1981) finds as best fit

$$C_{pa} = 152.456 + 7.122\,T, \tag{6.78}$$

with the heat capacity in $J\,kg^{-1}K^{-1}$ and the temperature in K. For temperatures below 150 K, the heat capacity may also be expressed as a linear function of temperature but with different coefficients (given in Yen, 1981).

The thermal conductivity, k, introduced as the constant of proportionality in Fourier's law for heat conduction (equation (6.11)) depends on density and temperature. For solid ice, a regression analysis of all available data suggests (Yen, 1981)

$$k_i = 9.828 \cdot e^{-0.0057T}, \tag{6.79}$$

with k_i in $W\,m^{-1}K^{-1}$ and T in K. For snow, the transfer of heat is a more complex process, involving conduction and convection, as well as diffusion of water vapor. Measurements of conductivity often include all these processes, and therefore k is often replaced by the effective thermal conductivity, k_e. In most experimental studies, the effective conductivity is correlated with snow density. Yen (1981) shows that all data can be reasonably represented by

$$k_e = 2.22362 \left(\frac{\rho_f}{1000} \right)^{1.885}, \tag{6.80}$$

where k_e is expressed in $Wm^{-1}K^{-1}$, and the density of firn, ρ_f, in kg/m^3. Pitman and Zuckerman (1967) investigated how temperature affects the effective conductivity, using vapor-grown ice crystals and conducting measurements at various temperatures and densities. Except for the lowest densities (about 100 kg/m^3), their data suggest the following relation (Yen, 1981):

$$k_e = 0.0688 \exp(0.0688\,T_c + 4.6682 \cdot 10^{-3}\,\rho_f), \tag{6.81}$$

where T_c represents the firn temperature in °C. The actual thermal conductivity is often described by Van Dusen's equation (Van Dusen, 1929):

$$k = 2.1 \cdot 10^{-2} + 4.2 \cdot 10^{-4}\,\rho_f + 2.2 \cdot 10^{-9}\,\rho_f^3. \tag{6.82}$$

Paterson suggests that this formula gives a lower limit for the conductivity; an upper limit can be found by the expression derived by Schwerdtfeger (1963) based on Maxwell's work on the electrical conductivity of heterogeneous media:

$$k = \frac{2\,k_i\,\rho_f}{3\rho_i - \rho_f}. \tag{6.83}$$

Here, k_i and ρ_i represent the conductivity and density of ice, respectively.

As a result of many processes, the firn density increases with depth. Density changes in snow are caused by (Yen, 1981): (1) compaction due to the weight of snow above, (2) destructive metamorphism (that is, sharp-edged freshly fallen snow crystals are smoothed by migration of water molecules, resulting in settling and increase in density), (3) constructive metamorphism (transfer of vapor within the snow cover due to the temperature gradient), and (4) melt metamorphism (the change in snow structure due to melt-freeze cycles and the change in crystals as a result of the presence of meltwater). Yen (1981) summarizes various models proposed to parameterize each of these processes.

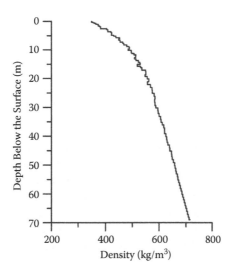

FIGURE 6.10   Density as a function of depth as measured at South Pole, Antarctica. (Based on data from Hogan, A. W., and A. J. Gow, *Geoph. Res.,* 102(D), 14021–14027, 1997.)

At present, most studies use empirical or semiempirical expressions. Figure 6.10 shows the depth–density profile measured at South Pole, Antarctica (Hogan and Gow, 1997). In good approximation, the profile to a depth of 65 m below the surface may be represented by

$$\rho_f(d) = 380.92 + 11.46d - 0.16d^2 + 8.85 \cdot 10^{-4} d^3, \qquad (6.84)$$

where d denotes the depth below the surface in m. Another empirical relation adopts an exponential increase with depth according to

$$\rho_f(d) = \rho - (\rho - \rho_s)e^{-Cd}, \qquad (6.85)$$

where $\rho = 917$ kg/m³ is the density of solid ice, ρ_s the density of surface snow, and C a site-specific constant (Schytt, 1958).

The implicit assumption behind density profiles (6.84) and (6.85) is that the firn density does not change with time at any location and depth. This is known as Sorge's law, which states that when snow accumulates steadily and there is no melting, at any particular location the density is always the same at any one depth below the surface. This law was suggested by Bader (1954) based on observations made by Ernst Sorge in Greenland during the winter of 1930–1931 in a 16 m deep hand-dug pit. For many modeling applications, temporal variations in firn density may be ignored and simple steady-state firn densification models are adequate. A notable exception is when considering interannual variability in snowfall and associated short-term elevation changes (Arthern and Wingham, 1998; Helsen et al., 2008).

Where surface melting occurs, the downward percolation of meltwater, as well as internal melting due to the absorption of penetrating short-wave radiation, dominates the densification process. Water movement through snow is a complex process that depends on the presence of vertical drainage channels, ice layers, the slope of the surface, and many other factors. A review of theories for water flow is presented by Male (1980), but none of these is readily implemented into a numerical model. In their numerical model, Greuell and Konzelmann (1994) apply a three-step procedure to simulate the downward percolation of meltwater. First, if the calculated temperature at any grid element exceeds the melting temperature, the temperature is set equal to the melting temperature and the excess heat used for melting and the amount of meltwater formed added to the water content of the grid element. Next, downward percolation and refreezing are simulated, starting at the surface (the uppermost grid element). Refreezing is limited by the condition that the temperature of the firn cannot exceed the melting temperature (refreezing releases latent heat, thus raising the firn temperature), and by the available meltwater and pore space. Some of the meltwater that remains after calculating the amount of refreezing may actually remain in the grid element (instead of percolating downward) as capillary water (that is, water retained against gravity in the pores of the firn). There is a maximum of water that can be retained this way, given by the maximum of the ratio of capillary water to pore volume, the so-called irreducible water amount. If the amount of meltwater present exceeds the irreducible water amount, the water content of the grid element is set equal to this maximum and the remainder is added to the next deeper grid element. This procedure may result in the collection of meltwater in the lowermost grid element. In reality, meltwater is likely to percolate downward until encountering an impermeable ice layer. Some of the meltwater may run off on top of this layer, while some of the meltwater may form a slush layer of snow saturated with water. Greuell and Konzelmann (1994) do not explicitly model the runoff; rather, meltwater collecting in the lowest grid element is used to saturate the snow, starting with the lowermost grid element and proceeding upward. The thickness of the slush layer is restricted, and the remainder of the meltwater is assumed to run off. The implicit assumption made in this procedure is that the model always extends to the depth of the first impermeable ice layer, and that no other layers form at shallower depths during the simulation. Greuell and Oerlemans (1986) use a simpler model in which meltwater is allowed to penetrate downward until a layer with a density larger than 800 kg/m^3 is reached and the water is assumed to run off.

Wakahama (1968) and Colbeck (1973) suggest that the rate of firn compaction increases when the water saturation is large. Except just above layers of impermeable ice, these conditions normally do not occur, and for high-density, wet snow with almost spherical grains, the rate of increase in density should be similar to that for dry snow (Yen, 1981). For freshly fallen, low-density snow, Yen (1981) argues that the presence of liquid water will accelerate the destructive metamorphism, thereby increasing the settling rate. Refreezing of meltwater may result in the formation of ice layers and lenses, thus providing a rapid mechanism for transforming snow into ice. A general expression for the change in density caused by melting or refreezing has not yet been derived, and most models incorporate the effects of melt-freeze cycles in an ad hoc way. For example,

Greuell and Oerlemans (1986) take the rate of densification to be a function of meltwater such that a melting rate of 2 cm/day increases the densification rate by a factor of about 7. Greuell and Konzelmann (1994) do not include the decrease in density when internal melting occurs because inclusion of this process led to numerical problems. Instead, these authors prescribed an unrealistically low value for the irreducible water amount to reduce the densification rate due to refreezing of meltwater.

6.8   CALCULATED NEAR-SURFACE SNOW TEMPERATURES AT SOUTH POLE STATION

By coupling the calculation of surface fluxes to the thermodynamic equation (6.17), the change in firn temperatures throughout the year or under altered climate conditions can be numerically simulated. As an example, the annual temperature cycle at South Pole Station is considered in this section. The purpose here is not to present the most accurate calculation of englacial temperatures but rather to illustrate the procedure and how model results can be validated against available data. The model used in this section is essentially the same as that used by Van der Veen and Jezek (1993) to study the seasonal firn temperature near Vostok Station.

The temperature equation (6.17) may be simplified by making the usual assumptions that horizontal advection and diffusion of heat may be neglected and that strain heating is small near the surface of the glacier. In an exception to the general convention used in this book, the vertical z-axis is taken positive *downward*, with z = 0 at the surface. The thermodynamic equation then becomes

$$\frac{\partial T}{\partial t} = \frac{\partial}{\partial z} K \frac{\partial T}{\partial z} + w \frac{\partial T}{\partial z} - \frac{1}{\rho C_p} \frac{\partial Q_a}{\partial z}. \qquad (6.86)$$

The last term on the right-hand side corresponds to the last term on the right-hand side of equation (6.17) and represents the heat added to a layer at depth z. Neglecting melting and refreezing of meltwater (as is appropriate for central East Antarctica), the only source of heat below the surface is short-wave solar radiation penetrating into the firn as described by equation (6.47). The amount of radiation absorbed in a layer at depth z below the surface is the difference between the radiation entering this layer at the top and exiting at the bottom to deeper layers. Thus,

$$\frac{\partial Q_a}{\partial z} = \beta \, I_s(h-z), \qquad (6.87)$$

with the radiation flux at depth, $I_s(h-z)$, given by equation (6.48).

Near the surface of the glacier, large temperature gradients may develop as the surface is being cooled or heated, while at greater depths, the gradients tend to become smaller and not changing as much throughout the year. This means that the grid spacing should be small near the surface (a few cm) but the spacing may increase with depth. The calculations discussed here apply to the upper 55 m of firn, and the spacing increases from 8 cm just below the surface, to 12 m between the lowest two model layers. Because of the variable grid spacing, vertical derivatives cannot be simply estimated from central differences. One possibility is to introduce a new coordinate, s = ln(z), and rewriting equation (6.86) in terms of this coordinate. By taking the spacing between adjacent s-layers constant, ds = 0.25, say, the physical layer thickness increases from 8 cm to 12 m over the model domain. Another possibility for calculating vertical derivatives at gridpoint I is to approximate the temperature profile by a third-order polynomial through the three neighboring grid points I − 1, I, and I + 1. Derivatives are then readily estimated from the coefficients of this polynomial. Results for both approaches are practically the same.

To solve equation (6.86) numerically, two boundary conditions are needed. At the lower boundary, the temperature is prescribed at the measured temperature and kept fixed. The surface temperature is calculated explicitly and changes in response to the heat exchange between the surface and the atmosphere, as described by the total energy flux Q_s, and as a result of downward heat conduction to deeper layers. Following Goddard (1975), the downward heat flux is parameterized as

$$Q_d = \frac{K}{d}(T_d - T_s), \tag{6.88}$$

where T_d represents the temperature at depth d below the surface, here taken to be the temperature of the first depth layer.

Vertical advection of heat (the second term on the right-hand side of equation (6.86)) can be estimated by taking the vertical velocity, w, to decrease linearly from the surface to the bed of the glacier. At the surface, w equals the accumulation rate (noting that z increases downward), while at the bed, w is zero. However, accumulation at South Pole is very small, and consequently, vertical advection has little effect on the calculated temperatures. Similarly, the heat flux associated with precipitation is very small and not discussed here.

As input for the model, the air temperature, wind speed, cloudiness, and air pressure need to be prescribed. For these, the multiyear monthly mean values given in Dalrymple (1966) and shown in Figure 6.11 are used, with the understanding that these values are representative for the midmonth climate. To obtain smoothly varying records of climate, these midmonth values are linearly interpolated. Preferably, a more continuous record of, say, daily observations should be used, but such long-term records are scarce.

The thermodynamic equation (6.86) is solved by numerical integration. At the start of the simulation, the model is run until a steady state is reached for the applicable climate (in this case, January 1). To adequately model the annual cycle, several years are simulated to minimize the effect of the initial temperature conditions. Results discussed below pertain to the fifth model year. Because of the small grid spacing near the surface, a small time step (1 minute) must be used to prevent numerical instabilities.

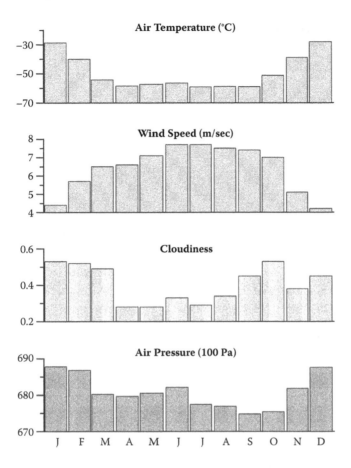

FIGURE 6.11 Monthly mean multiyear meteorological data used as input for the calculation of the firn temperature at South Pole Station. (Based on data from Dalrymple, P. C., in *Studies in Antarctic Meteorology,* American Geophysical Union, Washington, DC, 195–231, 1966.)

Calculated surface fluxes at local noon are shown in Figure 6.12. During the austral winter (mid-March to early September), no solar radiation reaches the surface, and incoming long-wave radiation is small because of the low air temperatures. As a result, the net radiative flux is directed toward the atmosphere, representing loss of heat for the snow surface. During the austral summer, incoming solar and long-wave radiation increase, reversing the direction of the net radiative heat flux. The latent heat flux is small compared with these radiative fluxes. The reason for this is that the air is very dry because of the low temperatures. The sensible heat flux is more important in heating the snow during the austral winter, when the air is slightly warmer than the snow surface (Figure 6.13). During the summer months, when incoming radiation raises the snow temperature above that of the air, the sensible heat flux is directed toward the atmosphere. Finally, heat conduction from the surface to the deeper layers (equation (6.88)) is slightly positive during the winter (indicating that

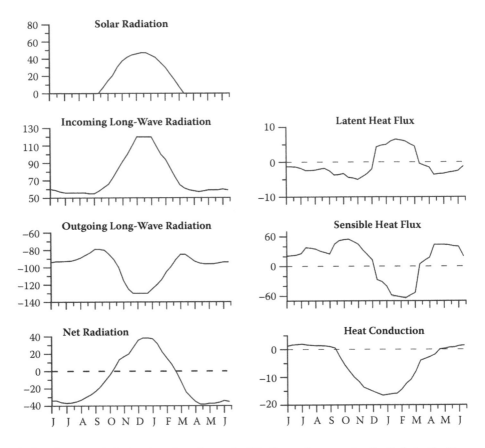

FIGURE 6.12 Calculated surface fluxes (in W/m^2) at solar noon for South Pole. Positive fluxes are directed downward and represent heat gain for the glacier surface. Note the difference in vertical scales for the three panels on the right.

heat is transported upward in the firn because the surface is coldest), but negative during the prolonged summer, when the surface is warmer than the snow below.

The calculated surface temperature agrees within a few degrees with monthly mean surface temperatures measured during the International Geophysical Year, or IGY (1957–1958), as shown in Figure 6.13. The measurements given in Dalrymple (1966) do not include the summer maximum (December–January). The comparison shows that calculated midwinter temperatures are too high. This may be because multiyear averages for the meteorological variables were used in the calculation, whereas the monthly mean surface temperatures shown in Figure 6.13 apply to one year only. In particular, the air temperature, and thus the incoming long-wave radiation and the sensible heat flux, may vary by several degrees from year to year (Schwerdtfeger, 1970), which could easily raise or lower the temperature of the snow surface by a few degrees. Another possibility for the discrepancy between model calculations and measurements is that the model is driven by monthly mean meteorological conditions, which are interpolated to obtain a continuous record. As a result,

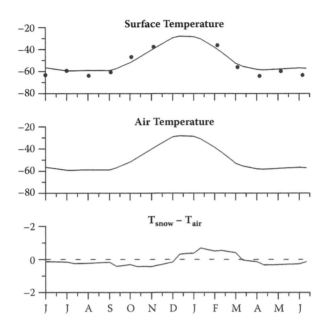

FIGURE 6.13 Comparison between the calculated surface temperature and measured monthly mean values (upper panel; note that the measurements from Dalrymple and others (1966) do not include December and January). The middle panel shows the air temperature (in °C) used as input, and the lower panel shows the calculated difference between the temperature of the glacier surface and the air temperature.

the midwinter minimum in air temperature, as well as the midsummer maximum, are apt to be underestimated. This points to the need of simultaneously obtaining accurate and continuous records of, most importantly, air temperature and wind speed, as well as the snow surface temperature. The calculated temperature of the snow surface represents the most critical model result that can be directly verified against measurements to test the model and especially the parameterizations of the heat exchange between the glacier and atmosphere.

Radiative fluxes can be measured directly and the net monthly mean radiative flux measured at South Pole is shown in Figure 6.14, as well as the monthly mean flux computed by the model (note that the measurements do not cover the months of December and January). Agreement is satisfactory given the uncertainty in measurements (Dalrymple et al., 1966) and the fact that the prescribed air temperature may be different from the actual air temperature during the period of measurements (the IGY), as noted above.

The turbulent fluxes of latent and sensible heat can also be measured directly by means of eddy-covariance sensors. However, this is expensive and requires significant instrument power (D. van As, pers. comm., 2012). Therefore, in most applications, these fluxes are calculated from measured vertical profiles of air temperature and wind speed, by using formulas similar to those given in Section 6.6. Fluxes calculated by Dalrymple et al. (1966) are shown in the lower two panels of Figure 6.14,

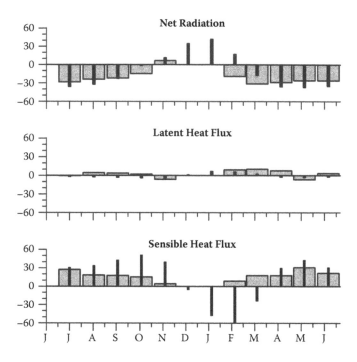

FIGURE 6.14 Comparison between monthly mean surface fluxes (in W/m²) calculated with the numerical model (black bars) and those inferred by Dalrymple (1966) (hatched boxes).

as well as the monthly mean fluxes calculated by the model. As discussed above, the latent heat flux is small, according to Dalrymple et al. (1966) and to the model. The difference between both estimates for the sensible heat flux is rather large, with the exception of the winter months. This is a disturbing result that most likely can be attributed to the parameterization used, as well as to the air temperature profile. For example, Stearns and Weidner (1993) give estimates for the sensible heat flux that differ substantially from those in Dalrymple et al. (1966).

Calculated englacial temperatures are shown in Figure 6.15. As expected, the annual variation affects only the upper 10 m of the firn (a posteriori justifying the lower boundary condition of fixed temperature). Because the austral winter is of longer duration than the austral summer, the annual cold wave penetrates somewhat deeper than its warm counterpart. The relative warming of the surface during the summer is slightly larger than the relative cooling during the winter. This general pattern agrees well with the firn temperatures measured at South Pole Station during 1957–1958 (Dalrymple, 1966), as shown in Figure 6.16. There are some differences, most notably in the timing of the winter minimum and summer maximum temperatures at depth. This may, however, be very well the result of differences between the actual climate conditions during this period and those prescribed for the model calculations.

The comparison between model results and measured data highlights one of the problems encountered most often, namely, incomplete or nonsimultaneous data sets.

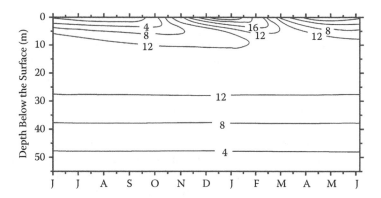

FIGURE 6.15 Calculated annual cycle in temperature at depth for South Pole Station. Contours show the difference between the calculated temperature at depth and the temperature at 66 m depth (kept fixed in the model).

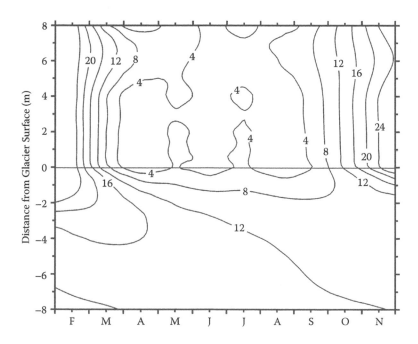

FIGURE 6.16 Temperatures in the firn and atmospheric boundary layer measured at South Pole Station during 1957–1958. To allow for a comparison with Figure 6.15, contours show the difference between the measured temperature and the temperature at 66 m depth. (Based on data from Dalrymple, P. C., in *Studies in Antarctic Meteorology,* American Geophysical Union, Washington, DC, 195–231, 1966.)

The calculations are based on multiyear monthly mean climate meteorological parameters, interpolated to obtain hourly values describing the annual cycle. However, comparable data against which the model results can be tested are not available, but only firn temperatures measured during the course of one year. This particular year may not be indicative of the longer-term climate, given the large variability in air temperatures observed at South Pole Station. To circumvent this problem, simultaneous meteorological observations and measurements in the firn should be conducted. Kuipers Munnike et al. (2009) present measurements of the surface energy balance at Summit, Greenland, and show that measured firn temperatures can be satisfactorily modeled with an energy balance model similar to the one used in this section.

Inspection of the thermodynamic equation (6.86) shows that the temperature at depth is determined by vertical conduction and vertical advection of heat, absorption of solar radiation by deeper layers, and the density-dependency of the thermal diffusivity. In addition, where firn temperatures are close to the melting temperature, meltwater penetration and refreezing may also be important. Greuell and Konzelmann (1994) conducted some sensitivity experiments to determine which of these processes is most important. They conclude that for the location of the ETH Camp, in East Greenland, the largest amount of heating of the upper firn layers is caused by refreezing of downward percolating meltwater. This is probably the least understood process included in their model. The second most important process is absorption of radiation at depth, as confirmed by Kuipers Munnike et al. (2009). The density-dependency of the diffusivity has only a small effect, while other processes in their model do not result in noticeable heating or cooling of the upper layers. Of course, these conclusions cannot be directly applied to other locations, such as central East Antarctica, where surface melting does not occur.

7 Subglacial Processes

7.1 INTRODUCTORY CONCEPTS

In a grounded ice sheet, three mechanisms may cause the ice to flow in the direction of decreasing surface elevation. The first mechanism is *internal deformation* as described by Glen's flow law discussed in Chapter 2. The ice deforms under its own weight to relieve the internal stresses. The associated deformational velocity is zero at the glacier base and reaches a maximum at the upper surface (Section 4.2), and is determined by the geometry—through the driving stress—and ice temperature. Typically, these velocities range from less than a few meters per year to a few hundred meters per year and change only slowly over time as the glacier geometry adjusts to mass imbalances. The second flow mechanism, *basal sliding*, becomes important where basal temperatures have reached the pressure melting temperature and subglacial water is present and is responsible for the dynamic behavior exhibited by outlet glaciers of the Greenland and Antarctic ice sheets. The transition from a frozen bed (cold-based glacier) to one that is lubricated by basal water (warm-based glacier) can lead to an increase in discharge by a factor of 10 or more. The third process that may contribute to glacier flow is *deformation of subglacial sediments*. As noted by Clarke (2005), processes that act in the layer extending a few meters above and below the ice-bed interface can have a greater impact on glacier dynamics than processes operating within the ice itself (such as fabric development and temperature changes), yet this layer remains poorly studied and understood.

Many complex theories of glacier sliding have been developed in attempts to derive a sliding relation, linking sliding speed to relevant quantities such as basal drag, bed roughness, water pressure, and many others. However, the range of possible basal substrates (hard rock, deformable sediment, etc.) and interactions between the various basal processes involved has impeded the development of a full theory. Similarly, obtaining suitable field measurements to test proposed sliding relations is hampered by the inaccessibility of the glacier bed. Consequently, most numerical models use an ad hoc sliding relation that is only partially supported by theory or observations.

Weertman (1979) presents an overview of what he termed, "the unsolved general glacier sliding problem." In short, this problem can be stated as, "the determination of the velocity, U, of a small block of ice that is within the glacier and just above the glacier bed" (p. 98). There are many factors that influence the value of this velocity, and these factors cannot be considered separately (although this is often done) because many feedbacks exist. For example, the sliding speed is believed to increase as the thickness of the interstitial water film increases, which leads to more energy being dissipated by the sliding process, allowing more basal ice to be melted, increasing the thickness of the water layer. Also, under the action of sliding, erosion may occur, which alters the characteristics of the bed surface. An overview of sub-glacial processes related to glacier dynamics can be found in Clarke (2005), which makes it clear that accounting for all these processes and interactions in a single sliding theory may be an unattainable goal and the search for finding *the* sliding relation an illusory quest. Nevertheless, much can be learned by studying the importance of individual processes, without accounting for all interactions.

Presenting a full overview of all theoretical work and field measurements pertaining to the glacier sliding problem would be a formidable task and probably require an entire book by itself. Therefore, after presenting the classic model for ice sliding over a hard bed, the discussion in the following sections is restricted to some of the main issues, not claiming to be in any way complete and exhaustive.

One of the first attempts at deriving a mathematical model for glacier sliding can be found in Koechlin (1944), who noted that both basal slip and internal deformation contribute to the observed surface velocity. The sliding component causes glacial erosion. By considering balance of forces at the bed of a slab of ice on an incline, he concluded that sliding can occur when the slope of the incline and the ice thickness are sufficiently large to overcome friction or drag, taken to be proportional to the bed roughness (Koechlin, 1944, p. 115). He tested his theory against observations made on a number of mountain glaciers, although he did not actually formulate a sliding law relating the sliding velocity to bed roughness and gravitational driving force.

The first complete theory of glacier sliding was proposed by Weertman (1957a, 1964), who noted that the bed of a glacier is usually irregular, containing many protuberances impeding the flow of ice and preventing sliding by rigid translation. To explain why sliding nevertheless occurs, Weertman proposed two mechanisms, namely, pressure melting (or regelation) and enhanced flow due to stress concentrations around the obstacles. To estimate the velocity associated with each process, Weertman made the assumption that the glacier bed consists of a smooth inclined plane, with cubical protuberances of dimensions L whose centers are separated by a distance λ (Figure 7.1). Otherwise, the bed is assumed to be smooth and unable to support a shear stress. This means that basal resistance must be due to horizontal normal stresses acting against the vertical faces of the cubicals, being compressive

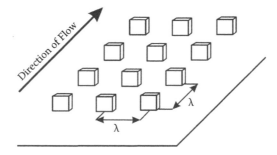

FIGURE 7.1   Idealized model for the glacier bed. Each cube has dimensions L × L × L, and these are separated from each other by a distance λ in both directions. (From Weertman, J., *J. Glaciol.,* 3, 33–38, 1957a. Reprinted from the *Journal of Glaciology* with permission of the International Glaciological Society and the author.)

on the upglacial side and tensile on the downglacial side (so called from drag). This stress difference causes the pressure melting temperature to be lower on the upstream side of the obstacle than on the downstream end. As a result, upstream melting and downstream refreezing occurs, allowing the ice to move around the cube (regelation). The latent heat released by the refreezing is conducted through the obstacle to the upstream side to maintain the temperature difference. Secondly, the stress difference across the obstacles results in enhanced creep of ice around the obstacles.

Consider first the regelation process. In each area of the bed measuring λ by λ, there is one cubic protuberance that provides all basal resistance for this area (frictional traction being zero). If τ_b denotes the average basal drag, then each obstacle must provide a resistive force equal to $\lambda \cdot \lambda \cdot \tau_b$. This resistance is due to an extra compressive stress on the stoss side of the protuberance and an extra tensile stress on the lee side (these normal stresses are in excess of the hydrostatic pressure). From symmetry it follows that the compressive and tensile stresses are equal but of opposite sign. The area upon which these normal forces act is $L \cdot L$ (the side of the obstacle) so that the magnitude of each normal stress is equal to $(\lambda^2 \tau_b / L^2)/2$. This stress difference results in a difference in the pressure melting temperature across the protuberance that is directly proportional to the stress difference. That is

$$\Delta T = \frac{C \tau_b}{R^2}, \tag{7.1}$$

where $R = L/\lambda$ is the bed-roughness parameter and C a constant. If the ice is taken to be at the pressure-melting temperature everywhere, a temperature gradient thus exists across the obstacle, given by (7.1). As a result, heat is conducted across the obstacle in the upglacial direction. Because the distance over which the temperature difference is present is L, the heat flux is given by

$$Q_u = k_b L \Delta T, \tag{7.2}$$

where k_b represents the thermal conductivity of the basal material. The assumption is made that there is no flow of heat through the ice. Under steady-state conditions, this upglacial heat flow must be compensated by the release of latent heat on the downglacial side of the obstacle, where meltwater refreezes. The volume of ice that is melted per unit time on the stoss side of the obstacle and refreezes on the lee side is then

$$V = \frac{Q_u}{\rho L_f}, \tag{7.3}$$

with ρ representing the density of glacier ice and L_f the specific latent heat of fusion. The sliding velocity associated with this regelation process follows from the above expressions and is given by

$$U_r = \frac{V}{L^2} =$$

$$= C_r \frac{\tau_b}{L} R^{-2}. \tag{7.4}$$

Here, C_r represents a constant that incorporates all the material parameters in equations (7.1)–(7.3).

The second process that allows the ice to flow around the basal obstructions is enhanced deformation due to the normal stresses induced by the obstacle. Because the basal resistance arises from these normal stresses and the shear stress parallel to the bed is zero, the constitutive relation (3.48) becomes

$$R_{xx} = 2B\dot{\varepsilon}_{xx}^{1/n},\tag{7.5}$$

if bridging effects and transverse strain rates may be neglected. As before, the compressive stress on the upglacial face of the obstacle is $(\tau_b R^{-2})/2$ and the strain rate is, from the constitutive relation (7.5),

$$\dot{\varepsilon}_{xx} = -\left(\frac{\tau_b}{4R^2 B}\right)^n,\tag{7.6}$$

where the minus sign indicates compression. Making the assumption that this strain rate acts over a distance L upstream of the obstacle, the velocity must decrease over this distance from the sliding value,

$$U_c = \left(\frac{\tau_b}{4R^2 B}\right)^n L,\tag{7.7}$$

to zero at the stoss side of the obstacle. Similarly, the tensile strain rate on the lee side results in an increase in velocity from zero to the value given by (7.7) at a distance L downstream of the obstacle.

Comparing the two expressions (7.4) and (7.7) for the sliding velocity associated with both processes shows that, for a given bed roughness (fixed value of R), the pressure-melting mechanism results in increased sliding as the obstacle size decreases, while enhanced creep results in larger sliding around larger obstacles. It may therefore be expected that an intermediate obstacle size exists that offers the most resistance to the flow of the glacier. This size of the controlling obstacle can be found by equating expressions (7.4) and (7.7) and is given by

$$L_c = C_s\left(\frac{\tau_b}{R^2}\right)^{(1-n)/2},\tag{7.8}$$

where C_s represents a sliding constant incorporating the material parameters and the rate factor. The associated sliding velocity is

$$U_s = C_r\left(\frac{\tau_b}{R^2}\right)^{(1+n)/2}.\tag{7.9}$$

The Weertman theory discussed above was expanded by Nye (1969a, 1970) and Kamb (1970) for arbitrary bed shapes (instead of the rather artificial cubical model).

Both theories are mathematically rigorous and more difficult to understand than the Weertman model but, perhaps surprisingly, yield similar results. The roughness of the bed is described by the spectral power density (Nye) or by a spectral roughness function (Kamb), which is essentially the Fourier series or integral of the basal topography. In both derivations, the roughness is assumed to be small, that is, the difference between the mean bed slope and the actual bed slope is small. Furthermore, both adopt a linear rheology for glacier ice (n = 1). Following Weertman, the basal ice is supposed to be free of debris and separated from the impermeable bed by a thin water film only, so that the form of the lower ice surface is identical to that of the bed. Instead of using a controlling obstacle size, Kamb and Nye introduce the transitional wavelength, λ_t. For basal roughness on scales less than this wavelength, sliding is dominantly by regelation, while for larger roughness scales, the principal sliding mechanism is enhanced creep. The resulting sliding relation is similar to expression (7.9) with n = 1. Kamb (1970) also discusses how the conventional non-linear constitutive relation can be approximated as a linear relation in the sliding theory. For a "white roughness" (that is, the variance of the basal topography is the same for all wavelengths), he arrives at a similar relation between sliding velocity and basal shear stress as did Weertman. For a "truncated roughness" (in which the short wavelength components are absent), the sliding velocity is proportional to the nth power of the basal shear stress.

As pointed out by Lliboutry (1968), a major weakness of the Weertman model is the unrealistically large stress acting on the basal obstructions. The compressive stress on the upglacial side is $(\tau_b R^{-2})/2$. For a basal drag $\tau_b = 100$ kPa and a typical bed roughness R = 0.1, this implies a compressive stress of 5000 kPa and an equally large tensile stress on the downglacier side of the obstacles. Such large tensile stresses would almost certainly lead to basal fracturing or formation of cavities. Moreover, according to equation (7.9), the only controls on the sliding velocity are drag at the glacier base and the roughness of the bed, both of which may be expected to change little over short time spans. However, measurements of surface velocity on alpine glaciers have revealed important short-term fluctuations but no corresponding changes in driving stress. For example, on White Glacier, Iken (1973) observed distinct diurnal variations in the surface velocity during the melt season. After heavy rainfalls, the rate of sliding increased strongly, indicating that variations in subglacial water pressure may be important. Such variations cannot be explained with the Weertman or Nye–Kamb theory. Therefore, the next step must be to investigate sliding when separation between the glacier sole and the bed occurs.

7.2   SLIDING WITH CAVITATION

Cavitation occurs where the basal ice loses contact with the bed, usually on the lee side of a protuberance (Figure 7.2). The first to point out that in addition to the regelation and enhanced creep processes introduced by Weertman (1957a), flow with cavity formation should be considered was Lliboutry (1958a, b). In a series of subsequent papers, he improved and modified the theory for sliding with cavitation (Lliboutry, 1959, 1968, 1979, 1987a, b), and that work has revealed some fundamental aspects that have aided in explaining observed sliding variations on glaciers.

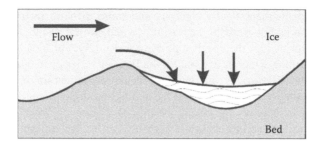

FIGURE 7.2 Formation of subglacial cavities in the lee of topographic irregularities. The smaller arrows indicate closure of the cavity due to advection of ice from upstream and due to vertical downward creep of the ice above. This closure is countered by melting of the cavity roof due to viscous heat dissipation.

The first step is to determine the pressure distribution around a basal obstacle to find an expression for basal drag. To do so, consider first the two mechanisms contributing to the basal resistance. The flow is assumed to be planar, so that from equation (3.20), the net basal resistance is given by

$$\tau_b = R_{xz}(b) - R_{xx}(b)\frac{\partial b}{\partial x}, \tag{7.10}$$

where $z = b$ corresponds to the elevation of the bed. To better understand the physical processes associated with basal drag, consider the components parallel and perpendicular to the bed. Using standard tensor transformation formulas (Section 1.1), the shear stress parallel to the bed is found to be

$$\sigma_{//} = (R_{zz}(b) - R_{xx}(b))\frac{\partial b}{\partial x} + R_{xz}(b), \tag{7.11}$$

while the full stress normal to the bed is given by

$$\sigma_{\perp} = R_{xx}(b)\left(\frac{\partial b}{\partial x}\right)^2 + 2R_{xx}(b)\frac{\partial b}{\partial x} + R_{zz}(b) - \rho g H, \tag{7.12}$$

when higher-order terms may be neglected. The basal resistance can now be written as

$$\tau_b = \sigma_{//} - R_{zz}(b)\frac{\partial b}{\partial x}. \tag{7.13}$$

The first term on the right-hand side represents skin friction and is commonly associated with vertical shear in velocity. If the bed is perfectly lubricated (as in the Weertman and Nye–Kamb theories) this contribution to the basal drag is zero. The second term corresponds to the form drag, which is the resistance due to pressure

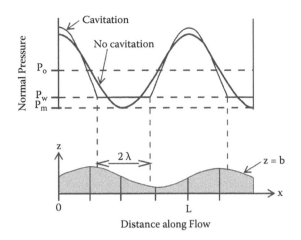

FIGURE 7.3 Bed geometry and normal pressure at the glacier bed with and without cavitation. The cavity is taken to be symmetric around the inflection point at x = L/2, and has a length 2λ. (From Schweizer, J., and A. Iken, *J. Glaciol.,* 38, 77–92, 1992. Reprinted from the *Journal of Glaciology* with permission of the International Glaciological Society and the authors.)

fluctuations generated by the flow over the obstacle. In most sliding theories, this is the only contribution to basal resistance.

Form drag may be important where bridging effects occur, that is, where the weight of an ice column is not supported from directly below and the vertical resistive stress, R_{zz}, is nonzero. To calculate the magnitude of the form drag, a simplified basal topography is considered, namely a sinusoidal bed (Figure 7.3)

$$b(x) = B_o \sin\left(\frac{2\pi x}{L}\right). \qquad (7.14)$$

If no separation occurs, the vertical resistive stress at the bed is taken to be out of phase with the bed topography in such a way that the pressure on the bed is largest on the stoss side of the obstacle and reaches a minimum on the lee side (Figure 7.3; Lliboutry, 1968). Then

$$R_{zz}(x) = R_o + \Delta R \cos\left(\frac{2\pi x}{L}\right). \qquad (7.15)$$

In this expression and in the following, stresses are referred to their value at the glacier sole.

Averaged over one wavelength, the weight of the ice must be supported by the bed, and hence

$$\int_0^L R_{zz}(x)\, dx = 0, \qquad (7.16)$$

from which it follows immediately that R_o must be zero. Furthermore, if the bed is perfectly lubricated, skin friction is zero. Then, if τ_b represents the average basal resistance, equation (7.13) for basal drag gives

$$\tau_b = -\frac{1}{L} \int_0^L R_{zz}(x) \frac{\partial b}{\partial x}\, dx, \qquad (7.17)$$

or

$$\Delta R = -\frac{\tau_b L}{B_o \pi}. \qquad (7.18)$$

Note that the vertical coordinate axis is taken positive upward, so that a negative value of R_{zz} corresponds to a pressure larger than lithostatic on the underlying bed. The pressure on the bed is equal to minus the full vertical normal stress in the ice, or

$$P_n = P_o + \frac{\tau_b L}{B_o \pi} \cos\left(\frac{2\pi x}{L}\right). \qquad (7.19)$$

Here $P_o = \rho g H$ represents the lithostatic pressure. The minimum pressure on the bed occurs at $x = L/2$ (the inflection point on the lee side of the obstacle; Figure 7.3) and is given by

$$P_m = P_o - \frac{\tau_b L}{B_o \pi}. \qquad (7.20)$$

Where the subglacial water pressure equals this minimum pressure, separation of the ice from the bed will occur.

Equation (7.20) relates basal drag to the basal pressure and reflects the assumption that basal resistance arises from compressive stresses acting on the upstream slopes of bed obstacles. As pointed out by Iken (1981), the consequence of this model is that an upper limit to basal drag exists. The minimum pressure cannot become less than the water pressure in the cavities without introducing nonzero acceleration terms in the force-balance equations, and

$$P_m \geq P_w. \qquad (7.21)$$

Then, from equation (7.20),

$$P_o - \frac{\tau_b L}{B_o \pi} \geq P_w, \qquad (7.22)$$

or

$$\frac{\tau_b L}{B_o \pi} \leq N, \qquad (7.23)$$

where $N = P_o - P_w$ represents the effective basal pressure. The upper bound on basal drag depends on the geometry of the bed through the amplitude, B_o, and wavelength, L, of the bed perturbations. This result is derived here for a simple sinusoidal bed, but the analysis by Schoof (2005) shows that an upper bound on basal drag exists for any arbitrary bed geometry and this upper bound depends on bed obstacles that have the steepest slopes. In essence, this upper bound exists because when cavitation occurs, it becomes more difficult to exert sufficient stress on the remaining undrowned bed obstacles (Schoof, 2005).

Although Iken (1981) noted the existence of an upper bound to basal drag where cavitation occurs or, equivalently, a critical water pressure above which sliding was said to become unstable, that study considered only the stable sliding regime with basal drag smaller than the upper bound. Fowler (1987a) argued that for real glaciers basal drag does not reach a maximum because there will always be bed obstacles large enough to offer sufficient form drag to allow basal drag to increase. However, as equation (7.23) shows, as well as the analysis of Schoof (2005), this is true only if the slope of the larger obstacles is greater than that of the obstacles that already have been drowned or cavitated.

The existence of an upper bound for basal drag has important consequences for modeling glacier flow. First, where the driving stress is larger than the maximum basal drag, other resistive stresses must balance part of the driving stress. Most ice-flow models are based on the lamellar flow assumption in which the driving stress is balanced entirely by drag at the glacier bed (Section 4.2), but this may not be appropriate for outlet glaciers approaching flotation ($N \to 0$). Second, it forces a different view of what constitutes a sliding law. The common view is that such a law relates the sliding velocity to basal drag and effective basal pressure. It is more appropriate to view a sliding relation as expressing basal drag as a function of sliding speed and effective basal pressure (Schoof, 2005).

Lliboutry has suggested a number of relations to link the sliding velocity to the stress distribution calculated above. For example, the velocity can be estimated by comparing the rate of basal melting to the rate of closure of the cavity or by considering the two components of velocity (in the horizontal and vertical directions) and requiring that the net velocity be directed along the lower surface of the ice. For an undulating random bumpy profile with extensive cavitation, Lliboutry (1987a) suggests the following local relation between basal drag and sliding velocity

$$\tau_b = f N + \frac{D U_s}{(f N)^{n-1}}. \tag{7.24}$$

Here f represents a constant that depends on the distribution of obstacle heights; D is a constant that includes the rate factor of the ice, bed roughness, and other parameters; n represents the exponent in Glen's flow law; and N represents the effective subglacial water pressure. The first term on the right-hand side represents Coulomb friction between solids, assumed to be negligible in most sliding theories.

The most common sliding relation used in numerical models is a generalization of the second term on the right-hand side of equation (7.24), and this is often referred to as modified Weertman sliding

$$\tau_b = B_s \, U_s^p \, N^q, \tag{7.25}$$

where B_s, P, and q are positive sliding parameters and U_s represents the sliding velocity (for example, Budd et al., 1979; Bindschadler, 1983; Fowler, 1987a). In the more common glaciological notation, this relation is written as

$$U_s = A_s \frac{\tau_b^m}{N^p}. \tag{7.26}$$

Bindschadler (1983) compared various sliding relations against observations and found a best fit for m = 3 and p = 1. In their model of the West Antarctic Ice Sheet, Budd and others (1984) adopted this sliding relation with m = 1 and p = 2.

Sliding relations of the form (7.25) do not allow for an upper limit to basal drag, irrespective of the value of the effective basal pressure. For this reason, Schoof (2005) suggested the following heuristic sliding law:

$$\frac{\tau_b}{N} = C \left(\frac{U_s}{U_s + D_o \, N^n} \right), \tag{7.27}$$

where D_o is a constant that depends on the wavelength of dominant obstacles, their slope, and the rate factor for glacier ice. Support for a sliding relation of this form comes from finite-element modeling of ice flow past bedrock obstacles (Gagliardini et al., 2007). Their modeling results suggest a sliding law of the form

$$\frac{\tau_b}{N} = C \left(\frac{\chi}{1 + \alpha \chi^q} \right)^{1/n}, \tag{7.28}$$

where

$$\chi = \frac{U_s}{C^n \, N^n \, A_s}, \tag{7.29}$$

and

$$\alpha = \frac{(q-1)^{q-1}}{q^q}. \tag{7.30}$$

In these expressions, C is the maximum value reached by τ_b/N (similar to equation (7.23)), and A_s the sliding parameter in the absence of cavitation. Figure 7.4 shows this sliding law for different values of the exponent q. The value of q controls the post-peak

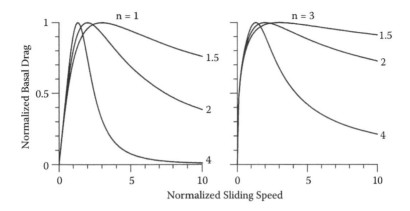

FIGURE 7.4 Normalized basal drag as a function of normalized sliding speed according to equation (7.28) for different values of the exponent, q (indicated by the labels), and for n = 1 (left panel) and n = 3 (right panel).

decrease in basal drag: the larger this exponent, the more rapidly basal drag decreases after peaking. The maximum basal drag corresponds to the sliding velocity

$$U_s = \frac{q}{q-1} A_s C^n N^n. \tag{7.31}$$

(Gagliardini et al., 2007).

Potentially, the double-valued sliding (7.28) can lead to flow instability. In the initial stages of cavitation and sliding, enough bed obstacles remain to provide resistance to flow. As cavitation progresses, and more smaller obstacles are drowned, basal resistance has to come from form drag offered by the remaining obstacles with ice contact. However, the maximum drag able to be generated from these obstacles depends on their slope; thus, unless larger obstacles also have steeper slopes, associated form drag will reach a maximum, irrespective of whether undrowned bed irregularities remain. Should further cavitation occur, basal drag will decrease while the sliding speed continues to increase.

Whether basal drag will reach the upper bound depends on the value of C in equation (7.28) and the effective basal pressure, N. For a sinusoidal bed, Gagliardini et al. (2007) found $C \approx 0.84(2\pi B_0/L) = 1.68\pi R$, where R represents the bed roughness. As derived in Section 7.4, the maximum effective basal pressure is given by the height above buoyancy, H_b (thickness of the ice in excess of the flotation thickness), corresponding to the situation where an easy connection exists between the subglacial drainage system and the proglacial water body. Under those conditions

$$\tau_b(max) = CN(max) =$$

$$= 1.68\pi R \rho g H_b = \tag{7.32}$$

$$\approx 47 R H_b,$$

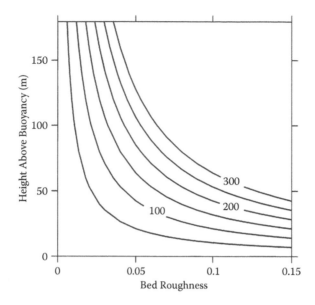

FIGURE 7.5 Contours of maximum basal drag (interval: 50 kPa) as a function of bed roughness and height above buoyancy. Contours for values greater than 300 kPa are not shown.

with basal drag in kPa. Figure 7.5 shows how this upper limit depends on bed roughness and height above buoyancy. For comparatively smooth beds or for glaciers that are close to flotation, the upper limit to basal drag can become less than 100 kPa, making it conceivable that ice flow enters the regime of rapid sliding with decreasing basal resistance.

All these theories contain some degree of simplification and may not be entirely realistic. Perhaps the most important objection is that the bed is assumed to be perfectly lubricated. That is, the tangential shear stress, or skin friction (equation (7.11)), is neglected and all basal resistance is attributed to the form drag resulting from the pressure fluctuations induced by the obstacle. Only in the very special case of clean basal ice may this be an appropriate simplification. In most cases, glaciers do not move frictionless over their bed, as can be inferred from erosive patterns such as glacial grooves and striae in regions formerly overlain by glaciers. Also, where observed, basal ice usually contains debris. As this dirty ice slides over its bed, friction may be expected. Thus, perhaps a more realistic approach is to include the effect of skin friction and to attempt to derive a relation between the sliding velocity and the skin friction (Morland, 1976; Schweizer and Iken, 1992).

The total basal drag, τ_b, is written as the sum of form drag, τ_f (the second term on the right-hand side of equation (7.13)) and tangential skin friction, τ_f, and

$$\tau_b = \tau_t + \tau_f. \tag{7.33}$$

Schweizer and Iken (1992) propose that the sliding velocity must be proportional to the nth power of the form drag, τ_f (which they term the "resultant effective shear stress"), while Morland (1976) considers a sliding relation that links the velocity to the skin friction, τ_t. These approaches probably represent the two limiting cases of form drag being most important and skin friction being dominant. Thus, a general sliding relation may look as follows:

$$U_s \propto C_1\tau_f^p + C_2\tau_t^q, \tag{7.34}$$

where C_1 and C_2 are constants that may depend on the effective water pressure, and other factors as well.

An entirely different approach to the problem of sliding with friction is taken by Boulton (1974), Hallet (1981), and Shoemaker (1986), among others. In these studies, the contact forces between debris particles in the ice and the underlying bed are calculated explicitly. Because this resistive force can be related to both the basal resistance and the velocity of the debris (and hence the velocity of the basal ice), a sliding relation can be derived. For example, Hallet (1981) suggests the following relation:

$$U_s = \frac{\tau_b}{\eta(R + \Omega\mu C_o)}, \tag{7.35}$$

where η represents the viscosity of the basal ice, R is a bed-roughness factor, Ω is a viscous drag term, μ describes the friction between the basal debris and the bed, and C_o denotes the concentration of debris in the basal ice layers. Most of these parameters are ill-constrained, thus limiting the practical use of this relation or variations on it.

7.3   GLACIER FLOW OVER A SOFT BED

In the models discussed so far in this chapter, the bed underneath the ice is taken to be hard and nondeformable. However, large areas previously covered by continental ice sheets are overlain with till, "the ubiquitous glacial deposit by which every former glacier is most surely traced" (Goldthwait, 1971, p. 3). Glacial geologists have long suggested that such sediments under glaciers may be actively deforming as a consequence of forces exerted by the moving ice. As early as 1894, W. J. McGee, in summarizing his work based on field observations in "the magnificent field of the southern Sierra" noted that "if a continuous sheet of comminuted debris intervene, the movement may be divided between its upper and lower surfaces; and if the intercalated sheet be thick, several planes of slip may exist within it and its own motion becomes differential" (McGee, 1894, p. 352). Where subglacial sediments are actively deforming, some fraction of the measured ice velocity will be due to the rate at which the sediment is deforming. Yet, the glaciological community was slow to realize the importance of ice moving over soft beds and potential implications for ice-sheet stability. For example, the 1981 second edition of Paterson's classic *The Physics of Glaciers* does not mention deformable beds. By the time the third edition was published in 1994, an entire chapter was devoted to "Deformation of Subglacial Till." What brought about this "paradigm shift in glaciology" (Boulton, 1986) was the discovery of a meters-thick layer of deforming sediments under Whillans Ice Stream in West Antarctica in the first half of the 1980s by scientists from the University of Wisconsin, Madison (Blankenship et al., 1986). Prior to this discovery, Geoffrey Boulton and colleagues had been trying to raise the awareness of the glaciological community to the potential importance of soft beds to glacier flow, primarily based on work conducted on Breidamerkurjökull in Iceland, but the observations on Whillans Ice Stream made the community realize the potential scale on which deforming sediments could act and, perhaps, affect the stability of large ice sheets.

Boulton and Jones (1979) pointed out that many glaciers rest on soft beds that consist of sediments that may be deforming. Regions that were formerly ice covered reveal a subglacially deposited till or other unlithified sediment, which appears to be very soft and with a high water content. Boulton and Jones (1979) proposed that if high water pressures develop in the sediments (due to basal melting and low transmissibility of the bed), the effective pressure may be sufficiently low to cause deformation of the subglacial sediments. In fact, Boulton and Hindmarsh (1987) argued that measurements of deformation within the till underlying Breidamerkurjökull in Iceland suggest that 88% of the ice movement is due to such deformation, compared with 12% by sliding at the glacier-bed interface. Similarly, the discovery of a layer of till under Whillans Ice Stream in West Antarctica (Blankenship et al., 1986, 1987) has led Alley and co-workers to advance the hypothesis that the high speeds measured on this ice stream, as well as the very small basal resistance (less than 20 kPa), are the result of the presence of a continuous layer of actively deforming till (Alley et al., 1986, 1987a).

Despite numerous research efforts over the last few decades, understanding of the role of subglacial sediments in modulating glacier flow remains frustratingly limited. How these sediments deform has been a contested topic in glaciology, and different constitutive relations have been proposed and incorporated into numerical ice flow models (c.f. Clarke, 2005). Even the more basic question concerning the extent of deforming sediments appears to be under debate (Boulton et al., 2001; Piotrowsky et al., 2001, 2002). In part, this may be due to the many faces of glacial till and its composition and the difficulties in firmly establishing origins of sediments left behind by glaciers. For example, Piotrowsky et al. (2002, p. 174) note that "the Norfolk Drift is as controversial as it is well known," with various researchers having proposed different origins of this sedimentary deposit. Moreover, the inaccessibility of sediments under glaciers effectively prevents controlled deformation experiments, and most flow laws for till are based on laboratory experiments. In the few experiments where subglacial sediments were accessed directly, observations suggest that till deformation may vary considerably on time scales from hours to days (for example, Kavanaugh and Clarke, 2006).

This section briefly discusses the most common models for till deformation in use by the glaciological community. It falls outside the scope of this book to discuss all relevant observations and theories. Indeed, this is a task better left to scientists with greater expertise in this area.

Boulton and Hindmarsh (1987) describe results of deformation experiments conducted under the margin of Breidamerkurjökull, a major outlet glacier of Vatnajökull in Iceland. Because this represents one of the few in situ measurements of till deformation, it is worthwhile to discuss these observations in some detail. During a series of experiments, tunnels were excavated within the basal ice and perpendicular to the margin (approximately in the direction of ice flow). Strain markers and piezometers were inserted into the subglacial sediment by drilling small holes, which were plugged after the markers and piezometers were in place. At the end of the experiment (136 to 244 hours), the sediment was excavated and the location of the strain markers noted. The markers buried in the till showed clear signs of shearing motion within the sediment. Boulton and Hindmarsh (1987) therefore conclude that the shear stress exerted by the ice on the sediment causes the sediment to deform, and this deformation is the major contributor to the ice velocity.

From a series of experiments conducted from 1977 to 1983, Boulton and Hindmarsh (1987) obtained seven triplets of values for strain rate in the subglacial till, basal drag, and effective normal pressure, and fitted two rheological models to these data. The first model represents a nonlinear Bingham fluid with yield strength τ_y. Deformation in such a material is zero if the shear stress exerted on it is smaller than the yield strength, while a larger shear stress, τ, results in deformation according to

$$\dot{\varepsilon} = C \frac{(\tau - \tau_y)^p}{N^q}. \tag{7.36}$$

The yield strength is determined by the Mohr–Coulomb failure criterion (for example, Vyalov, 1986, p. 101)

$$\tau_y = C_o + N \tan \phi, \tag{7.37}$$

with C_o the cohesion and ϕ the friction angle, both material properties of the till that may depend on porosity, water content, and other factors. The other model tested by Boulton and Hindmarsh (1987) is the nonlinear viscous fluid model, described by the following rheological relation:

$$\dot{\varepsilon} = D \frac{\tau^s}{N^t}. \tag{7.38}$$

Both rheological models fit the data equally well, and Boulton and Hindmarsh find only a slight rheological nonlinearity ($p = 1.3$, $q = 1.8$ in equation (7.36); $s = 0.6$, $t = 1.2$ in equation (7.38)). Given the paucity of data and the number of adjustable parameters, the model fitting should be viewed with skepticism. More importantly, the drag exerted by the moving ice on the underlying till is calculated from the glacier geometry. In itself, this is a dubious procedure, especially near the glacier terminus, where longitudinal stress gradients can be expected to be important; but moreover, no accurate surface elevation measurements appear to be available for this glacier. Basal drag varying between about 20 kPa and 105 kPa over short distances and time spans (as claimed by Boulton and Hindmarsh, 1987, Figure 7) appears highly improbable if these values are estimated from the driving stress (see also the discussion in Murray, 1997).

TABLE 7.1

Value of the Effective Viscosity for Various Materials

Material	Log (viscosity) in Pa s
Air [1]	−4.7
Water [1]	−2.9
Olive oil [1]	−1.0
Mud flow [2]	2.7
Soft clay [3]	9–10
Firm clay [3]	11–12
Stiff clay [3]	13–14
Hard clay [3]	14–16
Ice	13–14

Source: Data from (1) Turcotte, D. L., and G. Schubert, *Geodynamics* (2nd ed.), Cambridge, Cambridge University Press, 2002, p. 228; (2) Johnson, A. M., *Physical Processes in Geology*, San Francisco, Freeman, Cooper, & Co., 1970, p. 513; (3) Vyalov, S. S., *Rheological Fundamentals of Soil Mechanics*, Amsterdam, Elsevier, 1986, pp. 120–121.

In most coupled ice-flow/till-deformation models, a linear rheology for the subglacial till is adopted (for example, Alley et al., 1987b; Alley 1989b; MacAyeal, 1989). That is,

$$\tau = \eta\dot{\varepsilon}, \tag{7.39}$$

where the (effective) viscosity, η, may be dependent on the ambient effective pressure. Using the data from Boulton and Hindmarsh (1987), the viscosity of the till under Breidamerkurjökull ranges from $6.5 \cdot 10^{10}$ to $5.6 \cdot 10^{11}$ Pa s. For comparison, Table 7.1 lists viscosities of some other materials.

An entirely different rheological model for water-saturated till is suggested by Kamb (1991), based on experiments conducted on subglacial material recovered from the base of Whillans Ice Stream in West Antarctica. This fast-flowing ice stream is believed to be underlain by a meters-thick subglacial layer that is highly porous and saturated with water at high pore pressure, as suggested by seismic reflection studies (Blankenship et al., 1986, 1987). Subsequent drilling to the bed has confirmed these early inferences (Engelhardt et al., 1990). Samples of till recovered show a highly plastic but cohesive, sticky mud that contains an abundance of rock fragments of pebble size and smaller. Direct-shear tests on the freshly cored till suggest that the till behaves very much like a Coulomb-plastic material with a yield stress of about 2 kPa (Kamb, 1991). In terms of the nonlinear viscous rheology (equation (7.38)), this corresponds to a very large value of the exponent ($s \sim 100$; Kamb, 1991). Equivalently, perfect plasticity corresponds to a linear rheology (equation (7.39)), but with a very small viscosity, for shear stresses that are larger than the yield stress.

In numerical models, introducing a yield criterion with strain rate switching from zero to some large number when the yield stress is reached is likely to lead

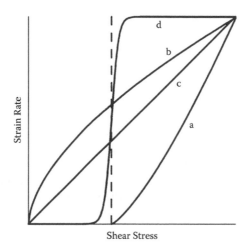

FIGURE 7.6   Relation between strain rate and applied shear stress for four rheological models for subglacial till. Curve a: equation (7.36) with p = 1.3; curve b: equation (7.38) with s = 0.6; curve c: equation (7.39); curve d: equation (7.40). The vertical dashed line indicates the yield strength. Scales are arbitrary.

to numerical instabilities or other problems. To avoid such numerical artifacts, Kavanaugh and Clarke (2006) propose the following relation to describe the rheological behavior of Coulomb-plastic till:

$$\dot{\varepsilon} = \frac{\dot{\varepsilon}_o}{2}\left[1 + \tanh\left(2\pi\frac{\tau - \tau_y}{\Delta\tau}\right)\right]. \tag{7.40}$$

By assigning a large value to the reference strain rate, $\dot{\varepsilon}_o$, and a small value to the failure range, $\Delta\tau$, this relation closely resembles stepwise behavior with yield strength τ_y.

A comparison of the four proposed flow relations for subglacial till is shown in Figure 7.6. Because Boulton and Hindmarsh (1987) found only small nonlinearity for the two relations tested, these are not significantly different from the linear relation, except, of course, that for relation (7.36) no deformation occurs until the yield strength is reached. Kavanaugh and Clarke (2006) measured simultaneously basal water pressure, pore water pressure, sediment deformation, glacier sliding, and sediment strength beneath Trapridge Glacier in Canada's Yukon Territory and, combined with hydromechanical modeling, tested these four flow laws. They conclude that for the duration of the five-day measurements, modeled instrument responses yield the best qualitative agreement with field measurements if Coulomb-plastic rheology is adopted. Iverson and Iverson (2001) use a numerical model to illustrate how deformation profiles such as those measured in the Breidamerkurjökull sediments can result from brief episodes of failure during which the stress locally at some depth exceeds the yield strength of the till at depth, instigating motion of a thin layer of sediment at that depth.

The perfect-plastic model and the (almost) linear-rheological model have important but very different consequences for the flow (and stability) of the overlying glacier. According to the first (Kamb) model, the shear stress on the subglacial layer is limited to the yield strength of the till, and if the till is moving, its flow is presumably driven by horizontal gradients in pressure. Also, in this model, basal drag experienced by the glacier must be small, a few kPa at most. Interpretation of measured surface velocities on Whillans Ice Stream suggest that, indeed, this may the case (Whillans and Van der Veen, 1997). If, on the other hand, the till obeys a linear rheology, as advocated by Alley and co-workers, the speed of the glacier is determined entirely by the viscous properties of the till. Equating the shear stress on the subglacial layer with basal drag, integration of expression (7.39) twice with respect to the vertical coordinate gives the velocity at the top of the till layer (and thus the velocity of the glacier):

$$U_t = \frac{2}{\eta} H_t \tau_b, \tag{7.41}$$

where H_t represents the thickness of the till layer and the velocity at the bottom of this layer is taken to be zero.

The controlling parameter that determines which of these two models applies is the effective viscosity of the subglacial till. To explore this issue somewhat further, consider the flow of a viscous material between two plates. The forces acting on a small element are shown in Figure 7.7. Assuming accelerations are negligible, the sum

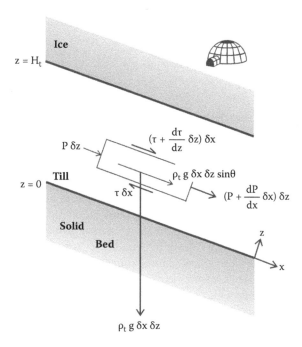

FIGURE 7.7 Forces acting on a small element (δx by δz, and unit width) of deformable material embedded between the glacier sole and a nondeforming solid bed.

of the forces acting in the direction of flow must be zero. Making this sum, and dividing by the volume of the element, gives the balance equation (compare Section 3.1),

$$\frac{\partial \tau}{\partial z} = \frac{\partial P}{\partial x} - \rho_i g \sin\theta,$$
(7.42)

where P represents the pressure and θ the slope of the lower boundary (taken positive when the bed slopes downward in the direction of flow, as shown in Figure 7.7). Adopting the linear-viscous model (equation (7.39)), the vertical shear in velocity is linked to the shear stress as

$$\frac{\partial u}{\partial z} = \frac{2}{\eta}\tau.$$
(7.43)

Combining equations (7.42) and (7.43) yields

$$\frac{\partial^2 u}{\partial z^2} = \frac{2}{\eta}\frac{\partial P}{\partial x} - \frac{2\rho_i g}{\eta} \sin\theta.$$
(7.44)

Integrating this expression twice with respect to the vertical z-coordinate gives the velocity profile in the viscous layer:

$$u(z) = \frac{1}{\eta}\frac{\partial P}{\partial z}z^2 - \frac{\rho_i g}{\eta} \sin\theta\, z^2 + Az + B.$$
(7.45)

The two integration constants, A and B, can be determined from the boundary conditions. At the lower boundary of the layer (z = 0), the velocity must be zero, and thus B = 0. Denoting the thickness of the layer by H_t and the velocity at the upper boundary by U_t, the second integration constant is

$$A = \frac{U_t}{H_t} - \frac{H_t}{\eta}\frac{\partial P}{\partial x} + \frac{\rho_i g H_t^2}{\eta} \sin\theta.$$
(7.46)

The velocity profile now becomes

$$u(s) = \frac{H_t^2}{\eta}\frac{\partial P}{\partial x}(s^2 - s) - \frac{\rho_i g H_t^2}{\eta} \sin\theta\, (s^2 - s) + U_t\, s,$$
(7.47)

where $s = z/H_t$ denotes a dimensionless vertical coordinate.

From the expression for the velocity profile and using equation (7.43), the shear stress at the top of the viscous layer, and thus the basal drag on the glacier, is found to be

$$\tau_b = \frac{1}{2}\frac{\partial P}{\partial x}H_t - \frac{1}{2}\rho_i g H_t \sin\theta + \frac{\eta}{2}\frac{U_t}{H_t}.$$
(7.48)

For very small values of the effective viscosity, η, the third term on the right-hand side may be neglected, and basal drag is determined by the gradient in subglacial pressure and slope of the bed. According to the Alley model, on the other hand, the first two terms are negligible and basal drag is controlled entirely by the viscosity of the till.

To quantify this analysis, consider again the experiments on Breidamerkurjökull, described in Boulton and Hindmarsh (1987). The ambient pressure gradient may, in first approximation, be set equal to the gradient in the ice overburden pressure (thus, gradients in the water pressure are neglected). Then

$$\frac{\partial P}{\partial x} \approx \rho g \frac{\partial H}{\partial x}, \tag{7.49}$$

where H represents the ice thickness. From Figure 1 in Boulton and Hindmarsh (1987) it may be concluded that the bed is near horizontal ($\theta = 0$) and the thickness gradient is equal to the surface slope (about -0.4). The thickness of the till layer is about 0.5 m, so that the contribution to basal drag from the pressure-gradient term (the first term on the right-hand side of equation (7.48)) is about -0.9 kPa. Note that this contribution to the basal drag is negative. This is because the viscous flowing material exerts a drag on the upper plate (i.e., the glacier sole) that is directed along the positive x-axis (the direction of flow). Basal drag on the glacier is taken positive when resisting the flow of the ice and directed along the negative x-axis. In principle, if the third term in equation (7.48) were zero and the bed horizontal or sloping downward in the direction of flow, basal drag on the glacier could thus be negative, with the ice being dragged along by the subglacial till driven by the horizontal pressure gradient. However, for realistic values of the pressure gradient and basal slope, the effect is small, a few kPa at most. The second term on the right-hand side of equation (7.48) is zero (no bed slope), while the third term can be calculated for different values of the effective viscosity. The results of a calculation using $U_t = 23$ m/yr (Boulton and Hindmarsh, 1987) are shown in Figure 7.8. The transition from pressure-driven flow to shearing flow occurs over a relatively narrow range of values of the effective viscosity, namely, about a factor of 10. Given the uncertainty in viscosity (c.f. Table 7.1), it thus seems possible that the till near the margin of Breidamerkurjökull is being squeezed out from underneath the glacier (as illustrated in Figure 7.9), rather than deforming under an applied shear stress.

The "squeezing" hypothesis was proposed by Price (1970) to explain the pattern of moraine ridges observed near the margin of the retreating Fjallsjökull, another glacier in Iceland. The former frontal position is clearly marked by moraine ridges, about 5 to 10 m high. Inside these larger moraines are series of smaller moraine ridges (about 1 to 4 m high) that appear to have formed proglacially, rather than being deposited in a subglacial cavity to be exposed by the retreating ice front. Also, the ridges follow the shape of the ice front, but no evidence appears that the glacier front acts as a bulldozer, building the moraine

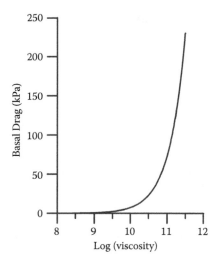

FIGURE 7.8 Basal drag on a glacier resting on a layer of deformable till, as a function of the effective viscosity of the till, calculated from equation (7.48), using $\partial P/\partial x = -4\mu Pa/m$, $H_t = 0.5$ m, $U_t = 23$ m/yr, and assuming zero bed slope.

ridges by plowing up the proglacial deposits. Price (1970) suggests that pressure-driven flow may be triggered by spring meltwater reaching the base of the glacier. The increased water content of the subglacial till would lower the bearing capacity of the till, until the semiliquid water-soaked material is being squeezed out from underneath the glacier.

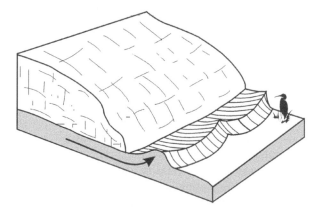

FIGURE 7.9 Illustration of the squeezing of till from underneath the glacier margin to produce proglacial moraine ridges. (From Price, R. J., *Arctic Alpine Res.*, 2, 27–42, 1970. Reproduced by permission of the Regents of the University of Colorado from *Arctic and Alpine Research*.)

7.4 SUBGLACIAL HYDRAULICS

The nature and efficiency of the subglacial drainage system is one of the primary controls on glacier sliding and deformation of subglacial sediments. Where drainage of meltwater is tortuous or impeded, one may expect high water pressures to build up, reducing the effective pressure at the glacier bed, resulting in increased basal sliding speeds (equation (7.26)) or reduced basal friction (equation (7.24)). Similarly, rheological properties of subglacial till depend strongly on the water content, and deformation occurs only when the till is water saturated. Consequently, any model aiming to predict how glaciers and ice sheets evolve over time must include a calculation that describes how the effective basal pressure changes in accordance with the glacier. This, in turn, requires a model for how meltwater at the glacier bed is routed underneath the glacier.

In 1972, Hans Weertman published a paper entitled, "General Theory of Water Flow at the Base of a Glacier or Ice Sheet," which started with the following comment: "There is at present no reasonably complete theory of the flow of water beneath glaciers and ice sheets. The theories that have been developed describe only limited parts of the problem" (Weertman, 1972, p. 287). Sad to say that 40 years later, this statement is perhaps even more true than when originally written. At the time, models for water flow at the bed consisted of sheet flow with variable thickness, and concentrated drainage through a system of tunnels either incised into the ice or into the bedrock underneath. Since then, the subglacial landscape has become even more muddied with the introduction of drainage through a linked-cavity system and the realization that many glaciers and ice sheets may be resting on soft sediments. As a result, despite a large number of theoretical analyses and increasingly more in situ observations, a general theory remains elusive. What is fairly certain from observations is that different drainage systems exist under different glaciers and, to complicate matters further, that one drainage system can morph into another, depending on supply of meltwater to the glacier bed, with consequences for glacier discharge. It has been suggested that rapid changes in glacier flow (such as surges) are associated with the switch from one drainage system to another (Fowler, 1987b; c.f. Section 10.4). Schoof (2010) proposes that faster flow on Greenland outlet glaciers is driven by variability in the input of water to the bed due to dynamic switching between channelized and linked-cavity drainage systems.

Before discussing the various drainage systems and their implications on glacier flow, it is instructive to first consider the range of possible water pressures underneath a glacier. This range can be determined independent of the water flux or nature of the drainage system and provides modelers with reasonable bounds on the effective basal pressure used in most sliding laws.

The maximum water pressure can be found by considering the effective basal pressure, N, defined as the difference between the weight-induced pressure of the ice above, P_o, and the water pressure, :

$$N = P_o - P_w =$$

$$= \rho g H - P_w. \tag{7.50}$$

The effective basal pressure cannot become negative, as this would correspond to a net upward force exerted on the base of the ice. Thus, the water pressure cannot become larger than the ice overburden pressure, and

$$P_w^{max} = \rho g H,$$
(7.51)

noting that P_w is taken positive when directed upward. If the water pressure equals this maximum value, the entire weight of the ice is carried by the subglacial water and the effective basal pressure is zero.

An expression for the minimum water pressure can be derived from the condition that subglacial water must be able to reach the glacier terminus. The discharge of water is driven by the downslope component of gravity and by the gradient in water pressure. Because the total pressure gradient is negative in the direction of flow, it is convenient to introduce Γ_t, defined as the negative of the total pressure gradient:

$$\Gamma_t = -\left(\rho_w g \frac{\partial b}{\partial x} - \Gamma_w \right),$$
(7.52)

where ρ_w denotes the density of (fresh) water, b the basal elevation (negative if below sea level), and Γ_w minus the gradient in water pressure. The first term between the brackets represents the component of gravity driving subglacial water in the basal downslope direction. If the water is to reach the glacier terminus, the expression between brackets must be smaller than zero, so

$$\Gamma_w \geq \rho_w g \frac{\partial b}{\partial x},$$
(7.53)

or

$$\frac{\partial P_w}{\partial x} \leq -\rho_w g \frac{\partial b}{\partial x}.$$
(7.54)

Integrating this expression with respect to the flow direction, x, gives the minimum water pressure needed to discharge water toward the terminus,

$$P_w^{min} = -\rho_w g b + C.$$
(7.55)

Note that the basal elevation, b, is negative when below sea level, so that in general, the minimum pressure is positive. The integration constant, C, is given by

$$C = g b_{gr} (\rho_w - \rho_s),$$
(7.56)

with b_{gr} the basal elevation at the glacier terminus (or the grounding line, if a peripheral ice shelf exists) and ρ_s the density of sea water. This constant takes into account the difference in density of fresh water and of sea water. If C were zero, fresh water would not be able to flow out from under the glacier and sea water would be injected under the glacier (Lingle and Brown, 1987). However, the value of C is small, and for the present discussion this constant may be neglected.

The minimum water pressure corresponds to the case of a subglacial aquifer with zero impedance or to water flowing in a film underneath the glacier (Weertman, 1972). The effective pressure is then

$$N^{max} = \rho g \left(H + \frac{\rho_w}{\rho} b \right),$$
(7.57)

which represents the infamous height above buoyancy, used in many modeling studies (for example, Van der Veen, 1987; Budd and Jenssen, 1987; Nick et al., 2010). This model applies if there is a full and easy water connection to the sea. However, in most cases, water flow may be expected to be obstructed to some extent, and larger water pressures are needed to overcome the resistance of the basal drainage system.

For a glacier resting on soft (perhaps deforming) sediment, the bed may be a water-saturated layer of porous material that acts as an aquifer for the discharge of the subglacial water (Lingle and Brown, 1987). The flow of water is driven by horizontal pressure gradients and described by Darcy's law. Accounting for the slope of the bed (z = b), the gradient in subglacial water pressure is given by (Lingle and Brown, 1987)

$$\frac{dP_w}{dx} = -\rho_w g \frac{db}{dx} - \frac{\rho_w g}{K} U_w,$$
(7.58)

where ρ_w denotes the density of the subglacial water, U_w the volume flux of water per unit of aquifer cross-sectional area, and K the hydraulic conductivity of the porous material. The first term on the right-hand side represents the pressure gradient needed to drive water along a sloping bed, and the second term represents the pressure gradient needed to overcome the impedance of the aquifer.

Lingle and Brown (1987) apply the aquifer model to Whillans Ice Stream in West Antarctica. Because the hydraulic conductivity of the subglacial layer is unknown, they adopt a modified Weertman sliding relation (equation (7.26)), so that the effective pressure (and hence the water pressure) can be calculated from the sliding velocity (calculated from mass-balance considerations) and basal drag (which is equated with driving stress). The inferred values for the hydraulic conductivity fall in the range of 0.019 to 0.064 m/s, which is orders of magnitude larger than the maximum hydraulic conductivity of water-saturated till (about $1.6 \cdot 10^{-6}$ m/s; Boulton and Jones, 1979). On the basis of this result, Lingle and Brown (1987) conclude that water flow through an aquifer is not likely to occur under Whillans Ice Stream, because for realistic conductivities, the pressure gradients needed to drive the subglacial water toward the grounding line become unrealistically large. Instead, they suggest that advection of the water and till mixture may be the mechanism responsible for the discharge of subglacial water. Alley (1989a), on the other hand, argues that because water flow through a subglacial aquifer is so ineffective under this ice stream, where

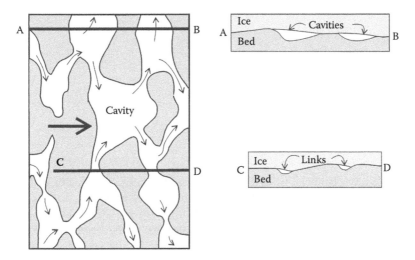

FIGURE 7.10   Schematic illustration of the linked-cavity system. The left panel presents a map view with shading indicating where the ice is in contact with the bed and small arrows showing the direction of water discharge through the network. The panels on the right give the vertical cross-sectional view along the lines AB and CD in the left panel; cavities form on the lee side of large obstacles in the bed, while the connecting links form behind smaller obstacles. (From Kamb, B., *J. Geoph. Res.,* 92(B), 9083–9100, 1987. Copyright by the American Geophysical Society.)

meltwater generation is large, the aquifer may be neglected entirely and water discharge occurs at the interface of basal ice and subglacial till.

A different model for the subglacial hydraulics is the linked-cavity system, schematically shown in Figure 7.10 (Walder, 1986; Fowler, 1987a; Kamb, 1987). As noted in Section 7.2, cavities may form on the lee side of basal obstructions. If there is an ample supply of meltwater, small channels, or links, may form between such cavities, probably on the lee side of lower bumps in the bed. Such a linked-cavity system will be stable if the closure of these links due to flow of ice from upglacier, as well as downward creep of ice from directly above, is compensated for by melting of the roof, due to viscous heat dissipation in the channel. Kamb (1987) argues that, provided the sliding velocity is sufficiently large and the water pressure not too high (but still high), the links will be stable. However, if the sliding velocity decreases or the water pressure increases, the rate of melting becomes too large for the ice flow and creep to keep the roof of the link in place. When this happens, the link may develop into a tunnel, and the glacier may become unstable.

To actually calculate the subglacial pressure in a linked-cavity system, an assumption about the dependency of the sliding velocity of the glacier on the effective pressure is needed and the shape of the conduits linking the cavities needs to be prescribed. The resulting relations are rather complex (Walder, 1986, equation (11); Kamb, 1987, equations (38) and (42)), but all show that the water pressure increases as the interstitial water flux increases. This result is fundamentally different from that derived by Röthlisberger (1972) for water flow through subglacial channels.

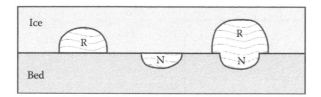

FIGURE 7.11   Cross-sectional view of a Röthlisberger-channel (left), a Nye-channel (middle), and an N-channel overlain by an R-channel (right).

Röthlisberger-channels, or R-channels, are at the base of a glacier and incised upward into the ice, as opposed to Nye-channels, or N-channels, that are incised downward into the glacier bed (Figure 7.11). Röthlisberger (1972) makes the assumptions that the circular conduits are in steady state (that is, melting of the roof is balanced by creep closure under the weight of the ice above), that no heat is transported by the water flowing through the channel, and that the stress in the ice above the channel is isotropic and homogeneous. Under these restrictions, a relation between pressure gradient and water flux can be derived, as shown in Section 7.5. According to the result, the water pressure, P_w, decreases as the discharge increases.

No consensus has been reached among glaciologists as to which model for the subglacial hydraulics is most appropriate. Weertman (1972) argues that channelized drainage is not very likely. A requisite for channelized water flow is that the conduits are able to capture subglacial meltwater that is formed under the entire glacier. For R-channels, and for N-channels that run approximately parallel to the ice-flow direction, the pressure gradient in the vicinity of the conduit is such that it drives water away from the channel rather than into it. Only a tributary N-channel may be able to capture the surrounding meltwater. However, Weertman and Birchfield (1982) doubt whether such channels can exist over extended periods of time. Erosion due to sliding of the glacier is likely to destroy the far ends of the N-channels faster than they can be deepened by the flowing water. Thus, the water flow through the channel decreases, thereby further shortening the conduit. On the other hand, Hooke (1989) argues that observational evidence favors a model in which a linked-cavity system is transected by a few broad, low conduits that drain the meltwater more directly. Most likely, the nature of the subglacial water flow varies from glacier to glacier and possibly throughout the year as well. As noted by Walder (2010), the details about how water is delivered to the main drainage paths (R-channels or linked cavities) may not be crucial and is likely governed by expressions similar to Darcy's law for water flow through porous media (as assumed in Flowers, 2008).

7.5   TUNNEL DRAINAGE

For temperate glaciers, meltwater may form at the surface and enter the glacier through a network of veins, crevasses, and englacial conduits, ultimately collecting at the glacier bed and discharged through the subglacial drainage system. The exact nature of the englacial drainage system remains a subject of debate, in particular how surface meltwater collects in englacial conduits. Irrespective of the details, it is generally assumed that at some depth below the surface the small veins and drainage tubes have evolved into an arborescent system of englacial tunnels (Shreve, 1972; Röthlisberger and Lang, 1987; Hooke, 1989). This is because the release of potential energy as water flows deeper into the glacier tends to favor the growth of larger veins at the expense of smaller ones (Shreve, 1972; Röthlisberger, 1972).

In many instances it is not very practical to discuss glacier drainage in terms of individual veins and tunnels. Rather, the overall water flow is commonly described in terms of the water pressure potential, defined as (Shreve, 1972; Lawson, 1993)

$$\phi = \phi_o + \rho_w \, gz + \rho g(h - z) + P(\dot{r}). \tag{7.59}$$

The flow of water is driven by the gradient in ϕ such that the water flows in the direction perpendicular to the surfaces on which the water potential is constant. In equation (7.59), ϕ_o represents a reference potential, taken constant, z the elevation above a horizontal reference surface, and $\dot{r}$ the rate of tunnel closure due to creep of ice. The second term on the right-hand side represents the potential energy of the water, while the third term corresponds to the water pressure, assumed to be equal to the weight of the overlying ice. The last term accounts for the difference between the water pressure in a conduit, and the ice overburden pressure.

Alternatively, retaining the water pressure as a variable, the potential is

$$\phi = \phi_o + \rho_w \, gz + P_w. \tag{7.60}$$

Applying this expression at the glacier base (z = b) and differentiating with respect to x gives the pressure gradient introduced in equation (7.52):

$$\Gamma_t = -\frac{\partial \phi}{\partial x} =$$

$$= -\left(\rho_w \, g \frac{\partial b}{\partial x} + \frac{\partial P_w}{\partial x} \right). \tag{7.61}$$

At this point, it may be useful to define some commonly used hydrological terms (indicated in Figure 7.12). The *piezometric height* is defined as the elevation difference between the level to which water would rise in a borehole connecting the conduit to the free surface (at atmospheric pressure) and the depth at which the conduit is located. Although the conduit may be located anywhere within the glacier, the piezometric height usually refers to the glacier bed. The surface connecting the piezometric elevations at various places in the glacier is called the *piezometric surface* or *hydraulic grade surface*. The piezometric height, H_p, is a direct measure of the water pressure in the conduit, and

$$\rho_w \, g H_p = P_w. \tag{7.62}$$

The piezometric height is also referred to as *pressure head*. If the height of the piezometric surface is referenced to a horizontal datum surface, the term *hydraulic head*

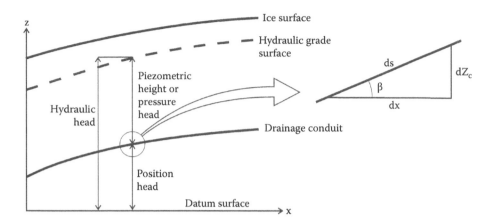

FIGURE 7.12 Definition of hydraulic terms and geometry used in the Röthlisberger model. The pressure head represents the height above the conduit of the piezometric level, that is, the level to which water would rise in a borehole connecting the conduit to the upper ice surface. The surface connecting the piezometric water levels at different points on the glacier is called the hydraulic grade surface. The height of the hydraulic grade surface above a horizontal reference (datum) surface is termed the hydraulic head. The height of the conduit above the datum surface is called the position head. Water flow in the conduit is driven by the slope of the hydraulic grade surface or, equivalently, by the gradient in the hydraulic head.

is used. The hydraulic head is the sum of the pressure head and the height of the conduit above the reference level, called the *position head*. The hydraulic head is equivalent to the water pressure potential.

Theoretical treatment of water flow in or under glaciers usually applies to steady conditions and constant water flux through an idealized tunnel or conduit. The tunnel is assumed to be circular, and its size is determined by the balance between melting of the tunnel walls due to frictional heating from the water passing through and creep closure due to the weight of the ice above. In this section, the model developed by Röthlisberger (1972) for drainage through tunnels is discussed in some more detail to illustrate the complexities involved in formulating a model for subglacial drainage. While this model is usually applied to estimate the water pressure at the glacier bed (for example, Bindschadler, 1983; Greuell, 1992), it may also be used to describe the water flow in englacial tunnels.

The tunnel or conduit is inclined at an angle β with the horizontal x-axis, such that $\tan \beta = dZ_c/dx$, with Z_c the elevation of the tunnel (defined with respect to a horizontal reference level; following the commonly used definition of the vertical z-axis, sea level is chosen as reference level, and $Z_c < 0$ if below sea level). For ease of discussion, the tunnel is assumed to be in the (x, z)-plane, so the length of a tunnel element is given by

$$ds = \frac{dx}{\cos\beta}. \tag{7.63}$$

The flux of water passing through a cross-section of tunnel per unit time is Q, assumed to be known and considered constant at any particular cross-section.

The change in water pressure, dp_w, over a horizontal distance dx can be expressed in terms of the water pressure potential in the tunnel, $\phi_c = \phi(Z_c)$, using equation (7.60) and is

$$dP_w = d\phi_c - \rho_w g \, dZ_c =$$
$$= d\phi_c - \rho_w g \tan\beta \, dx. \tag{7.64}$$

The second term on the right-hand side represents the change in pressure due to the change in potential head. That is, as water flows through an inclined tunnel, it gains or loses potential gravitational energy, depending on whether the tunnel is sloping downward or upward. The first term on the right-hand side of equation (7.64) represents the pressure loss associated with frictional heating.

To estimate the melting rate, it is assumed that the transfer of energy from the water to the surrounding ice occurs instantaneously, so that the heat produced in an element of the tunnel is used to melt the walls of the same element. If changes in the discharge velocity are neglected, the energy produced per unit time in the tunnel element is (Röthlisberger, 1972)

$$dE = Q \, d\phi_c. \tag{7.65}$$

Part of this energy is used to adjust the temperature of the water to the pressure melting temperature, T_p. The change of the pressure melting temperature with water pressure is given by $C_t = -7.4 \cdot 10^{-8}$ K/Pa, so for a change in water pressure equal to dP_w, the change in pressure melting temperature is

$$dT_p = C_t \, dP_w. \tag{7.66}$$

The amount of water that needs to be heated per unit time is Q, so the energy needed to adjust the water temperature is

$$dE_t = -C_t \, C_w \, \rho_w \, Q \, dP_w, \tag{7.67}$$

where $C_w = 4.217 \cdot 10^3$ Jkg^{-1}K^{-1} represents the specific heat capacity of water. The energy available for melting of the conduit walls is found by subtracting this expression from the energy production term (7.65) and is

$$dE_m = dE - dE_t =$$
$$= Q \, d\phi_c + C_t \, C_w \, \rho_w \, Q \, dP_w, \tag{7.68}$$

noting that $C_t < 0$. The volume of ice melted per unit time from the walls of the tunnel element is

$$dV_m = (\rho L_f)^{-1} \, dE_m, \tag{7.69}$$

in which $L_f = 334 \cdot 10^5$ J/kg represents the specific latent heat of fusion and ρ the density of ice.

Under steady-state conditions, the volume of ice melted balances the rate at which the tunnel walls contract due to creep closure. The closure rate is derived in Section 3.6 and is

$$S = \frac{A\,P_e^n}{(n-1)\,2^n},\tag{7.70}$$

in which A represents the rate factor of ice; n the exponent in the flow law; and P_e the effective pressure, that is, the difference between the ice overburden pressure, P_i, and the water pressure, P_w. Except for the constants in the denominator, this expression is the same as that derived by Nye (1953) and used by Röthlisberger (1972). Writing

$$\bar{A} = \frac{A}{(n-1)\,2^n},\tag{7.71}$$

the rate at which the radius, r, of the circular tunnel decreases due to creep closure is

$$\frac{dr}{dt} = \bar{A}\,P_e^n\,r.\tag{7.72}$$

Thus, for an element of length ds, the change in volume per unit time due to creep closure is

$$dV_c = 2\pi\,\frac{dr}{dt}\,ds =$$

$$= 2\pi\,\bar{A}\,P_e^n\,r^2\,ds.\tag{7.73}$$

Equilibrium conditions require that creep closure balances melting of the tunnel walls and $dV_m = dV_c$, or

$$2\pi\,\bar{A}\,P_e^n\,r^2\,ds = (\rho L_f)^{-1}\,dE_m.\tag{7.74}$$

The derivation is almost complete now. The energy available for melting is given by equation (7.68) and substitution in equation (7.74) yields the expression from which the water-pressure gradient can be obtained if the radius of the conduit is known.

The size of the tunnel can be eliminated from the above-derived equations by relating the radius, r, to the water flux, Q. If the flow in the tunnel is turbulent, the empirical Manning formula

$$Q = \pi r^2\,\frac{R^{2/3}}{m\sqrt{\rho_w\,g}}\left(\frac{d\phi_c}{ds}\right)^{1/2},\tag{7.75}$$

can be used to eliminate the tunnel radius (Röthlisberger, 1972). In equation (7.75), m represents the Manning roughness coefficient and R the hydraulic radius, that is, the cross-sectional area divided by the radius. For a circular tunnel, R = r/2. The

Manning roughness coefficient is of the order of 10^{-1} to 10^{-2} m$^{-1/3}$ s, depending on the roughness of the tunnel walls. Rearranging equation (7.75) gives

$$r^{8/3} = \frac{Q\,m\,2^{2/3}\,\sqrt{\rho_w\,g}}{\pi} \left(\frac{d\phi_c}{ds}\right)^{-1/2}. \tag{7.76}$$

Instead of the Manning formula (7.75), other expressions relating the water flux to the tunnel size have also been used (c.f. Lawson, 1993). However, whichever relation is used, the procedure remains the same, namely, eliminating the tunnel radius from the equilibrium condition (7.74). While the resulting expression for the pressure gradient may be different, the essential characteristics of the tunnel system appear to be the same as described by the Röthlisberger model.

The final expression is found by combining equations (7.68), (7.74), and (7.76), giving

$$\bar{A}\,P_e^n\,\pi^{1/4}\,2^{3/2}\,m^{3/4}\,Q^{3/4}\,(\rho_w g)^{3/8}\,\rho\,L_f \left(\frac{d\phi_c}{ds}\right)^{-3/8} ds =$$

$$= Q\,d\phi_c + C_t\,C_w\,\rho_w\,Q\,dP_w. \tag{7.77}$$

Dividing by $ds = dx/\cos\beta$ yields

$$Q\frac{d\phi_c}{dx} + C_t C_w \rho_w Q \frac{dP_w}{dx} = \bar{A}\,D(\cos\beta)^{-11/8}\,m^{3/4}\,Q^{3/4}\,P_e^n \left(\frac{d\phi_c}{dx}\right)^{-3/8}, \tag{7.78}$$

with

$$D = 2^{2/3}\,\pi^{1/4}\,L_f\,\rho(\rho_w g)^{3/8}. \tag{7.79}$$

To apply this expression to find the basal water pressure, the substitution

$$\phi_c = \phi_b =$$

$$= \phi_o + \rho_w\,g\,b + P_w, \tag{7.80}$$

is made, giving

$$\left(\frac{dP_w}{dx} + \rho_w g \tan\beta\right)^{11/8} + C_t C_w \rho_w \frac{dP_w}{dx}\left(\frac{dP_w}{dx} + \rho_w g \tan\beta\right)^{3/8} =$$

$$= \bar{A}\,D(\cos\beta)^{-11/8}\,m^{3/4}\,Q^{-1/4}\,(\rho g H - P_w)^n. \tag{7.81}$$

Obviously, this equation cannot be solved analytically, and the water pressure has to be determined by integrating equation (7.81) numerically. The usual procedure is to start the integration at the glacier terminus, where the water pressure is assumed to be known. If the glacier terminus is grounded on land above sea level, the usual boundary condition is that the water pressure equals the atmospheric pressure, while for termini grounded below sea level, the water pressure is assumed hydrostatic. However, these conditions may not always apply, and in some instances, water can emerge under pressure at the glacier terminus (Lawson, 1993).

Instead of solving equation (7.81) numerically, an approximate analytical solution can be found by neglecting terms that are small. Fowler (1987a) argues that, in first approximation, the gradient in water pressure may be neglected compared with the term involving the slope of the bed. That is,

$$\frac{dP_w}{dx} \ll \rho_w g \tan \beta, \tag{7.82}$$

leading to the approximate solution

$$P_w = \rho g H - \left(\frac{(\rho_w g \sin \beta)^{11/8}}{\overline{A} D m^{3/4}} Q^{1/4} \right)^{1/n}. \tag{7.83}$$

According to this solution, the water pressure is constant in a drainage tunnel, and it changes only where tributary tunnels join and the water flux changes. This approximation fails near the glacier terminus, where the water pressure is either hydrostatic or equal to the atmospheric pressure. If the gradient in water pressure were to be zero near the terminus, the water pressure would be equal to the atmospheric or hydrostatic pressure under the whole glacier. In addition, the approximate solution does not apply to tunnels that are horizontal or inclined upward.

Perhaps the most important result of the Röthlisberger model is that the water pressure is inversely related to the discharge flux (as can best seen from inspection of the approximate solution (7.83)). That is, if the discharge increases, the pressure gradient decreases. This is fundamentally different from the linked-cavity system (Kamb, 1987) in which the pressure increases as the water flux becomes larger. The theoretical description of other drainage systems essentially follows the analysis of Röthlisberger (1972), that is, steady-state conditions are adopted, with creep closure of the drainage conduits or cavities being balanced by melting of the ice.

The assumption of steady state may not be entirely realistic. There are important temporal changes in the amount of water added to the drainage system, for example, after a heavy rainfall event. Röthlisberger and Lang (1987) study the effect of changes in water flux with time by repeatedly calculating the steady-state water pressure using equation (7.81). For short-term fluctuations, with periods of a day or so, the size of the conduits cannot adjust rapidly enough, and the water pressure varies in phase with the discharge flux. For annual variations, it may be that the conduits adjust to the varying water input, so that steady-state conditions remain approximately valid. If so, variations in water pressure are out of phase with the water flux, and the seasonal decrease in discharge during the winter months may lead to higher water pressures. This means that the effective basal pressure is smallest during the winter season, resulting in maximum sliding speeds during that time of the year. On the other hand, observations on Columbia Glacier, Alaska, show a seasonal maximum in spring (May) and a minimum in fall (late September) (Krimmel, 1997). This suggests highest basal water pressures in the spring when water input to the glacier is largest. The high water pressures may be the result of the inability of the drainage system to accommodate increased discharge. Perhaps the drainage tunnels partly collapse during the winter when insufficient meltwater is available for countering creep closure. Starting in the spring and continuing through the summer, the tunnel system develops more fully and becomes more efficient in draining subglacial water. If correct, this model implies that even on a seasonal scale, time evolution of the drainage system must be incorporated into the theory.

Another reservation about the tunnel model is the assumption that the shape of the tunnels is circular. This may be reasonable for englacial tunnels entirely within the ice, but for tunnels at the glacier bed, the circular shape may be unrealistic. However, if the shape of the tunnel is (approximately) semicircular, the derivation presented above should also apply. More important is the question whether the tunnel will be completely filled with water. Paterson (1994) argues that for some distance upstream of the terminus, it is likely that the tunnel is only partially filled, in which case ice will be melted from the sides of the tunnel, but not from its roof. This would result in broad and low drainage tunnels. Given the many other approximations made in the theoretical descriptions, it is doubtful whether such a change has important consequences.

8 Fractures

8.1 SURFACE CREVASSES

Most of the topics and applications discussed in this book deal with ductile behavior of ice resulting in continuous deformation or creep under comparatively low stresses. This is the primary mode of glacier motion and most relevant to glacier modeling. However, most people traveling on glacier surfaces will be struck by physical manifestations of the brittle nature of ice, namely, crevasses. Crevasses are imposing chasms cutting through the surface of most glaciers that form primarily under tension when stretching cannot be accommodated by ductile flow and, instead, the ice behaves as a brittle material susceptible to rupturing. Most crevasses are confined to the upper few tens of meters and do not importantly affect glacier flow. There are, however, several reasons why crevasse initiation and propagation has received some attention in the glaciological literature. First, the orientation of crevasses—usually perpendicular to the direction of principal tensile stress—may contain information about the regional stress field (c.f. Van der Veen, 1999a). Second, fracture propagation is the primary mechanism by which icebergs break off from glacier termini, and better understanding and quantification of this process may aid in formulation a calving relation for numerical models (Van der Veen, 2002; Benn et al., 2007a,b). Third, rapid downward propagation of water-filled crevasses may be a mechanism by which meltwater ponds formed at the glacier surface can almost instantaneously penetrate the full ice thickness, thus providing a drainage route for surface water to reach the glacier bed and provide additional lubrication (Van der Veen, 2007; Das et al., 2008). For these reasons, it is worthwhile to present a discussion of crevasses on glaciers based on the review of Van der Veen (1999a). The following sections in this chapter introduce the mathematical tools for modeling crevasses and fracture propagation on glaciers with application to iceberg calving.

Perhaps the earliest explanation for crevasses was given by Scheuchzer (1723), who suggested, in what has been characterized as poor medieval Latin, that crevasses formed during the summer as a result of expansion of air bubbles in the ice (Walker and Waddington, 1988). A century later, Hugi (1830) proposed that thermal stresses would generate crevasses during warm spells. While these earlier theories were soon proven to be incorrect, they appropriately recognized crevasses to be the product of stresses, or forces, acting on the ice. It was not long thereafter that these forces became identified with differential speed across the glacier surface, and by the 1840s it was generally accepted that "the formation of crevasses betokens a local distending force" (Forbes, 1859, p. 151). Similarly, Hopkins (1862, p. 706) concluded that "when the maximum normal tension is the force to which the cohesive power of a glacial mass first gives way, the result, as above observed, must be an open fissure, or crevasse, the direction of which must manifestly be perpendicular to that of the tension producing it."

In a paper presented May 1, 1843, at the Cambridge Philosophical Society (Hopkins, 1844a), the principles of force balance were outlined and used to find the orientation of principal stress acting on the surface of a glacier. Hopkins's discussion is in terms of forces, but it is more appropriate to frame the discussion in terms of stresses. If A represents the normal stress directed along the horizontal x-axis (taken here to be the direction of flow, along the axis of the glacier), B the normal stress in the direction of the other horizontal y-axis, and F the tangential or shear stress in the xy-plane, the greatest normal tensile stress, P_1 is given by

$$P_1 = \frac{1}{2}\left\{ A + B + \sqrt{(A-B)^2 + 4F^2} \right\},$$ (8.1)

with its direction determined by

$$\tan 2\alpha = \frac{2F}{A - B},$$ (8.2)

where α represents the angle between the direction of P_1 and the along-flow x-axis. As concluded in a second memoir (Hopkins, 1844b, art. 9), "The maximum action here spoken of is the maximum *tension* at the proposed point, and since it is perpendicular to the corresponding line of separation, *there will manifestly be the greatest tendency to form a fissure along that line, and a fissure will be formed along it if the maximum tension be greater than the cohesive power at the proposed point*" (ital. original). In other words, if the tensile stress, P_1, exceeds a critical value depending on the strength of ice, a crevasse (or fissure) will open perpendicular to the direction of P_1. In a later paper, Hopkins (1862) applied the above formulas to explain the orientation of crevasses as frequently observed on glaciers in the Alps.

Consider first a glacier descending a valley with approximately parallel sides so that the normal stress in the direction perpendicular to the flow direction may be neglected (B = 0). Toward the margins the speed approaches zero and the dominant stress acting on the ice is the shear stress, F, representing lateral drag due to resistance offered by the rock walls. Setting A = 0 in equation (8.1) gives the principal tensile stress, P_1 = F, while equation (8.2) yields an angle of 45° with the axis of the glacier as the direction of principal tensile stress. Thus, near the margin, crevasses will strike at 45°, as in region I in Figure 8.1. Near the centerline of the glacier, on the other hand, the effect of drag from the margins is negligible (F = 0), and the stress responsible for crevasse opening is the normal tensile stress, A, acting along the flow axis so that crevasses open perpendicular to this axis (region II in Figure 8.1). Where the valley walls diverge, longitudinal or splaying crevasses may form. From continuity arguments it follows that the speed along the x-axis must decrease as the ice spreads laterally, and A becomes negative (compressive) while the other normal stress, B, is positive (tension). Where this tensile stress becomes sufficiently large, fracturing in the direction of the glacier axis will occur. The diverging flow causes

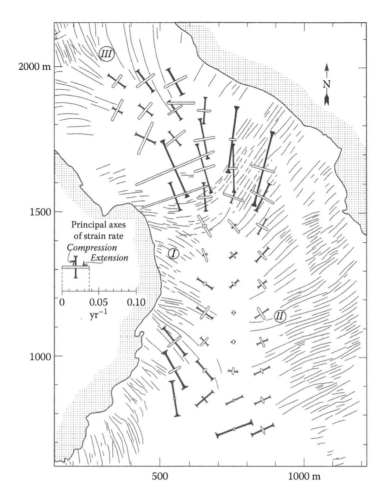

FIGURE 8.1 Principal surface strain rates and crevasse orientation on Blue Glacier, Olympic Mountains, Washington state. Ice flow is toward the top of the figure. (From Meier, M. F. et al., *J. Glaciol.*, 13, 187–212, 1974. Reprinted from the *Journal of Glaciology* with permission of the International Glaciological Society and the authors.)

these crevasses to splay radially, resulting in the pattern near the top in Figure 8.1 (region III).

The patterns predicted by Hopkins's model conform in many instances with those observed on glaciers. There are exceptions, however, in particular where crevasses meet glacier margins. For example, on Blue Glacier shown in Figure 8.1 near the margin opposite region I (below the North arrow), crevasses strike the margin at angles approaching 90°. Without considering in detail the stress distribution in relation to crevasse orientation, the causes for such deviations from predictions cannot be identified. Careful testing of the Hopkins model is possible for those glaciers on which surface strain rates (velocity gradients) were measured and crevasse patterns mapped. Several such studies are summarized in Van der Veen (1999a). Generally, crevasses appear to form in a direction close to perpendicular to the direction of principal tensile strain but there are exceptions (for example, Zumberge et al., 1960; Kehle, 1964; Hambrey and Müller, 1978; Whillans et al., 1993). Some authors disregard differences between prediction and observation, based either on measurement uncertainties, on the fact that observed crevasses may have formed upglacier and rotated by the ice flow, or on the suggestion that not all crevasses are visible. In other words, the Hopkins model for crevasse formation has not yet been unambiguously tested, but it continues to be accepted. Those studies in which quantitative estimates of crevasse rotation are made conclude that these estimates do not agree with observations. Further, some crevasses show signs of strike-slip faulting, in which the two walls move parallel to each other, with displacements that can be much larger than the opening rate (for example, Kehle, 1964; Hambrey, 1976). Finally, crevasse traces persist over much greater distances and well into the ablation area, instead of being melted away completely, as would be expected for crevasse depths of several tens of meters.

A difficulty inherent in comparing theoretical crevasse orientation with observations is that until recently, no attempts were made to measure stresses on glaciers directly (first results of direct measurements of stress at depth in Worthington Glacier, Alaska, are described in Pfeffer et al., 2000). Instead, crevasse orientations are compared to the directions of principal strain or strain rate, because these quantities are readily measured on the surface of a glacier. That is, crevasses are assumed to form in the direction of the principal tensile strain rate. However, this direction need not coincide with the direction of principal tensile stress. As pointed out by Ambach (1968), strain rates may not be the best indicator for crevasse studies, as the principal stresses are not simply scaled to the principal strain rates. If $\dot{\varepsilon}_1$ and $\dot{\varepsilon}_2$ represent the two principal components of strain rate, the principal components of the full stress are (Nye, 1959, p. 416)

$$\sigma_1 = 2\eta\dot{\varepsilon}_1 + \eta\dot{\varepsilon}_2, \tag{8.3}$$

$$\sigma_2 = \eta\dot{\varepsilon}_1 + 2\eta\dot{\varepsilon}_2, \tag{8.4}$$

with

$$\eta = B\dot{\varepsilon}_e^{1/n-1}, \tag{8.5}$$

the effective viscosity in the flow law for glacier ice linking stresses to strain rates. In this last expression, B represents the viscosity parameter, n = 3 the flow-law exponent, and $\dot{\varepsilon}_e$ the effective strain rate (the second invariant of the strain-rate tensor). These relations show that even if one of the principal strain rates is tensile, the corresponding principal stress does not need to be tensile also (namely, when the value of the compressive strain rate is twice that of the tensile strain rate, $\dot{\varepsilon}_2 = -2\dot{\varepsilon}_1$). The directions of the principal full stress and those of the principal strain rates are co-aligned so the orientation of crevasses is not affected by considering strain rates, rather the observational criterion for when crevassing first occurs.

Almost certainly, some criterion exists for crevasse initiation on glaciers, but there is disagreement in the literature as to what this criterion is. Meier (1958) and Vornberger and Whillans (1990) argue that crevasses form when the tensile principal *strain rate* reaches a threshold value. Other authors have proposed that crevasses are formed where the tensile *stress* reaches a critical value. Vaughan (1993) considered crevassing on 17 polar and alpine glaciers for which strain rates were available. Converting these strain rates to stresses using the constitutive relation for glacier ice, he derived failure envelopes suggesting that the tensile stress required for crevasses to form varies between 90 and 320 kPa. This range in required tensile stress is not caused by differences in ice temperatures but may reflect variations in ice properties that affect the strength of ice or, perhaps, by different crevasse spacings.

Early theories to estimate crevasse depth (Nye, 1955, 1957) made the assumption that crevasses penetrate to the depth at which the net longitudinal stress is zero, that is, where the longitudinal tensile stress equals the compressive ice overburden pressure. This model assumes zero cohesive strength of the ice and that crevasses can exist for all positive values of the tensile stress. The tensile stress tending to open a crevasse is the resistive stress, R_{xx}, defined in Section 3.1 as the full stress minus the weight-induced lithostatic stress, L. The depth, d, of the surface crevasse then follows from the requirement that

$$R_{xx}(d) = \rho g d, \tag{8.6}$$

where the right-hand side represents the lithostatic stress. Taking the stretching stress constant with depth, and assuming constant density, the crevasse depth can be estimated from

$$d = \frac{R_{xx}}{\rho g}. \tag{8.7}$$

For a typical stress of 150 kPa, this expression predicts a crevasse depth of about 17 m, using a constant density $\rho = 917$ kg/m^3 corresponding to solid ice. Generally, the density of the upper firn layer is considerably smaller than this value. Accounting for lower near-surface density is straightforward and, for a given stretching stress, would result in greater crevasse depth than that predicted by equation (8.7).

Nye's zero-stress model for estimating crevasse depth does not account for stress concentrations that may exist at the tip of the crevasse. This is a reasonable approximation for closely spaced crevasses because the tensile stress is nonexistent in the slabs separating neighboring crevasses. However, for single crevasses or those that are spaced sufficiently far apart so as to not affect stresses around neighboring crevasses, stress concentrations must be taken into account. Weertman (1973b) presents a model based on dislocation theory and finds that the depth of an isolated crevasse is greater than that predicted by equation (8.7) by a factor $\pi/2$. Again, a constant ice density is assumed in that model.

An alternative approach that can be applied to estimate the penetration depth of both single crevasses and of closely spaced crevasses is based on fracture mechanics in which the strength of ice is described through the fracture toughness, or resistance to crack propagation. The application of fracture mechanics to crevasses on glaciers is the focus of the next section.

8.2 FRACTURE MECHANICS

Crevasse propagation on glaciers may, in good approximation, be described using procedures of linear elastic fracture mechanics (LEFM) (Smith, 1976, 1978; Rist et al., 1996; Van der Veen, 1998a). This approach is based on the assumption that solid materials contain small flaws and cracks that affect their load-bearing capacity. These cracks can grow to larger fractures if stress concentrations near the crack tip become large enough to overcome the material strength or cohesion of the material (Broek, 1986). LEFM applies only to materials that are brittle and for which inelastic deformation near the crack tip can be ignored. Because glacier ice is a viscous material that deforms under stress, for slow loading conditions, ductile processes acting at crevasse tips should be accounted for when modeling fracture growth. A formal treatment of ductile processes can be included, using the J-integral approach (Hutchinson, 1979) or path-independent C* integrals (Riedel and Rice, 1980), but this would make the discussion much more complicated. In view of other uncertainties and approximations, a more complex theory of crevasse formation and propagation is not warranted for the present purposes. Because viscous properties of glacier ice make the strength of ice appear to be greater than if it were a true brittle material (Sanderson, 1988, p. 90), viscous effects can be accounted for by assigning a greater value to the fracture toughness. The LEFM approach is applicable provided the region near the crack tip where plastic deformation occurs is small compared with the depth of the crevasse (Gdoutos, 1993, p. 57). Following Broek (1986, p. 14), Van der Veen (1998a) estimates this plastic region to extend about 0.1–2.5 m from the crevasse tip, which is much less than the typical depth to which crevasses penetrate.

The fundamental assumption in the LEFM approach is that small starter cracks are present. Under suitable conditions, these cracks can grow into deeper fractures or crevasses. The initial crack size depends on the applied tensile stress. If the tensile stress is less than ~30 kPa, no crevasses can develop, while for stresses in the range of 40–60 kPa, an initial crack of 1–2 m depth is required. This initial depth decreases rapidly to a few cm for tensile stresses more common on glaciers and ice sheets (Van der Veen, 1999a). This requirement that a macroscopic initial crack must be present for crevasses to develop is a major weakness of the LEFM approach. On the scale of individual crystals, cracks could nucleate from dislocations or grain boundaries, and fracturing may result from crack nucleation or crack propagation; but on these microscopic scales, the concepts of stress intensity factor and fracture toughness may have limited applicability in the context as used here (Ingraffea, 1987). Modeling these microscopic processes is well beyond the scope of studies investigating crevasse propagation, and instead, the assumption is usually made that macroscopic cracks or flaws exist that can develop into full-fledged crevasses. In this respect, it should be noted that the fracture toughness of low-density snow and firn is orders of magnitude smaller than that of solid ice (Schulson and Duval, 2009). This opens the possibility that required starter cracks form from comparatively small

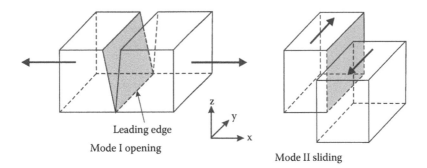

Leading edge

Mode I opening

Mode II sliding

FIGURE 8.2 Two principal modes of crack propagation. (Reprinted from Van der Veen, C. J., *Cold Regions Sci. Techn.*, 27, 31–47, 1998a. With permission from Elsevier.)

"flaws" in the firn (such as melt lenses or planes of structural weakness). Because of the smaller fracture toughness of firn, these flaws may propagate deeper than they could in solid ice and grow to deeper starter cracks that allow further downward propagation into more solid glacier ice.

Only one-dimensional Mode I fracturing is considered, with crevasses propagating downward perpendicular to a tensional stress (Figure 8.2). Other possibilities are the *sliding mode* (Mode II), resulting from shear in the plane of the fracture, and the *tearing mode* (Mode III), in which the applied shear stress is parallel to the leading edge of the crevasse. Two-dimensional crevasse propagation is discussed in Section 8.3.

To account for stress concentrations near the crevasse tip, the stress intensity factor is introduced. This factor can be estimated from the geometry of the fracture and the net longitudinal stress (for example, Sih, 1973; Broek, 1986; Gdoutos, 1993). For the Mode I geometry shown in Figure 8.2, the stress intensity factor is (Broek, 1986, pp. 10–11)

$$K_I = \beta \sigma_t \sqrt{\pi d}, \tag{8.8}$$

where β is a dimensionless factor that depends on the geometry (crevasse spacing, height-to-width ratio, etc.), d the depth of the crevasse, and σ_t the external or remote tensile (or compressive) stress. If the stress intensity factor is greater than a critical value, K_{Ic}, fracture growth will occur until the stress intensity factor becomes equal to or less than K_{Ic}. This critical value is called the fracture toughness and is independent of the crevasse geometry. A review of experiments conducted on ice to determine its fracture toughness is given by Schulson and Duval (2009), and these results suggest $K_{Ic} \approx 0.1$ to 0.4 MPa m$^{1/2}$. For low-density snow and firn, the critical stress intensity factor is considerably smaller, estimated to fall in the range of 0.05 to 2.0 kPa m$^{1/2}$ (Schulson and Duval, 2009).

In the LEFM approach, crevasse growth results from an externally applied tensile stress. This growth is impeded by the ice overburden pressure, which increases with depth. Thus, as the crevasse penetrates deeper, the compressive stress at the crevasse tip becomes greater, effectively lowering the stress intensity factor. When considering single-mode fracturing (as done here for Mode I), the stress intensity factor resulting from a combination of applied stresses can be obtained by addition

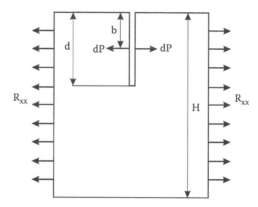

FIGURE 8.3 Geometry and notation used for calculating the stress intensity factor for a single crevasse. dP represents the lithostatic force at depth b below the surface acting on a segment db of the crevasse wall. (Reprinted from Van der Veen, C. J., *Cold Regions Sci. Techn.*, 27, 31–47, 1998a. With permission from Elsevier.)

of stress intensity factors corresponding to each stress (Broek, 1986, p. 84). Thus, for crevasse propagation on glaciers, crevasse opening associated with a tensile stress and crevasse closure due to the lithostatic stress may first be considered separately; addition then gives the combined effect, allowing crevasse depth to be estimated. For a single crevasse, the geometry is shown in Figure 8.3.

Consider first opening of a single crevasse resulting from a tensile stress, R_{xx}, assumed to be constant throughout the ice thickness (depth variation can be included but makes the solution more complex). The corresponding stress intensity factor is (Broek, 1986, p. 85)

$$K_I^{(1)} = F(\lambda) R_{xx} \sqrt{\pi d}, \tag{8.9}$$

where

$$F(\lambda) = 1.12 - 0.23\lambda + 10.55\lambda^2 - 21.72\lambda^3 + 30.39\lambda^4, \tag{8.10}$$

with $\lambda = d/H$ the ratio of crevasse depth to ice thickness. For shallow crevasses ($\lambda \to 0$), this expression reduces to that used by Smith (1976, 1978),

$$K_I^{(1)} = 1.12 R_{xx} \sqrt{\pi d}. \tag{8.11}$$

Figure 8.4 shows the stress intensity factor as given by equation (8.9) as a function of crevasse depth, for different values of the ice thickness. The approximate solution (8.11), which assumes infinite ice thickness, is indistinguishable from the H = 2000 m solution. For crevasses penetrating more than about 10% of the ice thickness, accounting for the finite thickness becomes important, and the stress intensity factor becomes greater than the approximate solution.

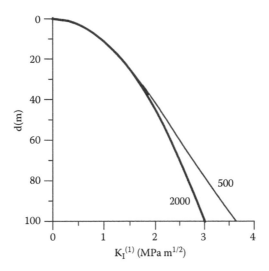

FIGURE 8.4 Stress intensity factor at depth d below the surface for a tensile stress of 100 kPa, and ice thickness of 500 m and 2000 m.

The second process to consider is crevasse closure resulting from the compressive lithostatic stress. Because this stress cannot be considered constant, calculating the corresponding stress intensity factor is somewhat more involved than for the case of constant tensile stress. Following Rist and others (1996) and Van der Veen (1998a), the near-surface density profile is described by the following polynomial

$$\rho_f(h - z) = \rho - (\rho - \rho_s)e^{-C(h-z)}, \tag{8.12}$$

where $\rho = 917$ kg/m^3 represents the density of ice, $\rho_s = 350$ kg/m^3 the density of surface snow, and $C = 0.02$ m^{-1}. The lithostatic stress at depth $(h - z)$ below the surface is then

$$L(h-z) = -\int_{h-z}^{z} \rho_f(\bar{z})g\,d\bar{z} =$$

$$= -\rho g(h - z) + \frac{\rho - \rho_s}{C}g\left(1 - e^{-C(h-z)}\right). \tag{8.13}$$

To find the stress intensity factor associated with this lithostatic stress, the procedure of Hartranft and Sih (1973) is followed. Let $b = h - z$ denote the depth below the surface. The lithostatic force acting over a small depth interval db is

$$dP(b) = L(b)\,db, \tag{8.14}$$

and the corresponding stress intensity factor is (Tada et al., 1973, p. 2.27)

$$dK_I^{(2)} = \frac{2L(b)\,db}{\sqrt{\pi d}}\,G(\gamma,\lambda).$$ (8.15)

The function $G(\gamma, \lambda)$ is given by

$$G(\gamma,\lambda) = \frac{3.52(1-\gamma)}{(1-\lambda)^{3/2}} - \frac{4.35 - 5.28\gamma}{(1-\lambda)^{1/2}} +$$

$$+ \left[\frac{1.30 - 0.30\gamma^{3/2}}{(1-\gamma^2)^{1/2}} + 0.83 - 1.76\gamma\right](1 - (1-\gamma)\lambda),$$ (8.16)

where $\gamma = b/d$ and $\lambda = d/H$ as before.

The net stress intensity factor at the crevasse tip is found by integrating the contributions acting at different depths (equation (8.15))

$$K_I^{(2)} = \int_0^d dK_I^{(2)}.$$ (8.17)

Substituting equation (8.13) gives

$$K_I^{(2)} = \frac{2\rho g}{\sqrt{\pi d}} \int_0^d \left[-b + \frac{\rho - \rho_s}{\rho C}\left(1 - e^{-Cb}\right)\right] G(\gamma,\lambda)\,db.$$ (8.18)

While an analytical expression for the integral is not readily obtained, the integral can be evaluated using standard numerical techniques. Note that the value of $K_I^{(2)}$ is negative to indicate that this stress intensity factor is associated with the lithostatic stress tending to close a crevasse. Figure 8.5 shows the stress intensity factor calculated from equation (8.18) as a function of the crevasse depth, d, for a 2000 m thick glacier. Also shown is the constant density solution (light curve, obtained by setting $\rho_s = \rho$ in equation (8.12)), which significantly overestimates the stress intensity factor.

Combining the two stress intensity factors given by equations (8.11) and (8.18) allows the penetration depth of a crevasse to be estimated. Figure 8.6 shows the net stress intensity factor for two values of the tensile stress. The vertical dashed lines indicate the lower and upper values of the estimated range for fracture toughness of glacial ice (0.1 – 0.4 MPa m$^{1/2}$; Schulson and Duval, 2009). Consider the curve corresponding to R_{xx} = 100 kPa. Adopting the upper limit for the fracture toughness, in the upper 5 m the stress intensity factor is less than K_{Ic}, and crevasses shallower than 5.2 m cannot exist. However, if a crevasse has reached a depth of 5.2 m (for example, because it formed upstream under a greater tensile stress, or because it formed in low-density firn with a smaller fracture toughness), it has the potential

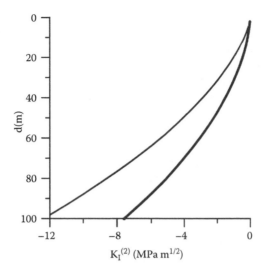

FIGURE 8.5 Stress intensity factor at depth d below the surface corresponding to the lithostatic stress for 2000 m thick glacier. The heavy curve corresponds to the calculation including low-density firn at the surface, and the thin curve corresponds to the constant density solution.

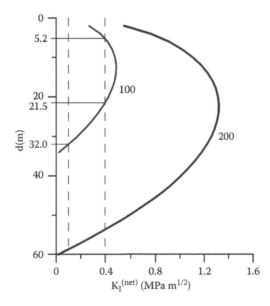

FIGURE 8.6 Net stress intensity factor at depth d below the surface for two values of the stretching rate (100 kPa and 200 kPa). The dashed vertical lines correspond to the estimated lower and upper limits of fracture toughness of glacial ice.

to grow deeper because the stress intensity factor initially increases with depth. At some depth below the surface, the stress intensity factor reaches a maximum before decreasing with depth due to the effect of the lithostatic stress. For this example, the stress intensity factor becomes less than 0.4 MPa m$^{1/2}$ at 21.5 m below the surface and further crevasse propagation is arrested. If the lower limit is adopted for the fracture toughness, the crevasse can penetrate to a depth of 32 m. Doubling the tensile stress to 200 kPa allows the crevasse to extend to a depth of 55–58 m, depending on the value for the fracture toughness.

Estimating crevasse depth using the LEFM approach is more involved than for the zero-stress model, and requires the stress intensity factor to be evaluated at depth for a range of model parameters to solve the expression

$$K_I^{(1)}(R_{xx},d) + K_I^{(2)}(d) = K_{Ic}, \tag{8.19}$$

for the crevasse depth, d. Figure 8.7 shows crevasse depth as a function of the tensile stress for two values of the fracture toughness. Depending on whether a low or high value is adopted for K_{Ic}, a tensile stress of 30–80 kPa is required for crevasses to be able to exist. Also shown in Figure 8.7 is the crevasse depth derived by Weertman (1973b),

$$d = \frac{\pi}{2\rho g}R_{xx}. \tag{8.20}$$

This solution agrees with the constant-density LEFM solution, except for small values of the tensile stress. This difference arises because the Weertman solution (8.20)

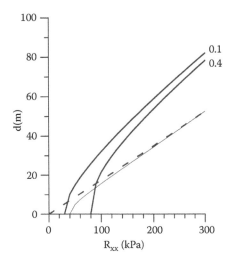

FIGURE 8.7 Penetration depth of a water-free crevasse on a 2500 m thick glacier. Labels indicate the value of the fracture toughness used in MPa m$^{1/2}$. The dashed line represents the Weertman (1973) solution and the thin solid line the constant density solution. (Reprinted from Van der Veen, C. J., *Cold Regions Sci. Techn.*, 27, 31–47, 1998a. With permission from Elsevier.)

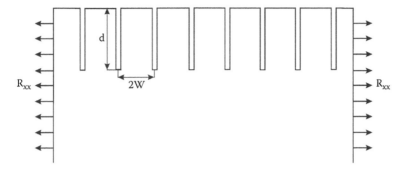

FIGURE 8.8 Geometry and notation used for equally spaced crevasses under tension. The depth of all crevasses equals d, while the distance between neighboring crevasses equals 2W. (Reprinted from Van der Veen, C. J., *Cold Regions Sci. Techn.*, 27, 31–47, 1998a. With permission from Elsevier.)

does not account for the strength of ice, and crevasses can exist for all values of the tensile stress. More importantly, assuming a constant density significantly underestimates the depth to which a crevasse can penetrate.

The discussion so far has focused on penetration depth of individual crevasses. Fields of closely spaced crevasses are probably more common on glaciers, and the "blunting" or "shadowing" effect will affect the stresses at the crevasse tips. In the thin slabs separating neighboring crevasses, no tensile stress exists, thus reducing stress concentrations (Weertman, 1973b). This only affects the stress intensity factor associated with the tensile stress; the lithostatic stress is the same for an individual crevasse and in a field of crevasses. For the geometry shown in Figure 8.8, the stress intensity factor is given by (Benthem and Koiter, 1973)

$$K_I^{(1)} = D(S) R_x \sqrt{\pi d S}, \qquad (8.21)$$

where

$$D(S) = \frac{1}{\sqrt{\pi}} \left[1 + \frac{1}{2}S + \frac{3}{8}S^2 + \frac{5}{16}S^3 + \frac{35}{128}S^4 + \frac{63}{256}S^5 + \frac{231}{1024}S^6 \right] +$$
$$+ 22.501 S^7 - 63.502 S^8 + 58.045 S^9 - 17.577 S^{10}, \qquad (8.22)$$

and

$$S = \frac{W}{W + d}. \qquad (8.23)$$

The assumption is made that the crevasses are evenly spaced at distance 2W and that all crevasses have the same depth, d. This solution for the stress intensity factor applies to fractures in a semi-infinite half plane (d ≪ H).

For crevasses that are widely spaced (W ≫ d), S ≈ 1 and D(S) ≈ 1.12, and the stress intensity factor equals the value estimated from equation (8.11) for an individual

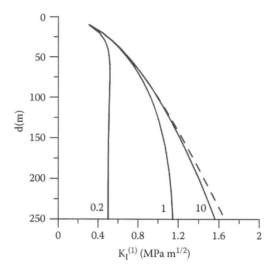

FIGURE 8.9 Stress intensity factor associated with a tensile stress at depth d below the surface, for different values of crevasse spacing. Labels indicate crevasse spacing in km. The dashed line corresponds to the solution for a single crevasse.

crevasse. As crevasses become more closely spaced, the stress intensity factor decreases (Figure 8.9), and a larger tensile stress is needed for crevasses in a field to occur, in excess of 300 kPa for very closely spaced crevasses (Van der Veen, 1998a).

For tensile stresses typically found on glaciers and ice sheets, surface crevasses penetrate a few tens of meters downward, and their presence may be expected to have little effect on glacier flow and dynamics. That situation can change, however, when the crevasses become filled with surface meltwater. Because the density of water is greater than that of firn and ice, the pressure of water in the crevasse compensates part or all of the lithostatic stress, allowing the crevasse to penetrate deeper. This effect can be included by adding a stress intensity factor corresponding to the water pressure. The procedure is similar to that followed to arrive at equation (8.18) for the stress intensity factor associated with the lithostatic stress. The depth of the crevasse is denoted by d, and the water level in the crevasse is at a distance below the surface. At depth $b > a$ the water pressure is

$$P_w(b) = \rho_w \, g(b-a), \tag{8.24}$$

where ρ_w represents the density of fresh water. Analogous to equation (8.18) the corresponding stress intensity factor is

$$K_I^{(3)} = \frac{2\rho_w \, g}{\sqrt{\pi d}} \int_a^d (b-a) \, G(\gamma, \lambda) \, db, \tag{8.25}$$

with $G(\gamma, \lambda)$ given by equation (8.16).

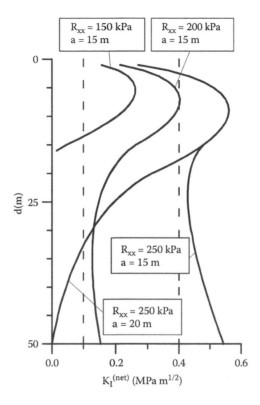

FIGURE 8.10 Net stress intensity factor at depth d below the surface of any crevasse in a field of crevasses spaced at 500 m, for different combinations of model parameters. The vertical dashed lines represent the estimated lower and upper fracture toughness of glacier ice. (Reprinted from Van der Veen, C. J., *Cold Regions Sci. Techn.*, 27, 31–47, 1998a. With permission from Elsevier.)

The important parameter that determines whether water-filled crevasses can penetrate the full ice thickness is the depth of the water level below the surface, a, and, secondly, the tensile stress. Some net stress intensity factors for any crevasse in a field of crevasses spaced at 500 m are shown in Figure 8.10 for different combinations of model parameters. If the water level is sufficiently close to the surface (a ≈ 10 m; Van der Veen, 1998a) and the tensile stress is greater than 100–150 kPa, a water-filled crevasse can reach the base of the glacier. This is the mechanism proposed by Van der Veen (2007) for how surface meltwater lakes can drain quickly and meltwater is transferred to the subglacial drainage system. In essence, hydrofracturing through the full thickness can occur provided the inflow of water into the crevasse is high enough to maintain a water level close to the surface. Generally, on cold glaciers, this will require abundant meltwater ponding on the glacier surface to initiate and sustain full thickness fracturing before refreezing at depth occurs. Das et al. (2008) report on rapid drainage (within less than two hours) of a supraglacial lake in West Greenland through 980 m of ice to the bed and argue that this drainage was initiated by water-driven fracturing that evolved into moulin drainage. Concurrent ground-based measurements showed that drainage coincided with increased seismicity, a temporary increase in ice velocity and surface uplift, and followed by slowing down and subsidence over the next 24 hours.

The model described above is essentially one-dimensional, and only the vertical direction is considered. This is equivalent to making the assumption that crevasses extend infinitely into the transverse direction. Based on reported crevasse depths, which seldom exceed ~30 m (Holdsworth, 1968), compared with widths visible on the glacier surface, which are typically a few hundred meters to several kilometers, this assumption appears reasonable. However, the direct consequence is that the model cannot predict the crevasse orientation as seen on the surface in relation to the applied stress. To do so requires a two-dimensional model in which crevasse propagation in the transverse direction as well as in the vertical is considered. As shown in the next section, stress intensity factors can be calculated for elliptical cracks intersecting the surface, and these solutions can be used to evaluate crevasse propagation in the plane perpendicular to the longitudinal stress.

8.3   TWO-DIMENSIONAL CREVASSE PROPAGATION

The following discussion of map view propagation of crevasses is from Van der Veen (1999a) and applies to quasistatic crevasse propagation.

For an elliptical fracture in a uniform stress field, the stress intensity factor was determined by Newman and Raju (1981) using results of three-dimensional finite-element analysis. Referring to Figure 8.11 for the notation, and making the assumption that the fracture depth is small compared with the total thickness, the stress intensity factor is given by

$$K_I = \sigma \sqrt{\pi d} \; [1.13 - 0.09\lambda] \; [1 + 1.464\lambda^{1.65}]^{-1/2}$$
$$[\lambda^2 \cos^2 \varphi + \sin^2 \varphi]^{1/4} [1 + 0.1(1 - \sin \varphi)^2], \tag{8.26}$$

with $\lambda = d/w$ the ratio of fracture depth and half-length.

At $\varphi = \pi/2$ (the deepest point of the fracture front), the stress intensity factor is

$$K_I(A) = \sigma \sqrt{\pi d} [1.13 - 0.09\lambda][1 + 1.464\lambda^{1.65}]^{-1/2}, \tag{8.27}$$

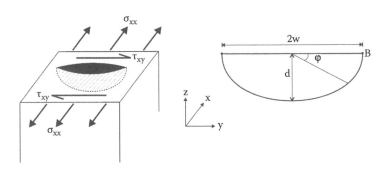

FIGURE 8.11   Geometry of an elliptical surface crack. (Modified from Van der Veen, C. J., *Polar Geog.,* 23, 213–245, 1999a.)

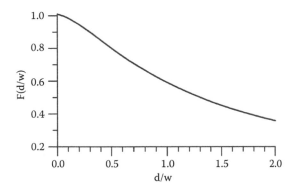

FIGURE 8.12 Correction factor F(d/w) as a function of the depth-to-width ratio, d/w. (Modified from Van der Veen, C. J., *Polar Geog.*, 23, 213–245, 1999a.)

which can also be written as

$$K_I(A) = F(\lambda)K_I^{edge}, \tag{8.28}$$

where

$$K_I^{edge} = 1.12\sigma\sqrt{\pi d}, \tag{8.29}$$

represents the stress intensity factor for an edge fracture of depth d subject to a constant tensile stress, σ (Broek, 1986, p. 10). The function $F(\lambda)$ is defined as

$$F(\lambda) = \frac{[1.13 - 0.09\lambda]}{1.12}[1 + 1.464\lambda^{1.65}]^{-1/2}, \tag{8.30}$$

and constitutes a correction factor to account for the two-dimensional structure of the elliptical fracture compared with the one-dimensional edge fracture. In Figure 8.12 this correction factor is shown as a function of the ration $\lambda = d/w$. Crevasses on glaciers are usually much longer than they are deep and $\lambda \to 0$ and $F(\lambda) \approx 1$. In other words, the depth of a crevasse on a glacier can, in good approximation, be estimated using the one-dimensional approach described in the previous section.

After the depth of the fracture is known, the length, 2w, can be estimated by considering the stress intensity factor at the crevasse tips at the surface

$$K_I(B) = \sigma\sqrt{\pi d}\,[1.13 - 0.09\lambda][1 + 1.464\lambda^{1.65}]^{-1/2}\,1.1\sqrt{\lambda}, \tag{8.31}$$

or

$$K_I(B) = 1.1\lambda F(\lambda)K_I^{plate}, \tag{8.32}$$

with

$$K_I^{plate} = 1.12\sigma\sqrt{\pi w},\qquad(8.33)$$

the stress intensity factor of a through-the-thickness crack with length 2w, in a plate subject to a tensile stress, σ (Broek, 1986, p. 10). Thus, the stress intensity factor at the surface equals that of a fracture penetrating the entire ice thickness multiplied by a correction factor to account for the partial penetration of the fracture. The length of the elliptical fracture can be estimated from the condition that the stress intensity factor at the tips (equation (8.32)) equals the fracture toughness.

Of interest here is the orientation of a crevasse as observed on the glacier surface. In a uniaxial stress field, the orientation of a fracture is perpendicular to the principal tensile stress, and fracture growth is in the plane of the original fracture such that, at the surface, the fracture remains straight. On glaciers, however, crevasses may be subject to combined loading of longitudinal tension and lateral shear. This fracturing can be described as a combination of Mode I (opening mode) and Mode II (sliding mode), and fracture propagation usually takes place at an angle with respect to the original fracture.

There are two approaches to estimating the propagation direction. According to the maximum principal stress criterion, fracture growth occurs in the direction perpendicular to the maximum principal stress at the fracture tip (Erdogan and Sih, 1963; Broek, 1986, p. 375). The strain energy density criterion postulates that a fracture propagates in the direction of minimum strain energy density (Sih, 1974; Broek, 1986, p. 377). Differences between orientations calculated from these methods are small (Broek, 1986, p. 380). Here, the maximum principal stress criterion is used because of its more clear physical meaning.

The stress intensity factor describes the stress distribution at the tip of a fracture. Because the problem is linear, stresses associated with Mode I and Mode II fracturing can be evaluated separately and added to yield the net stress. Consider first Mode I, in which the fracture opens in the direction of applied tensile stress. Denoting the corresponding stress intensity factor by K_I and using polar coordinates (r, θ) (Figure 8.13), the stresses at the tip are (Broek, 1986, p. 96)

$$\sigma_\theta^{(I)} = \frac{K_I}{\sqrt{2\pi r}}\cos\frac{\theta}{2}\left[1 - \sin^2\frac{\theta}{2}\right],\qquad(8.34)$$

$$\tau_{r\theta}^{(I)} = \frac{K_I}{\sqrt{2\pi r}}\sin\frac{\theta}{2}\cos^2\frac{\theta}{2}.\qquad(8.35)$$

Similarly, the stresses associated with Mode II (resulting from shear in the plane of the fracture) are (Broek, 1986, p. 96)

$$\sigma_\theta^{(II)} = \frac{K_{II}}{\sqrt{2\pi r}}\left[-\frac{3}{4}\sin\frac{\theta}{2} - \frac{3}{4}\sin\frac{3\theta}{2}\right],\qquad(8.36)$$

$$\tau_{r\theta}^{(II)} = \frac{K_{II}}{\sqrt{2\pi r}}\left[\frac{1}{4}\cos\frac{\theta}{2} + \frac{3}{4}\cos\frac{3\theta}{2}\right].\qquad(8.37)$$

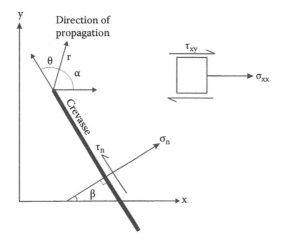

FIGURE 8.13 Definition of angles used in deriving the direction of crevasse propagation. (Modified from Van der Veen, C. J., *Polar Geog.*, 23, 213–245, 1999a.)

Adding the two contributions gives the net stress at the tip of the fracture (Broek, 1986, p. 375)

$$\sigma_\theta = \frac{1}{\sqrt{2\pi r}}\cos\frac{\theta}{2}\left[K_I \cos^2\frac{\theta}{2} - \frac{3}{2}K_{II}\sin\theta\right], \tag{8.38}$$

$$\tau_{r\theta} = \frac{1}{2\sqrt{2\pi r}}\cos\frac{\theta}{2}[K_I \sin\theta + K_{II}(3\cos\theta - 1)]. \tag{8.39}$$

It is now postulated that crevasse extension is in the direction for which σ_θ is a principal stress or, equivalently, the direction for which $\tau_{r\theta} = 0$. This direction, represented by the angle θ_m, is found by setting equation (8.39) to zero

$$K_I \sin\theta_m + K_{II}(3\cos\theta_m - 1) = 0, \tag{8.40}$$

from which it follows that

$$\tan\frac{\theta_m}{2} = \frac{1}{4}\frac{K_I}{K_{II}} \pm \frac{1}{4}\sqrt{\left(\frac{K_I}{K_{II}}\right)^2 + 8}. \tag{8.41}$$

The direction of crevasse propagation, θ_m, is defined with respect to the orientation of the crevasse, as illustrated in Figure 8.13. How the crevasse is oriented with respect to the external stresses, that is, with respect to a coordinate system fixed to the direction of glacier flow, determines the normal stress and shear stress acting on the crevasse, and hence the two stress intensity factors in equation (8.41).

Let β denote the angle between the crevasse orientation and the x-axis, the latter taken to be in the mean direction of ice flow. The tensile stress in the along-flow direction is σ_{xx} and the shear stress is τ_{xy}. The stress acting perpendicular to the crevasse walls is then (Jaeger and Cook, 1976, p. 13)

$$\sigma_n = \sigma_{xx} \cos^2 \beta + 2\tau_{xy} \sin \beta \cos \beta, \tag{8.42}$$

while the shear stress parallel to the crevasse walls is given by

$$\tau_n = -\frac{1}{2}\sigma_{xx} \sin 2\beta + \tau_{xy} \cos 2\beta. \tag{8.43}$$

From these expressions, the two stress intensity factors at the tips of a crevasse of length 2w can be estimated as

$$K_I = \sigma_n \sqrt{\pi w}, \tag{8.44}$$

$$K_{II} = \tau_n \sqrt{\pi w}. \tag{8.45}$$

The growth direction relative to the x-axis equals

$$\alpha = \frac{\pi}{2} + \beta - \theta_m, \tag{8.46}$$

and can be calculated as a function of the crevasse orientation using the formulas given above. However, as the crevasse propagates, the fracture orientation, and thus the angle β, changes. To account for this, the above expressions must be applied repeatedly to calculate the angle of propagation as the fracture grows in map view. In the case of only a far-field tensile stress, σ_{xx}, the orientation of the starter crack is unimportant. If the starter crack is misaligned from the tensile stress, the crevasse orients itself and the direction of growth is within a few degrees of the expected direction perpendicular to the tensile stress (Van der Veen, 1999a, Figure 10). For more complex stress configurations, crevasse propagation is not necessarily in the direction perpendicular to the principal tensile stress.

Consider the case of a tensile stress and a shear stress acting jointly on a crevasse. Such a stress configuration characterizes most valley glaciers and ice streams that are subject to lateral drag arising from friction between the moving ice and stagnant rock walls or more slowly moving interstream ridge ice. Making the assumption that crevasses form perpendicular to the principal tensile stress, the optimum angle, β_m, follows from the condition that the shear stress parallel to the crevasse walls, τ_n, is zero, and

$$\tan 2\beta_m = \frac{2\tau_{xy}}{\sigma_{xx}}, \tag{8.47}$$

which is the expression derived earlier by William Hopkins (equation (8.2)). Figure 8.14 shows the crevasse orientation as a function of the stress ratio τ_{xy}/σ_{xx} as predicted by the Hopkins model. For tension only, the crevasse is perpendicular to the direction of tensile stress, while in simple shear, the crevasse strikes at an angle

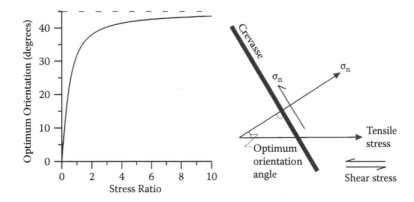

FIGURE 8.14 Theoretically predicted orientation (defined with respect to the tensile axis) of a crevasse subject to tension and lateral shear, as a function of the ratio between shear stress and tensile stress. (Modified from Van der Veen, C. J., *Polar Geog.*, 23, 213–245, 1999a.)

of 45° to the x-axis. For the LEFM model, the direction of crevasse propagation can be evaluated as a function of the crevasse orientation. Results for different stress ratios are shown in Figure 8.15; the horizontal dashed lines correspond to the optimum direction calculated from equation (8.47).

In the LEFM approach, the reason that crevasse propagation is not necessarily in the direction perpendicular to the principal tensile stress is that combined Mode I and

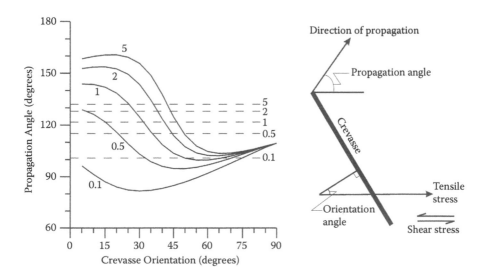

FIGURE 8.15 Angle of propagation (defined with respect to the tensile axis) as a function of crevasse orientation, for different values of the ratio of shear stress and tensile stress (indicated by the labels). The horizontal dashed lines correspond to the direction of principal tensile stress. (Modified from Van der Veen, C. J., *Polar Geog.*, 23, 213–245, 1999a.)

Mode II fracturing is considered here. Mode I fracturing corresponds to the case of a fracture opening perpendicular to the (principal) tensile stress, while Mode II fracturing corresponds to strike-slip faulting, with the two faces of the fracture moving parallel to each other. If a crevasse were always to be perpendicular to the direction of principal tensile stress, fracturing would be described entirely by Mode I, as there would be no shear stress parallel to the crevasse walls. By allowing for strike-slip faulting (Mode II fracturing), this shear stress need not be zero, and consequently, the crevasse need not be oriented in the direction predicted by equation (8.47). This does not contradict the assumption that crevasse propagation is in the direction perpendicular to the principal tensile stress at the crevasse tip (equation (8.40)). Each fracturing mode leads to stress concentrations near the crevasse tip that are not simply scaled to the macroscopic stress field but, instead, are directionally dependent (equations (8.34)–(8.37)). Inspection of these equations shows that the direction of principal tensile stress at the crevasse tip is not necessarily the same as that of the macroscopic principal tensile stress.

In the discussion so far, no consideration has been given to the probability that a starter crack oriented at an angle β to the tensile axis develops into a mature crevasse. This point is investigated by considering the stress intensity factor at the crevasse tips.

For pure Mode I at fracture, the principal tensile stress at the tip of the fracture is

$$\sigma_1 = \frac{K_{Ic}}{\sqrt{2\pi r}}, \tag{8.48}$$

where K_{Ic} represents the fracture toughness of the material. For mixed-mode fracturing, it is postulated that crack extension takes place if the principal tensile stress at the crevasse tip has the same value as at fracture in an equivalent Mode I case (Broek, 1986, p. 376). That is, σ_θ evaluated at $\theta = \theta_m$ using equation (8.38) is set equal to σ_1 from equation (8.48) to give

$$K_{Ic} = \cos\frac{\theta_m}{2}\left[K_I \cos^2\frac{\theta_m}{2} - \frac{3}{2}K_{II}\sin\frac{\theta_m}{2}\right], \tag{8.49}$$

or

$$K_{Ic} = K_I \cos^3\frac{\theta_m}{2} - 3K_{II}\cos^2\frac{\theta_m}{2}\sin\frac{\theta_m}{2}. \tag{8.50}$$

The direction of propagation, θ_m, as well as the two stress intensity factors, K_I and K_{II}, depend on the orientation of the crevasse through the angle β.

Figure 8.16 shows how the equivalent stress intensity factor varies as a function of crevasse orientation and ratio of shear stress to tensile stress. The equivalent stress intensity factor is normalized with the stress intensity factor for optimal crevasse orientation, that is, the direction for which the shear stress parallel to the crevasse walls vanishes (equation (8.47); these orientations are indicated by the vertical dashed lines). The curves in Figure 8.16 clearly show that the maximum equivalent stress intensity factor occurs for crevasse orientations that are different from the optimal

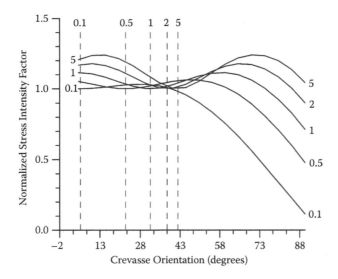

FIGURE 8.16 Normalized equivalent stress intensity factor at the surface tips of an elliptical crevasse subject to tension and shear, as a function of crevasse orientation. Labels indicate the value of the ratio between lateral and tensile stress. Vertical dashed lines correspond to the optimal crevasse orientation with zero shear stress acting along the crevasse walls. (Modified from Van der Veen, C. J., *Polar Geog.*, 23, 213–245, 1999a.)

orientation perpendicular to the macroscopic principal tensile stress. The reason for this difference is that the Mode II stress intensity factor also contributes to the effective stress intensity factor, allowing this quantity to exceed the value calculated for Mode I fracturing only and used for normalization (equation (8.49)). This finding has important consequences for relating crevasse patterns to surface strain or stress. Fracture growth occurs when the (equivalent) stress intensity factor exceeds a threshold value, the fracture toughness of the material subject to fracturing. Now consider the curve labeled 2 in Figure 8.16, which shows two maxima where the effective stress intensity factor is about 1.2 times that corresponding to Mode I fracturing in the direction perpendicular to the principal tensile stress. Thus, initial cracks oriented at the corresponding angles are most likely to grow further, albeit in directions that are closer to the optimum orientation than the orientation of these starter cracks.

The model for crevasse initiation and propagation along the glacier surface discussed in this section (from Van der Veen, 1999a) assumes the glacier to be homogeneous and subject to a uniform stress field. Under those conditions, fractures will continue to propagate laterally. This is not very realistic; crevasses observed on glaciers have finite lengths, and propagation is often arrested at structural boundaries such as suture zones between ice originating from adjacent tributaries. This is clearly visible in the feature map of the Ronne Ice Shelf shown in Figure 8.17. This map was digitized by Hulbe and LeDoux (2011) from the MODIS (Moderate Resolution Imaging Spectroradiometer) Mosaic of Antarctica, a composite of images taken between November 20, 2003, and February 29, 2004, with a resolution of 125 m

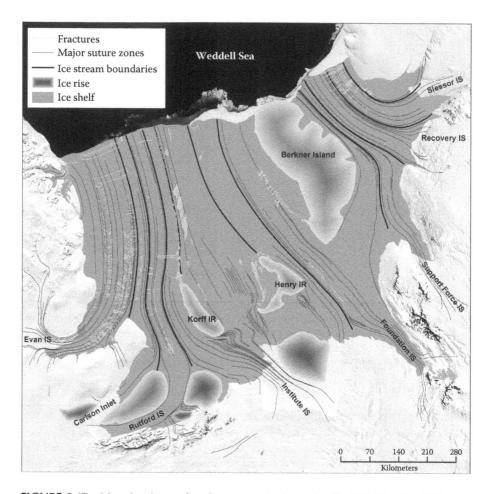

FIGURE 8.17 Map showing surface features on the Ronne Ice Shelf digitized by Hulbe and LeDoux (2011) from the MODIS Mosaic of Antarctica shown as the background image. Major ice streams entering the ice shelf are shown, and their lateral margins can be traced across the ice shelf (dark lines). Light gray lines indicate sutures that appear to arrest lateral growth of crevasses. (Redrawn by Kyle Purdon using data files from Hulbe and LeDoux, 2011.)

(Scambos et al., 2007). A complete discussion of the features is given by Hulbe et al. (2010), but in short, it shows how fractures along the lateral boundaries of outlet streams feeding in the ice shelf are advected downstream and form long rifts. As the crevasses advect through the ice shelf, they propagate laterally with orientations dictated by the stress field. Over most of the ice shelf, the length of these fractures is constrained with locations of crack tip arrest coinciding with suture zones and other structural boundaries. Hulbe et al. (2010) use the displacement discontinuity boundary element method to model horizontal propagation of crevasses subject to Mode I and Mode II fracturing and show that, in the absence of suture zones, fractures would continue to propagate laterally. Thus, observed crevasse patterns indicate that growth is arrested by the presence of these suture zones.

A more heuristic approach to modeling fracture fields is adopted by Albrecht and Levermann (2012). Defining the two-dimensional fracture density field, ϕ, as the mean area density of fractured ice with $\phi = 0$ corresponding to crevasse-free regions, the assumption is made that the field is horizontally advected with the ice velocity, U. For one-dimensional flow along a central flowline, this gives

$$\frac{\partial \phi}{\partial t} + U \frac{\partial \phi}{\partial x} = f_s + f_h, \tag{8.51}$$

where f_s represents a local production term and f_h the rate at which crevasses are closing. Albrecht and Levermann (2012) make the assumption that the production term depends on the stretching rate and fracture density as

$$f_s = \gamma \dot{\varepsilon}_{xx} (1 - \phi), \tag{8.52}$$

with γ a dimensionless parameter. The term in parenthesis reduces the local production proportional to crevasse density and reflects interactions among existing crevasses that tend to reduce the local stress.

Ignoring the healing term, f_h, the steady-state crevasse density is found by substituting equation (8.52) in the right-hand side of the advection equation (8.51) and setting the time derivative to zero. After some rearranging this gives

$$\frac{1}{1 - \phi} \frac{\partial \phi}{\partial x} = -\frac{\gamma}{U} \frac{\partial U}{\partial x}, \tag{8.53}$$

which can be integrated to give

$$\phi = 1 - (CU)^{-\gamma}. \tag{8.54}$$

The integration constant, C, can be found from the boundary condition that at the start of the flowline (x = 0) the velocity equals U_o and the fracture density equals ϕ_o. This yields the following solution for ϕ:

$$\phi(x) = 1 - (1 - \phi_o) \left(\frac{U(x)}{U_o} \right)^{-\gamma}. \tag{8.55}$$

Albrecht and Levermann (2012) first apply their model to an idealized ice-shelf geometry. For an ice shelf that is sufficiently wide that lateral drag and lateral spreading may be ignored, the steady-state profiles derived in Section 5.5 can be used to find an analytical expression for $\phi(x)$. To keep the algebra simple, consider the profile corresponding to zero net mass balance (equation (5.71)). The corresponding velocity profile is

$$U(x) = H_o U_o \left[\frac{(n + 1)C}{H_o U_o} x + H_o^{-(n+1)} \right]^{-1/(n+1)}, \tag{8.56}$$

with

$$C = \left(\frac{\rho g (\rho_w - \rho)}{4 B \rho_w} \right)^n, \tag{8.57}$$

where ρ and ρ_w represent the density of ice and sea water, respectively, and B the viscosity parameter. Taking n = 3, the fracture density is then given by

$$\phi(x) = 1 - (1 - \phi_o) \left(\frac{4 C H_o^3}{U_o} x + 1 \right)^{-\gamma/4}. \tag{8.58}$$

This steady-state solution is shown in Figure 8.18 calculated using the following parameter values: $\gamma = 0.5$, $C = 7.73 \times 10^{-11}$ m^{-3} yr^{-1}, $\phi_o = 0.1$, $H_o = 600$ m, and $U_o = 300$ m/yr (Albrecht and Levermann, 2012). The fracture density increases rapidly just beyond the grounding line, where the stretching rate is largest and ϕ relatively small. As $\dot{\epsilon}_{xx}$ decreases and ϕ increases, the local production rate decreases and the fracture density becomes almost constant.

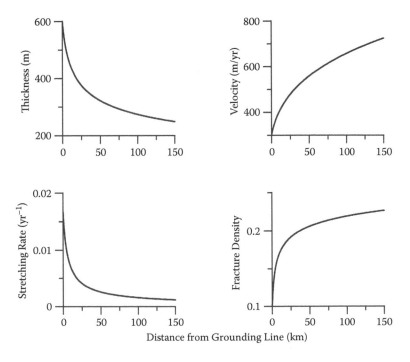

FIGURE 8.18 Modeled fracture density on a free-floating ice shelf. The upper left panel shows ice thickness (equation (5.71)) and the upper right panel the velocity (equation (8.56)). The lower panels show the stretching rate on the left and fracture density (equation (8.58)) on the right.

For more realistic ice-shelf geometries, Albrecht and Levermann (2012) combine the fracture propagation model with a map-view numerical ice-shelf model solving for the stress balance using the shallow-shelf approximation. With appropriate choice of model parameters, simulated fracture density agrees qualitatively with observed fracturing on the Ronne–Filchner Ice Shelf (shown in Figure 8.17). As noted by the authors, the heuristic fracture model may provide an avenue to incorporate a more realistic calving boundary condition into map view models of ice shelves and floating ice tongues.

8.4 BASAL CREVASSES

Basal or bottom crevasses extend from the base upward into the ice. Their existence was suggested by Weertman (1973b) based on the analogy with a water-filled surface crevasse. Because bottom crevasses on floating glaciers are filled with sea water, the water pressure in the crevasse partially counters the weight-induced lithostatic stress, tending to close the crevasse. Using radar sounding, bottom crevasses have been detected on several ice shelves and floating ice tongues including the Ross Ice Shelf (Jezek et al., 1979; Anandakrishnan et al., 2007) and the Larsen Ice Shelf (Swithinbank, 1977; McGrath et al., 2012), as well as on grounded glaciers (for example, Harper et al., 2010).

For a free-floating ice shelf spreading in the x-direction only, the resistive stress, R_{xx}, is proportional to the ice thickness (equation (4.58)). For that case, Weertman (1973b, 1980) derived the following expression for the height of a basal crevasse

$$h = \frac{\pi H}{4}, \tag{8.59}$$

where H is the ice thickness. This crevasse height is greater than the height at which the net longitudinal stress is zero by a factor $\pi/2$ as a result of stress concentrations at the crevasse tip. Where multiple closely spaced basal crevasses occur, these stress concentrations may be neglected (as in the case of fields of surface crevasses; Weertman, 1973b), and the crevasse height equals half the ice thickness. Jezek (1984) investigated height of basal crevasses on the Ross Ice Shelf and found that measured crevasse heights are considerably less than predicted by equation (8.59), indicating that the spreading stress is reduced compared with that for a free-floating ice shelf as a result of a compressive back stress.

A model for basal crevasses based on the LEFM approach is presented by Van der Veen (1998b), who investigated the conditions under which basal crevasses can occur. The model is very similar to that described in Section 8.2 for surface crevasses and is summarized in this section. The geometry is shown in Figure 8.19. A glacier of thickness H is considered, with z = 0 at sea level, the ice surface at z = h, and the base of the glacier at z = b (negative if below sea level). The crevasse height above the ice base is d_b. As in the case of surface crevasses, three stresses contribute to the net longitudinal stress at depth, namely, the lithostatic stress (tending to close the crevasse), the water pressure in the bottom crevasse, and the tensile resistive stress, R_{xx}, both of which tend to open the crevasse.

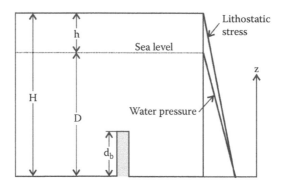

FIGURE 8.19 Geometry used to calculate the penetration height of basal crevasses. (Reprinted from Van der Veen, C. J., *Cold Regions Sci. Techn.*, 27, 213–223, 1998b. With permission from Elsevier.)

Consider first the lithostatic stress. As in Section 8.2, allowance is made for the presence of low-density firn at the surface by adopting the density profile (8.12), and the lithostatic stress at depth is given by equation (8.13).

To estimate the opening stress associated with water filling the basal crevasse, the water pressure at the base of the glacier must be known. As shown in Section 7.4, this pressure must fall in the range

$$-\rho_w \, g \, b \;\le\; P_w(b) \;\le\; \bar{\rho} \, g \, H. \tag{8.60}$$

In this expression, $\bar{\rho}$ represents the depth-averaged density. Note that for a floating ice shelf, the minimum and maximum water pressures are equal. Following Van der Veen (1998b), the water pressure at the bed is expressed in terms of the piezometric height, H_p, which equals the height above the base to which water in a borehole to the bed will rise. That is,

$$P_w(b) \;=\; \rho_w \, g \, H_p. \tag{8.61}$$

Introducing the height above buoyancy (or thickness in excess of flotation)

$$H_{ab} \;=\; H + \frac{\rho_w}{\bar{\rho}} \, b, \tag{8.62}$$

the inequality (8.60) can be rewritten as

$$0 \;\le\; H_p + b \;\le\; \frac{\bar{\rho}}{\rho_w} \, H_{ab}. \tag{8.63}$$

The water pressure in the crevasse is then given by

$$P_w(z) \;=\; \rho_w \, g \, (b + H_p - z), \tag{8.64}$$

at depths below the piezometric surface, and $P_w(z) = 0$ above that surface.

The third stress to be considered is the tensile stress associated with glacier flow. In first approximation, this stress may be linked to the stretching rate as (compare with equation (4.62))

$$R_{xx}(z) = 2B(T)\dot{\varepsilon}_{xx}^{1/3},\qquad (8.65)$$

where $B(T)$ is the temperature-dependent viscosity parameter (equation (2.15)) and $n = 3$ in the flow law. The assumption is made that the stretching rate, $\dot{\varepsilon}_{xx}$, is constant with depth, and depth variation in the stretching stress is solely due to temperature variations. Van der Veen (1998b) adopts a simple linear temperature profile,

$$T(z) = \frac{h-z}{H}T_b + T_s,\qquad (8.66)$$

where T_s and T_b represent the ice temperature at the surface and bed, respectively. Note that basal crevassing requires the presence of pressurized water at the glacier base so that the basal temperature equals the pressure melting temperature. In their application to the Larsen C Ice Shelf, McGrath and others (2012) adopt a parabolic temperature profile to estimate the tensile stress from measured surface strain rates.

In Section 8.2 stress intensity factors corresponding to each of the three stresses controlling crevasse propagation were estimated separately. This was done to discuss and illustrate how each stress contributes to the combined stress intensity factor. However, the combined stress intensity factor, $K_I^{(b)}$, can be found directly by integrating the net stress acting on the crevasse walls over the height of the crevasse, similar as the procedure leading to equation (8.18). Adding the three stress contributions, the net longitudinal stress at depth is given by

$$\sigma_n(z) = -\rho g(h-z) + \frac{\rho - \rho_s}{C}g\left[1 - e^{-C(h-z)}\right] + \rho_w g(H_p + b - z) + R_{xx}(z).$$

$$(8.67)$$

For a single basal crevasse, the net stress intensity factor is found by (numerical) integration

$$K_I = \int\limits_{b}^{b+d_b} \frac{2\sigma_n(z)}{\sqrt{\pi d_b}}\, G(\gamma,\lambda)\, dz,\qquad (8.68)$$

with the function $G(\gamma, \lambda)$ given by expression (8.16) (Hartranft and Sih, 1973).

Consider first a single bottom crevasse on a floating ice shelf on which the height above flotation is zero and the piezometric height corresponds to sea level. In that case, there are three model parameters that determine the penetration height, namely, ice thickness, stretching rate, and surface temperature. Of these, the ice thickness has the smallest effect, with crevasse height decreasing by a few meters only for an increase in thickness from 500 to 2000 m (Van der Veen, 1998b, Figure 6). The other two parameters have a much stronger effect, as shown in Figures 8.20 and 8.21.

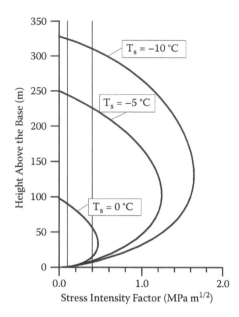

FIGURE 8.20   Effect of surface temperature on the stress intensity factor for a floating ice shelf (H = 500 m) and depth-independent stretching rate ($\dot{\varepsilon}_{xx}$ = 0.01 yr^{-1}). (Reprinted from Van der Veen, C. J., *Cold Regions Sci. Techn.*, 27, 213–223, 1998b. With permission from Elsevier.)

It may also be noted that accounting for the low-firn density layer through the density profile (8.12) has only a minor effect on the calculated stress intensity factor. This is because the density varies only near the surface (and thus affects penetration depth of surface crevasses), but for the lower parts of the glacier and over the height of bottom crevasses, the density is nearly constant.

    Figure 8.20 shows the calculated stress intensity factor for three temperature profiles. In all three cases, the bottom temperature is set at 0°C. Close to the glacier base, the lithostatic stress and the water pressure nearly balance (at the base they are the same), and crevasse propagation is determined by the stretching stress, R_{xx} and, in first approximation, the stress intensify factor increases as the square root of crevasse height. Thus, as the crevasse grows, the stress intensity factor increases, allowing the crevasse to penetrate further. But as the crevasse extends farther into the ice, the water pressure decreases more rapidly than does the lithostatic stress, as schematically indicated in Figure 8.19. This causes the net longitudinal stress to decrease, and after reaching a maximum, which depends on the magnitude of the stretching stress, the stress intensity factor starts to decrease as the crevasse grows upward. The two vertical lines correspond to the lower and upper values for the fracture toughness of glacier ice given by Rist and others (1996), and crevasse height is given by the intersection of these lines with the curves of the stress intensity factor. While there are two solutions, the solution corresponding to a shallow crevasse is unstable and a small perturbation will cause the crevasse to either heal or grow to

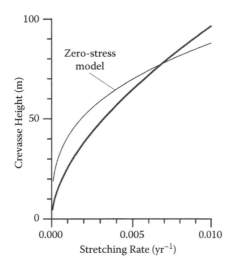

FIGURE 8.21 Height of a bottom crevasse on an isothermal floating ice shelf (H = 500 m; $T_s = 0°C$) as a function of stretching rate, using $K_{Ic} = 0.1$ MPa m$^{1/2}$ for the fracture toughness. The thin curve represents the zero-stress solution. (Reprinted from Van der Veen, C. J., *Cold Regions Sci. Techn.*, 27, 213–223, 1998b. With permission from Elsevier.)

the second, stable solution. As shown, the colder the surface ice, the higher a basal crevasse will penetrate upward. This is because, for constant stretching rate, the corresponding stress increases as the temperature decreases. Similarly, increasing the stretching rate allows the crevasse to penetrate farther into the ice (Figure 8.21). For the isothermal case shown in this figure, the crevasse height estimated from the zero-stress model in which the crevasse grows upward to the level where the net longitudinal stress is zero agrees well with the full solution.

McGrath and others (2012) identified a series of basal crevasses along a 31 km long transect across the northern sector of the Larsen C Ice Shelf in the Antarctic Peninsula. These crevasses show as clear refractions in echograms obtained from ground-penetrating radar. Near the lower end of the transect, individual crevasses penetrating 70–134 m into the ice and that are widely spaced (0.5–2.0 km) are observed. The crevasse opening width (at the base of the ice shelf) ranges from 20–70 m at the upstream end to 150–240 m on the last 15 km of the transect. Apparently, the more widely spaced crevasses transitioned from a region of multiple shallow basal fractures. Applying a model similar to that described in this section and using measured surface strain rates to estimate the stretching stress, these authors find that the LEFM model accurately predicts the height to which the basal crevasses penetrate. As would be expected for floating ice, surface troughs observed in visible imagery overlie the major basal crevasses. These surface depressions are likely to result in ponding of surface meltwater, which would create the reservoir needed for meltwater-driven hydrofracturing (Section 8.2) and, because these depression exist in conjunction with basal crevasses, could promote full-thickness fracturing (McGrath et al., 2012; c.f. Section 8.5).

For grounded glaciers, the most important parameter that determines whether basal crevasses can occur is the height of the piezometric head above the glacier base, which in turn depends on the basal water pressure. Rather than discussing the results in terms of height above buoyancy, the concept of flotation level is used here. This level is defined as

$$H_f = \frac{\bar{\rho}}{\rho_w} H, \tag{8.69}$$

and the difference between the piezometric head and this level is used to estimate crevasse propagation. This difference, $\Delta H_p = H_f - H_p$, describes how close the grounded glacier effectively is to flotation, irrespective of the height above buoyancy; where this difference is zero, the basal water pressure equals the ice overburden pressure and the effective pressure at the bed is zero. Note that because meltwater under glaciers is fresh, the density for fresh water should be used in these expressions.

The contour map in Figure 8.22 shows how the crevasse height depends on stretching rate and depth of the piezometric head below the flotation level. Even a small drop of a few meters below the flotation level can inhibit the growth of basal crevasses. Thus, basal crevasses can only occur where the basal water pressure is close to the ice overburden pressure and the glacier is effectively near flotation.

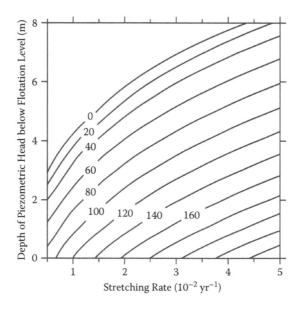

FIGURE 8.22 Contour map of basal crevasse height on a grounded glacier as a function of the depth of the piezometric head below the flotation level and stretching rate ($T_s = 0°C$). (Reprinted from Van der Veen, C. J., *Cold Regions Sci. Techn.*, 27, 213–223, 1998b. With permission from Elsevier.)

8.5 ICEBERG CALVING

Calving occurs where a floating or grounded glacier terminus meets the ocean or a proglacial lake, and blocks of ice detach from the glacier to form icebergs ranging in size from small bits and pieces to large tabular icebergs. It is the only mechanism by which ice is rapidly transferred from grounded glaciers and ice sheets to the world's oceans. Calving into proglacial lakes and embayments has been implicated as the process responsible for the rapid disappearance of large Northern Hemisphere ice sheets following the last glacial maximum (for example, Pollard, 1984; Stokes and Clark, 2004), with recurrent episodes of armadas of icebergs invading the North Atlantic (Heinrich Events) documented in ocean sediment cores (Heinrich, 1988; Bond and Lotti, 1995). More recently, retreat of floating calving termini in Greenland and Antarctica occurred concurrently, with large changes in discharge from the grounded feeder glaciers (for example, De Angelis and Skvarca, 2003; Joughin et al., 2004) and have raised concerns about accelerated contributions from the polar ice sheets to global sea level rise as climate continues to warm. At the present, iceberg calving accounts for almost all of the mass loss of the Antarctic Ice Sheet (Rignot et al., 2008) and perhaps as much as half the mass loss of the Greenland Ice Sheet (Van den Broeke et al., 2009). Yet the controls on calving rate or terminus position remain the topic of debate.

A wide range of calving environments exists, producing icebergs the size of small bergy bits to large tabular icebergs detaching from floating Antarctic ice shelves. As a start, two main factors can be identified as determining the type of calving, namely, whether the glacier is cold or temperate, and whether the terminus is grounded or floating. This gives four possible calving termini, of which three are known to exist. Temperate glaciers usually do not form floating ice tongues, although the terminus of Columbia Glacier in Alaska became ungrounded in 2007 as the glacier continued its prolonged retreat (Walter et al., 2010). Within the warm, grounded category a further distinction can be made between glaciers terminating in freshwater lakes and those terminating in sea water. To differentiate between these types of calving, the term *tidewater calving* is applied to calving from termini grounded in sea water, while the term *lacustrine calving* refers to calving from grounded glaciers that terminate in freshwater proglacial lakes. For comparable geometries—in particular water depth—calving rates on lacustrine glaciers are an order of magnitude smaller than those on tidewater glaciers (Funk and Röthlisberger, 1989; Warren et al., 1995; Warren and Aniya, 1999).

The type of iceberg produced by calving glaciers depends on the type of glacier terminus. Tidewater glaciers usually produce icebergs that are relatively small compared with the size of the glaciers, with dimensions of a few hundred meters at most. The rate at which these smaller bergs are produced can be large, in particular during rapid retreat, jamming the adjacent fjord waters with floating icebergs. Greenland tidal outlet glaciers usually produce larger bergs, with horizontal dimensions up to a few km. Calving from Antarctic ice shelves is more infrequent, with long periods of little or no calving activity punctuated by the detachment of large tabular icebergs that appear to originate as a result of giant rifts cutting through these ice shelves (Lazzara et al., 1999).

Given the range of calving environments and the essentially stochastic nature of iceberg detachment, one could question whether a deterministic universal calving relation exists. Calving involves propagation of fractures through the ice, which depends on the local stress regime, presence of pre-existing cracks (surface crevasses), strength of the ice, availability of surface meltwater, and other factors. Accounting for all these effects in a universally applicable calving relation may be an illusory quest. Nevertheless, it may be possible to formulate a "bulk calving relation" that can be implemented in numerical models and that reflects temporally averaged behavior of calving glaciers. That is, no attempt is made to model production of individual icebergs, but rather, an average rate of iceberg production is predicted. In this section, several models that have been proposed are reviewed.

Traditionally, most systematic calving observations derive from tidewater glaciers. A tidewater glacier is a glacier that terminates in the sea, where it usually ends in an ice cliff from which icebergs are discharged. In the polar regions, the terminus may be floating; but temperate tidewater glaciers, such as those found in Alaska, are usually grounded up to the calving ice front although parts of the terminal region may incidentally become afloat. Floating tidewater glaciers are essentially ice shelves that have formed in fjords or valleys.

The position of the terminus of tidewater glaciers appear to go through cycles of slow advance, typically at a rate of several tens of meters per year, lasting perhaps as long as 1000 yr, followed by much shorter periods of rapid retreat (Meier and Post, 1987). Typical retreat rates range from a few hundred meters per year to a few kilometers per year. The cyclic behavior of tidewater glaciers is illustrated in Figure 8.23, starting with the terminus at its most advanced position (panel a). The front of the

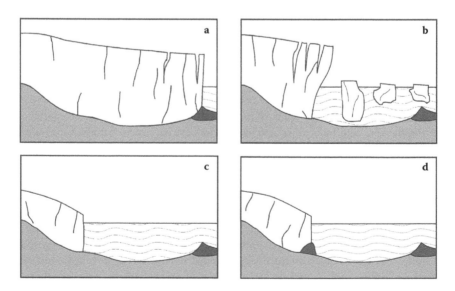

FIGURE 8.23 Illustrating the life cycle of tidewater glaciers from the most advanced position (panel a) through rapid retreat (panel b) to the most retracted position (panel c), followed by slow advance of the glacier terminus (panel d).

glacier rests on a terminal moraine shoal at the end of the fjord. A large portion of the advanced glacier lies in the ablation area, where surface melting during the summer exceeds winter snowfall. This may render the terminus position more susceptible to small perturbations in climate or ablation rate. While the glacier front cannot readily advance, a moderate thinning could cause the terminus to retreat into deeper water, resulting in increased calving rates and accelerated retreat (panel b). Retreat of the terminus is halted where the water depth becomes sufficiently shallow to reduce the rate of iceberg production (panel c). Subglacial erosion and transport of sediment results in the formation of a new moraine at the foot of the retracted terminus position (panel d). This moraine reduces the water depth at the terminus, allowing the front to advance as the moraine is being recycled in a conveyor-belt fashion. The rate of advance is slow, determined mainly by how fast material is eroded from the upglacial side of the moraine and deposited on the seaward side. Advance is halted when the terminus reaches the end of the fjord or where the fjord widens.

There is no obvious climatic control on the change in terminus position. Rather, the behavior of a tidewater glacier may be largely controlled by the phase of the cycle and by the balance between calving rate and glacier speed. Neighboring glaciers are known to behave markedly differently. For example, Harvard and Yale Glaciers both terminate in College Fjord, which feeds into Prince William Sound, Alaska, and derive from the same snowfields as Columbia Glacier. Harvard Glacier has been advancing at a rate of about 20 m/yr since 1899, while its immediate neighbor, Yale Glacier, appears to have retreated since the early 19th century. Initially, this retreat was at a modest rate of about 7 to 11 m/yr, but since 1957, the rate of retreat has increased considerably, reaching 345 m/yr between 1974 and 1978. More recently, the retreat of Yale Glacier has slowed and the ice front may have reached its stable retracted position (Sturm et al., 1991). Despite these different behaviors of neighboring calving glaciers, there is some suggestion that the initiation of retreat may be associated with climate. During several decades prior to the onset of retreat, the net surface mass balance of Columbia Glacier was negative (Tangborn, 1997). This may have caused the terminal region to thin and become unstable, leading to the rapid retreat of the calving front. Perhaps prolonged periods of warmer-than-usual climate predispose a tidewater glacier in its most advanced position to collapse. Once initiated, however, the course of retreat appears to be largely controlled by internal dynamics rather than climate factors.

The scenario for slow advance followed by rapid retreat is based largely on the linear relation between water depth and calving rate derived by Brown et al. (1982), who conducted a statistical analysis of data for 12 Alaskan tidewater glaciers and found that the annual calving rate, U_c (in km/yr), is correlated with the water depth, D (in m), as

$$U_c = 0.024 \ D. \tag{8.70}$$

Similar relations have been found for other glaciers (for example, Pelto and Warren, 1991; Kennett et al., 1997), but a number of questions remain that cast doubt on the water-depth relation as a viable explanatory and predictive model for iceberg calving (Van der Veen, 1996, 2002a).

A first objection against the water-depth relation is that this model applies to annual-averaged calving rates only (Brown et al., 1982) and breaks down when seasonal calving rates are considered (Sikonia, 1982; Krimmel, 1997; Meier, 1997). Moreover, even if annual-mean calving rates are considered, the water-depth model does not explain all of the variations in calving rate observed on Columbia Glacier. From late 1984 to early 1989, the calving rate on this glacier remained more or less constant, yet the terminus continued to retreat into deeper water, with the water depth increasing from around 200 m to around 300 m. Since then, the calving rate doubled, despite the terminus remaining grounded at a water depth of around 350 m (Van der Veen, 2002a, Figure 6). Meier and Post (1987) propose that the linear calving relation may apply to glaciers whose terminus is close to steady state, but that another calving relation becomes important during the period of accelerated retreat of the terminus. These authors also point out that the terminus appears to retreat to the point where the effective basal pressure approaches zero.

A second observation that casts doubt on the water-depth model is that, during its rapid retreat, the speed of Columbia Glacier increased almost as much as did the calving rate, and an excellent correlation exists between these two quantities. A similar linear relation between glacier speed and calving rate exists for the other Alaskan glaciers considered by Brown et al. (1982), as well as for lacustrine calving glaciers and for 12 floating tidal glaciers in northern Greenland (Van der Veen, 2002a). If, indeed, the calving rate is determined by geometrical factors such as water depth at the terminus, there is no physical explanation why these factors would cooperate with physical processes affecting glacier speed that both would vary in unison on both grounded tidewater glaciers and floating ice tongues. It may be noted that on short time scales (on the order of days), the correlation between calving rate and ice speed breaks down. O'Neel et al. (2003) measured ice motion and terminus positions at 2 to 8 hour intervals nearly continuously between May 2 and June 4, 1999, on LeConte Glacier, Alaska. Over this 30-day period they found no significant variation in ice flux to the terminus, but large variations in the calving flux. As noted earlier, calving is a highly stochastic process and, as a result, the terminus position can fluctuate significantly over short time scales, for example, after a major calving event. These individual events need not directly correlate with ice velocity and are likely controlled by other factors such as tidal forcing, subaerial collapse of seracs, forward toppling of overhanging cliffs, notch melting at the water line, existing planes of structural weakness, and so forth. However, when averaged over time periods sufficiently long to reduce the importance of individual calving events (about 10 days or so; O'Neel et al., 2003, Figure 5), the calving flux appears to be mostly constant over the time period considered. The goal of any calving criterion discussed in this section is not to predict individual calving events, but rather, to predict the average calving rate or change in terminus position. Thus, the observations of O'Neel et al. (2003) do not necessarily negate the arguments presented by Van der Veen (1996, 2002a).

A different mechanism for iceberg calving is suggested by Van der Veen (1996), based on the observation that during the retreat of Columbia Glacier, the thickness at the terminus remained at about 50 m in excess of the flotation thickness. This suggests that calving occurs whenever the terminal thickness in excess of flotation becomes less than some critical value, given by

$$H_c = \frac{\rho_w}{\rho} D + H_o, \tag{8.71}$$

where D represents water depth, ρ and ρ_w the densities of sea water and ice, respectively, and H_o the thickness at the terminus in excess of flotation ($H_o \approx 50$ m for Columbia Glacier). Vieli et al. (2001) proposed a slightly modified relation where the minimum height above buoyancy is replaced by a small fraction, q, of the flotation thickness at the terminus and

$$H_c = \frac{\rho_w}{\rho} (1 + q) D. \tag{8.72}$$

The height-above-buoyancy model is essentially different from previous models because the *position* of the terminus is determined by the local water depth and ice thickness, whereas in earlier models, the *calving rate* is linked to these parameters. According to Van der Veen (1996), retreat is initiated when the thickness at the terminus becomes too small to maintain contact with the bed, and the rate of retreat is controlled by the thinning rate of the snout and by the ice speed. Consequently, if the glacier speed increases (as it did during the retreat of Columbia Glacier), the calving rate increases simultaneously to maintain the rate of terminus retreat. The implication of this scenario is that the rapid retreat of tidewater glaciers is not the result of increased calving as the glacier retreated into deeper water. Instead, retreat is initiated and maintained by thinning of the glacier, probably associated with an increase in ice discharge and perhaps the preceding period of negative net surface mass balance. Thus, retreat may be expected to continue as long as large ice speeds are maintained.

Vieli et al. (2001) incorporate the calving relation (8.72) into a numerical model simulating advance and retreat of a tidewater glacier. In their model, the calving rate is a derived quantity obtained from the difference between ice velocity at the terminus and change in terminus position. Interestingly, a linear relationship between calving rate and water depth is found for the case of a slowly advancing or retreating terminus, but not for rapid terminus changes. Rapid retreat or advance of the terminus is controlled by basal topography and occurs preferentially where the bed slopes upward in the downglacier direction. During retreat through a basal depression, the ice velocity near the terminus increases because the sliding velocity is taken inversely proportional to the effective basal pressure. Whether or not the terminus is able to advance across a bed overdeepening depends on the water depth and the fraction, q, in the calving criterion (8.71). If the maximum water depth is too great, readvance cannot occur because mass loss by calving remains too large. A similar result was obtained by Nick et al. (2007), who employed a similar flowline model to simulate advance and retreat of Columbia Glacier. That study implemented both the water-depth calving criterion (8.70) and the height-above-buoyancy criterion

(8.72) and found that both yielded similar glacier behavior. Once initiated, retreat is rapid and continues until shallower water depth at the head of the fjord is reached. Irrespective of accumulation rate and calving criteria, the terminus does not advance into deeper water (>300 m water depth) unless sedimentation at the glacier front is included. By incorporating "conveyor belt" recycling of subglacial sediment and the formation of a sediment bank at the terminus, advance across basal troughs becomes possible, with the rate of advance determined by the rate of proglacial sediment deposition.

As noted by Benn et al. (2007a), a fundamental shortcoming of the height-above-buoyancy model is that it precludes the formation of ice shelves and floating ice tongues because the terminus position is always at a location where flotation has not been reached yet. This behavior contrasts with the observation that Columbia Glacier became ungrounded in 2007 (Walter et al., 2010). Further, many of the outlet glaciers in Greenland appear to form small floating ice tongues from which icebergs are discharged. To overcome this limitation of the calving criterion (8.71) or (8.72), Benn et al. (2007b) propose a criterion based on downward propagation of surface crevasses: calving occurs where the depth of surface crevasses equals the height above sea level or, equivalently, where the surface crevasse reaches water level. Nick et al. (2010) propose a modification of this calving model and posit that calving will occur where surface crevasses reach the depth to which basal crevasses penetrate upward into the ice (Figure 8.24).

In a field of closely spaced crevasses, there are no large stress concentrations near the tips of crevasses (Weertman, 1973b), and the penetration depth of surface crevasses may be estimated based on the assumption that crevasses will penetrate to the

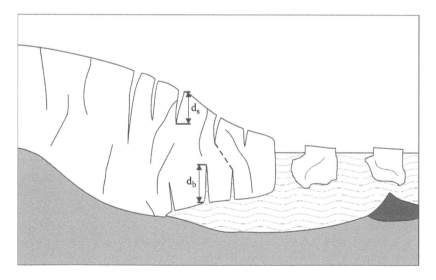

FIGURE 8.24 Geometry used in the crevasse-penetration calving model. The dashed line connecting the surface and bottom crevasse represents the intact part of the ice thickness along which shear failure may occur.

depth at which the longitudinal stress equals the compressive overburden pressure (equation (8.7)). If the crevasse is partially filled with water, the additional opening stress associated with the weight of the water allows deeper penetration, and the crevasse depth is given by (Benn et al., 2007b)

$$d_s = \frac{R_{xx}}{\rho g} + \frac{\rho_{ws}}{\rho} d_w. \tag{8.73}$$

In this expression, ρ_{ws} represents the density of surface meltwater filling the crevasse and d_w the water height in the crevasse so that the water level is at a depth $d_s - d_w$ below the ice surface.

Similarly, the propagation height of basal crevasses can be estimated from the condition that at this height, the net longitudinal stress must be zero. This stress is derived in Section 8.4 and given by equation (8.67). Ignoring the low-density firn layer (second term on the right-hand side) and assuming an easy connection between the subglacial drainage system and the adjacent ocean or lake (so that the piezometric height equals the depth, D, of the glacier base below sea or lake level), the net longitudinal stress is given by

$$\sigma_n(z) = -\rho g(h - z) + \rho_{wb} g(D - z) + R_{xx}. \tag{8.74}$$

The tensile stress, R_{xx} is taken constant throughout the ice thickness; ρ_{wb} denotes the density of sea water (tidewater calving) or fresh water (lacustrine calving). Setting $\sigma_n(z)$ equal to zero yields

$$z = \frac{\rho}{\rho_{wb} - \rho} \left[h - \frac{\rho_{wb}}{\rho} D + \frac{R_{xx}}{\rho g} \right]. \tag{8.75}$$

After some rearranging, the penetration height is found to be

$$d_b = \frac{\rho}{\rho_{wb} - \rho} \left(\frac{R_{xx}}{\rho g} - H_{ab} \right), \tag{8.76}$$

where H_{ab} represents the height above buoyancy, defined as

$$H_{ab} = H - \frac{\rho_{wb}}{\rho} D. \tag{8.77}$$

Nick et al. (2010) present results from the numerical flow-band model described in Section 9.4 and compared two calving criteria based on crevasse propagation. Model behavior is very similar, whether the position of the calving front is defined as the location where surface crevasses reach the waterline or whether this position is where surface and basal crevasses penetrate the full ice thickness. More important is the markedly different glacier behavior as compared to earlier studies in which calving is included through the water-depth relation or height-above-buoyancy criterion. The latter two calving models do not allow a floating ice tongue to develop

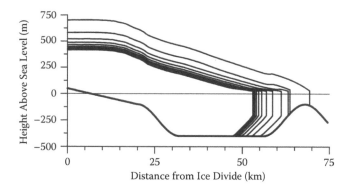

FIGURE 8.25 Simulated glacier profiles following glacier retreat from the most advanced position with the terminus grounded on the terminal moraine (at 70 km). Profiles are shown every 100 years. (From Nick, F. M. et al., *J. Glaciol.*, 56, 781–794, 2010. Reprinted from the *Journal of Glaciology* with permission of the International Glaciological Society and the authors.)

and, further, result in continued terminus retreat across basal overdeepenings when retreat from the terminal shoal is initiated. In contrast, a crevasse-based calving model allows stable grounding-line positions to be reached, even on a reverse bed slope. Figure 8.25 illustrates this behavior. Starting with a steady-state geometry with the terminus grounded on a terminal moraine (at 70 km), retreat was initiated by decreasing the surface mass balance along the entire glacier length. Terminus retreat is stopped well before the grounding line reaches shallower water, and a new steady state is attained, characterized by the presence of a small floating ice shelf. The formation of this shelf allows the grounding line to advance across basal overdeepenings. Figure 8.26 shows results from a model experiment starting with a

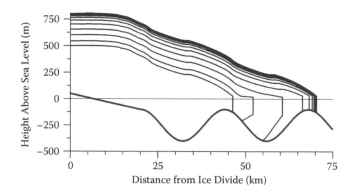

FIGURE 8.26 Simulated glacier profiles following glacier advance starting with the retracted position with the terminus grounded in shallow water (at 45 km). Profiles are shown every 50 years. (From Nick, F. M. et al., *J. Glaciol.*, 56, 781–794, 2010. Reprinted from the *Journal of Glaciology* with permission of the International Glaciological Society and the authors.)

steady-state profile with the terminus grounded in shallow water (at 45 km). Advance is initiated by increasing the accumulation rate, and a new steady state is reached with the terminus grounded at the terminal moraine (at 70 km).

The results of Nick et al. (2010) demonstrate that the choice of calving model incorporated into time-evolving numerical models determines behavior and stability of the model glacier. Earlier calving models (water-depth and height-above-buoyancy) always produce unstable glacier behavior with retreat of the terminus into deeper water, resulting in continued retreat until shallower water is reached. According to these models, advance across deeper water is possible only if proglacial sedimentation reduces water depth at the grounding line. In the crevasse-penetration models, the calving flux is not directly linked to water depth, and grounding-line dynamics appear to be less sensitive to basal topography (Nick et al., 2010).

A second difference between the crevasse-penetration model and earlier calving models is the ability to reproduce seasonal advance and retreat of the terminus, as observed on Greenland outlet glaciers (for example, Howat et al., 2008; Joughin et al., 2008). Because downward propagation of surface crevasses is directly linked to the amount of water in these crevasses, the rate of iceberg calving increases when surface melting occurs. Consequently, calving rates are greatest during late spring and summer and smallest during the winter, resulting in seasonal retreat and advance of the terminus (Nick et al., 2010). This strong dependency of calving rate on the water level in crevasses was also found by Cook et al. (2012), who noted that this introduces another difficulty in modeling iceberg calving because the water level in crevasses is difficult to determine either through in situ measurements or by using surface mass balance modeling.

The crevasse-penetration calving model applies only to temperate glaciers subject to surface melting. In the absence of water-filled surface crevasses, full thickness fracturing will not occur, nor will surface crevasses be able to reach the waterline. This can be readily shown to be the case by considering a floating ice tongue for which the height above buoyancy, H_{ab}, equals zero. The combined height of surface and basal crevasses is then

$$d_{tot} = d_s + d_b = \frac{R_{xx}}{\rho g} + \frac{\rho}{\rho_w - \rho} \frac{R_{xx}}{\rho g} = \frac{\rho_w}{\rho_w - \rho} \frac{R_{xx}}{\rho g}. \tag{8.78}$$

For a free-floating ice shelf, the stretching stress is proportional to the ice thickness, and substituting equation (4.58) into equation (8.78) gives

$$d_{tot} = \frac{1}{2} H. \tag{8.79}$$

Thus, an additional calving mechanism must be invoked when considering cold Antarctic ice shelves. Bassis and Walker (2012) propose that shear failure may occur on the intact part of the ice thickness separating surface and bottom crevasses (indicated by the dashed line in Figure 8.24). Their analysis suggests that a floating ice tongue can exist only when the ice entering the terminus region is relatively intact with few pre-existing crevasses. Another possible mechanism, suggested by Jeremy Bassis (pers. comm., 2012), is that calving may not be determined by crevasse depth estimated from the local stress field, but may be related to deeper crevasses that formed upstream in thicker ice and that are advected to the terminus. As the crevasse moves downglacier, the ratio of crevasse depth to ice thickness increases because the ice thins, while the fracture does not heal or adjust to the altering stress field. Such crevasses could become detachment boundaries for tabular icebergs.

Cuffey and Paterson (2010, p. 127) state that "simulations of calving fluxes in glacier and ice sheet models remain unconvincing." It is not entirely clear to what models they are referring—their discussion does not include the crevasse penetration model proposed by Benn et al. (2007b) and implemented in a numerical flowline model by Nick et al. (2010). Indeed, a solid theoretical and observational basis for formulating a calving relation is currently still lacking. (Of course, the same can be, and has been, said about the commonly used sliding relation; Fowler et al., 2001.) With regard to the crevasse penetration model, it could be argued that it requires crevasses to remain filled with surface meltwater. Where meltwater is abundant, there is ample evidence that englacial drainage efficiently routes this water to the bed. Moreover, there is little evidence from radar or other measurements that surface and bottom crevasses are co-located or that these can intersect. Finally, as noted above, the crevasse penetration model does not apply where surface melting is mostly absent yet iceberg calving is observed to take place. The main argument in favor of the crevasse penetration model is more heuristic or circumstantial: this model does predict seasonal variations in calving rate and terminus position that appear to be correlated with the onset and intensity of surface melting (as on Columbia Glacier; Tangborn, 1997), and it is based on plausible physical mechanisms.

Establishing a solid foundation for any model of iceberg calving—whether that be the crevasse penetration model or some modification to it, or an entirely different model—remains one of the grand challenges in glaciology. Considering that the choice of calving relation adopted in a numerical model can have a great impact on the behavior and stability of the modeled glacier (Nick et al., 2010), the lack of fundamental understanding of the calving process should be of concern to the glaciological community.

9 Numerical Ice-Sheet Models

9.1 INTRODUCTORY REMARKS

The word *model* has a wide range of meanings, but in the context of this chapter, the meaning is restricted to computer code written and developed to simulate flow of glaciers and ice sheets. The primary motivation for developing numerical models of glaciers is to better understand the behavior of glaciers and how they may react to changes in external forcing. These models are commonly based on some fundamental laws or assumptions thought to describe glacier flow. However, these simplifications do not necessarily apply to real ice sheets, and it is therefore of fundamental importance that models are thoroughly evaluated.

When it comes to evaluating model performance, the terminology used by different researchers can be confusing with words such as *verification*, *validation*, and *testing* often used interchangeably. In an attempt to bring some uniformity in terminology, Van der Veen (1999c) and Van der Veen and Payne (2003) propose the four terms *verification*, *validation*, *calibration*, and *confirmation* to describe the different levels of the process of model evaluation. This terminology is to a large extent based on similar discussions in the hydrological sciences (for example, Anderson and Bates, 2001).

Models cannot be verified. The term *verification* implies that the truth of the model has been demonstrated (Oreskes et al., 1994). More formal definitions are given by Ayer (1953) and by Fetzer (1988), who concludes that verification of models that describe open natural systems is "not even a theoretical possibility" (p. 1048). To illustrate this point, consider that ice-sheet models are based on concepts of continuum mechanics with Glen's flow law providing the relation between applied stress and resulting rate of deformation. This relation is based on laboratory experiments on single ice crystals and small blocks of ice, conducted over time scales that are insignificant compared with the residence time of ice in glaciers and ice sheets. Theoretical considerations provide some constraints on permissible relations between the stress and strain-rate tensors but do not conclusively constrain the flow law. Verification of the flow law, and ice-sheet models in general, would require experimental observations of processes acting on the scale of individual crystals throughout entire ice sheets and throughout their entire history, and without interfering with the flow of the glacier. This is, of course, not a feasible approach. A pragmatic approach to evaluate the veracity of a model is to adopt a two-step procedure involving validation and confirmation, in order to demonstrate that some level of confidence can be placed in model predictions.

Van der Veen and Payne (2003) propose that the term *validation* should be restricted to assessing only the numerical model and to demonstrate that the model does not contain any obvious or detectable flaws. This means that when the conceptual model for the system to be modeled is accepted, predictions of the numerical program should be consistent with this conceptual model. For example, if isothermal lamellar flow along a flowline on a horizontal bed is considered, the numerical model should predict a steady-state profile corresponding to the Vialov profile. Bueler et al. (2005) derive several analytical solutions for the isothermal shallow ice equations and propose a specific suite of tests to validate numerical models (note that these authors use the term *verification* for the process of validation). Another possibility for validating models is conduct model intercomparisons, in which similar models developed by different groups are subjected to benchmark experiments (for example, Pattyn et al., 2008, 2012). Such intercomparisons can bring to light inconsistencies among models that arise from numerical details such as grid spacing and numerical schemes used to solve the governing partial differential equations. For example, Vieli and Payne (2005) compared the ability of different models to simulate grounding-line migration and found that in models using a fixed grid size, grounding-line dynamics are strongly influenced by the discretization scheme used. The implication of this finding is that little confidence can be placed in results obtained with these models. Again, it is important to note here that a successful validation does not, and cannot, make any claims regarding how appropriate the conceptual model is for describing actual processes controlling glacier flow (Van der Veen, 1999c).

After ensuring that the numerical code reliably solves the mathematical equations describing the conceptual physical model, the next step in the evaluation process is to compare predictions of the model with actual observations. Most models contain unknown or poorly constrained parameters whose values can be adjusted to improve or optimize the match between observations and model predictions. This process of parameter tuning is called *calibration* and serves to demonstrate that the model is capable of reproducing observations within a certain range or parameter uncertainty (Anderson and Woessner, 1992). It is important to keep in mind that calibration does not provide any information about how appropriate the model is in describing processes occurring in nature. Rather, it shows that models usually contain enough adjustable parameters to make model predictions match observations.

The final step in the evaluation process is that of *confirmation*, or demonstrating that the model constitutes a realistic description of the physical processes being modeled. This again involves comparing model predictions with additional observations that were not used in the calibration process. Of course, models can never be absolutely confirmed, as this would imply verification of the model. However, following Strahler (1987), a "ladder of excellence" can be envisioned, with each rung representing some level of confidence that can be placed in the model's ability to describe processes in nature. As the model is confirmed by increasingly more observations, it rises up the ladder. It is important in this respect that model predictions to be tested against observations are not trite or obvious ("if snowfall increases, the ice sheet will grow"). In the words of Popper (1963, p. 112), "Confirmation should only count if they are the result of *risky predictions*."

More often than not, insufficient data are available to determine whether the conceptual framework on which a particular model is based is correct or not. To illustrate this, consider the study of Nick et al. (2007), who applied a flowline model to simulate advance and retreat of Columbia Glacier, an Alaskan tidewater glacier undergoing rapid retreat since the early 1980s. The conventional view is that retreat will continue until the terminus reaches the head of its fjord, where the bed rises above sea level. Subsequent terminus advance down the fjord where the bed is below sea level is made possible by "conveyor-style" sediment recycling and the formation of a sediment bank at the glacier terminus (Section 8.5). Nick et al. (2007) include two calving criteria, the height-above-buoyancy model and the water-depth relation (c.f. Section 8.5) and find that irrespective of which calving relation is implemented, the model glacier cannot advance into deeper water (~300 m water depth or more) unless sedimentation at the glacier front is included. A later study (Nick et al., 2010) incorporates a different calving criterion (based on crevasse depth; Section 8.5) and finds that this model does allow terminus advance across basal overdeepenings without having to invoke sedimentation near the terminus. In the absence of additional observations or information on, for example, sedimentation rates near the terminus, or calving rates, it is not possible to assess which of these two modeling studies best encapsulates the physical processes controlling advance and retreat of tidewater glaciers.

Models can be separated into two categories, namely, diagnostic and prognostic. A *diagnostic* model describes a certain process such as parameterizations of the surface mass balance, or the relation between depth-averaged ice velocity and driving stress for lamellar flow. A *prognostic* model predicts how a process or quantity evolves with time. The most widely used prognostic equations are the continuity equation from which changes in ice-sheet geometry are calculated, and the thermodynamic equation that allows temperature changes to be estimated. When doing so, the implicit assumption is that the diagnostic equations remain valid as the ice sheet changes shape or climate alters significantly. Because many diagnostic models are tuned to present-day observations only, it is not a priori certain that they may be applied under vastly different conditions. The modeler should therefore conduct sensitivity experiments to investigate how small changes or uncertainties in the diagnostic equations affect the behavior of the model ice sheet. Failing to do so renders any model predictions rather useless. Of course, if the model turns out to be very sensitive to certain parameterizations that are ill tested, predictions are not very useful either, other than in identifying key processes that need to be understood better. Such sensitivity studies are all the more important for complex models that cannot be easily reproduced by others.

Numerical models are used in many different ways. One application is to describe in detail the flow of a glacier, adjusting the model parameters and including different physical processes, to better match model predictions with observations. Another application of numerical models is to test the sensitivity of the model glacier to certain processes and how small changes may affect the model behavior. Such studies usually do not pertain to one glacier in particular but are used to identify and study the nature of possible feedback mechanisms. Finally, numerical models can be used to predict past and future evolution of real ice sheets. While such studies are valuable in providing some idea about possible responses, it should be realized that very few of these models have been adequately evaluated and confirmed. The usual procedure is to run the numerical model to a steady state that agrees reasonably well with the current observed geometry of the Greenland or Antarctic ice sheets. If agreement is not satisfactory, model parameters (such as the illustrious enhancement factor) are adjusted. In other words, the observations are used to tune the model, which is entirely different from actually testing the model.

The following sections in this chapter introduce the basic concepts of numerical ice-sheet models. The objective is not to discuss the most advanced models in great detail, but rather to provide the necessary background to construct relatively simple models for ice flow that can be used to investigate glacier dynamics.

9.2   NUMERICAL METHODS

Solving prognostic equations numerically is a science in itself. Many pitfalls exist that may lead to spurious results or errors becoming infinitely large. This is because continuous functions are discretized on a numerical grid, and derivatives are calculated from these grid values. This procedure introduces rounding errors that may adversely affect the solution. While unstable integration schemes are readily discovered, there are other, more subtle, errors that go undetected. Many excellent textbooks exist on the numerical solution of partial differential equations (for example, Smith, 1985). In this section only a few of the most important issues applying to finite-difference modeling are briefly discussed.

Only single-valued, finite and continuous functions of horizontal distance, x, and time, t, are considered here. In a numerical model, the function values are known at discrete gridpoints, with coordinates $k \cdot \Delta x$ and $n \cdot \Delta t$, where k and n represent (positive) integers. Although not necessary, the horizontal grid spacing, Δx, and time step, Δt, are considered constant. To arrive at expressions for horizontal differences, Taylor series expansion is used. For the ice thickness this expansion gives

$$H(x + \Delta x, t) = H(x,t) + \Delta x \frac{\partial H}{\partial x}(x,t) + \frac{(\Delta x)^2}{2} \frac{\partial^2 H}{\partial x^2} + O((\Delta x)^3), \quad (9.1)$$

and

$$H(x - \Delta x, t) = H(x,t) - \Delta x \frac{\partial H}{\partial x}(x,t) + \frac{(\Delta x)^2}{2} \frac{\partial^2 H}{\partial x^2} + O((\Delta x)^3), \quad (9.2)$$

where $O((\Delta x)^3)$ represents terms containing the third and higher powers of the horizontal grid spacing. Subtracting both expressions and neglecting higher-order terms gives the first horizontal derivative

$$\frac{\partial H}{\partial x}(x,t) = \frac{H(x + \Delta x, t) - H(x - \Delta x, t)}{2 \Delta x}. \quad (9.3)$$

Similarly, adding (9.1) and (9.2) yields an expression for the second derivative

$$\frac{\partial^2 H}{\partial x^2}(x,t) = \frac{H(x+\Delta x,t) - 2H(x,t) + H(x-\Delta x,t)}{(\Delta x)^2}. \tag{9.4}$$

These discretizations of derivatives are so-called central differences, because the derivative at a gridpoint is calculated using values at both neighboring gridpoints. For spatial derivatives, this is the most common discretization. For time integrations, a frequently used scheme is forward differencing

$$\frac{\partial H}{\partial t}(x,t) = \frac{H(x,t+\Delta t) - H(x,t)}{\Delta t}. \tag{9.5}$$

The errors involved in discretizing continuous derivatives can, at least in principle, be reduced to an acceptable level by decreasing the grid spacing, Δx, and time step, Δt. This does not necessarily mean that the numerical scheme will converge on the correct solution. This can best be seen by considering the advection equation.

The continuity equation derived in Section 5.2 is an example of an advection equation. For clarity of the following discussion, the advection velocity, U, is taken constant in the flow direction, and source terms are neglected (no surface accumulation). This is not a very realistic model for glaciers, but it demonstrates clearly some of the difficulties involved in numerically solving differential equations. Under these assumptions

$$\frac{\partial H}{\partial t} + U\frac{\partial H}{\partial x} = 0, \tag{9.6}$$

describes how the thickness changes as ice is being advected in the x-direction. The solution can be written as

$$H(x, t) = F(x - Ut), \tag{9.7}$$

subject to the initial condition $H(x, 0) = F(x)$. In other words, the solution is constant along lines $x - Ut = $ constant. Referring to Figure 9.1, the value of H at point A is equal to the initial value $H(x_0, t_0)$. Now consider the discrete form of the advection equation, using central differencing for the horizontal derivative, and forward differencing for the time integration

$$\frac{H_k^{n+1} - H_k^n}{\Delta t} + U\frac{H_{k+1}^n - H_{k-1}^n}{2\Delta x} = 0. \tag{9.8}$$

For brevity, the following notation is used here

$$H_k^n = H(k \cdot \Delta x, n \cdot \Delta t). \tag{9.9}$$

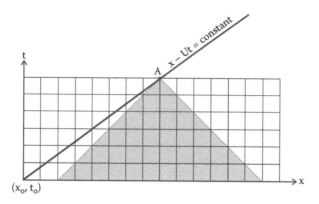

FIGURE 9.1 Divergence of the numerical solution. The shaded area includes gridpoints whose value affects the calculated result at gridpoint A.

Rewriting equation (9.8) and solving for the new ice thickness gives

$$H_k^{n+1} = H_k^n - \frac{U\Delta t}{2\Delta x}\left(H_{k+1}^n - H_{k-1}^n\right). \tag{9.10}$$

The numerical solution at point A as computed from this equation is determined by the gridpoints falling within the shaded area in Figure 9.1. The gridpoint (x_o, t_o) falls outside this region and consequently does not affect the calculated value at point A, even though $H(A) = H(x_o, t_o)$. Consequently, there is no reason to believe that the numerical solution will converge toward the "true" solution. To ensure convergence, the gridpoint (x_o, t_o) must fall within the region of points that determine the numerical solution at point A. This is the case if the following criterion is met

$$U\Delta t \le \Delta x. \tag{9.11}$$

This requirement is the Courant–Friedrichs–Lewy condition, or CFL condition for short.

Another source of error may occur when the numerical scheme is unstable and the numerical solution increasingly diverges from the "true" solution as time integration progresses. This stability can be investigated using the Von Neumann method. The assumption is made that the solution may be written as a Fourier series of harmonic functions and that the stability of each individual Fourier component may be considered separately. Although each harmonic function can be written in terms of sines and cosines, it is more convenient to use the complex exponential formulation. That is

$$H_\omega^n\, e^{-i\omega k\Delta x}, \tag{9.12}$$

with $i = \sqrt{-1}$, represents the Fourier component with frequency ω. The change in amplitude of the Fourier component may be written as

$$H_\omega^{n+1} = \lambda H_\omega^n. \tag{9.13}$$

Substituting (9.12) and (9.13) in the discrete form (9.8) of the advection equation gives

$$\lambda - 1 + i\frac{U\,\Delta t}{2\,\Delta x}\sin\Delta x = 0, \tag{9.14}$$

from which the magnitude of the amplification factor, λ, follows

$$|\lambda|^2 = 1 + \left(\frac{U\,\Delta t}{\Delta x}\right)^2 \sin^2\Delta x. \tag{9.15}$$

Thus, $|\lambda|$ is always larger than 1 and the amplitude of the harmonic component will continue to increase with time. The frequency of the Fourier component does not appear in equation (9.15), so that all components continue to grow. In other words, the numerical scheme (9.8) or (9.10) is unstable in the sense that it does not allow a steady solution to be reached.

A numerical scheme that is stable under certain conditions is one in which the time derivative is approximated by forward differencing, and the horizontal derivative by upward (or upstream) differencing

$$\frac{H_k^{n+1} - H_k^n}{\Delta t} + U\frac{H_k^n - H_{k-1}^n}{\Delta x} = 0, \tag{9.16}$$

or

$$H_k^{n+1} = H_k^n - \frac{U\,\Delta t}{\Delta x}\left(H_k^n - H_{k-1}^n\right). \tag{9.17}$$

To evaluate the stability of this numerical scheme, consider again the harmonic component (9.12) with the increase in amplitude given by (9.13). Substitution in equation (9.16) now leads to

$$|\lambda|^2 = 1 - 2\frac{U\,\Delta t}{\Delta x}\left(1 - \frac{U\,\Delta t}{\Delta x}\right)(1 - \cos(\omega\,\Delta x)). \tag{9.18}$$

For the solution to be stable, the amplification factor, $|\lambda|$, must be one or less. From (9.18) it follows that this criterion is met if

$$\frac{U\,\Delta t}{\Delta x} \leq 1, \tag{9.19}$$

which is the CFL condition derived earlier.

Having arrived at a numerical scheme that is both convergent and stable, one could proceed with numerically solving the advection equation (9.6). There are more problems looming on the horizon, however. According to equation (9.18), the damping of each harmonic Fourier component depends on its wavelength. This is illustrated in Figure 9.2, which shows $|\lambda|$ as a function of $\mu = U\Delta t/\Delta x$. The decrease in amplitude is largest for $\mu = 0.5$ and increases as the wavelength

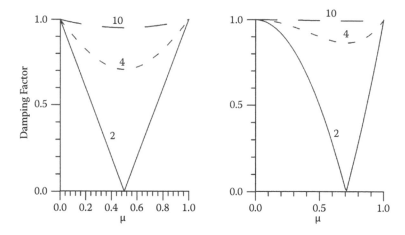

FIGURE 9.2 Damping factor in the upstream scheme (left panel) and in the Lax–Wendroff scheme (right panel) for harmonic Fourier components. Labels give the wavelength of the components, nondimensionalized with the horizontal grid spacing.

becomes smaller. For $\mu = 1$, the solution is neutral, with $|\lambda| = 1$ (the same is true for $\mu = 0$, but this corresponds to $\Delta t = 0$, which obviously is not an option for numerical time integration). In general, it may therefore be expected that after a sufficiently long time integration, small-scale features will have disappeared from the numerical solution. Figure 9.3 shows the calculated advection of two sine-waves (using numerical scheme (9.17)), as well as the analytical solution,

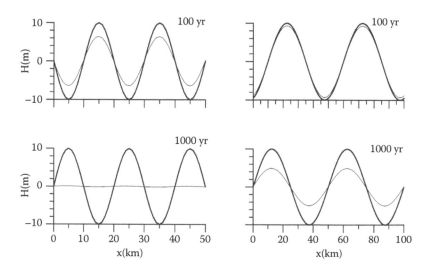

FIGURE 9.3 Damping of two sine waves (left panels: wavelength equals 20 times the horizontal grid spacing; right panels: wavelength equals 50 times the horizontal grid spacing) for the upstream/forward-in-time approximation of the advection equation. The light curves represent the solution calculated numerically and the heavy curves the analytical solution.

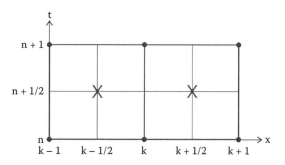

FIGURE 9.4   Grid geometry for the Lax–Wendroff scheme.

obtained by simply moving the initial wave along the grid. As boundary conditions, the grid is assumed to be cyclic, with gridpoint $x = -\Delta x$ corresponding to the last gridpoint, $x = k_{max} \cdot \Delta x$. To minimize unwanted damping, the horizontal grid spacing, Δx, can be decreased. However, to satisfy the CFL condition for stability, the time step, Δt, must be decreased at the same time, and this may not always be practical.

A numerical scheme that performs better than (9.17) is the Lax–Wendroff scheme. The function H is known at the gridpoints $k \cdot \Delta x$, at time $n \cdot \Delta t$; in Figure 9.4 these gridpoints are shown as black dots. Values of H at "intermediate" gridpoints (indicated by an X in Figure 9.4) are introduced to discretise the advection equation using central differencing for the x-derivative and the forward scheme for the time integration:

$$\frac{H_{k+1/2}^{n+1/2} - \frac{1}{2}\left(H_{k+1}^{n} + H_{k}^{n}\right)}{\frac{1}{2}\Delta t} = -U \frac{H_{k+1}^{n} - H_{k}^{n}}{\Delta x}. \tag{9.20}$$

Similarly

$$\frac{H_{k-1/2}^{n+1/2} - \frac{1}{2}\left(H_{k}^{n} + H_{k-1}^{n}\right)}{\frac{1}{2}\Delta t} = -U \frac{H_{k}^{n} - H_{k-1}^{n}}{\Delta x}. \tag{9.21}$$

The next step is to use central differencing, centered around $(n + 1/2) \cdot \Delta t$, to approximate the time derivative

$$\frac{H_{k}^{n+1} - H_{k}^{n}}{\Delta t} = -U \frac{H_{k+1/2}^{n+1/2} - H_{k-1/2}^{n+1/2}}{\Delta x}. \tag{9.22}$$

Using equations (9.20) and (9.21), the "intermediate" values on the right-hand side of (9.22) can be eliminated. The result is

$$\frac{H_{k}^{n+1} - H_{k}^{n}}{\Delta t} = -U \frac{H_{k+1}^{n} - H_{k-1}^{n}}{2\,\Delta x} + \frac{1}{2}U^{2}\,\Delta t \frac{H_{k+1}^{n} - 2H_{k}^{n} + H_{k-1}^{n}}{(\Delta t)^{2}}, \tag{9.23}$$

or

$$H_k^{n+1} = \frac{1}{2}\mu(1+\mu)H_{k-1}^n + (1-\mu^2)H_k^n - \frac{1}{2}\mu(1-\mu)H_{k+1}^n, \qquad (9.24)$$

with $\mu = U\,\Delta t/\Delta x$. Following the same procedure as above, it can be shown that this scheme is stable for $\mu \leq 1$, which is, once again, the CFL condition. The amplification factor is

$$|\lambda|^2 = 1 - 4\mu^2(1-\mu^2)\sin^4\left(\frac{\omega\Delta x}{2}\right). \qquad (9.25)$$

Comparing this expression with the damping factor for the previously discussed numerical scheme (9.17) shows that damping is somewhat less of a problem in the Lax–Wendroff scheme (Figure 9.2, right panel). Applying the Lax–Wendroff scheme to the advection of sine-waves, as done in Figure 9.3 for the upstream/forward-in-time scheme, shows no significant difference between the numerically determined solution and the analytical solution.

The numerical schemes discussed above are examples of explicit schemes, in which the new value of H at a particular gridpoint is calculated independently of the other gridpoints. Such schemes have the great advantage of being fairly simple to construct, and solving them is straightforward. Their biggest disadvantage, however, is that the time step must be small to satisfy the CFL condition. This restriction can be avoided by using an implicit scheme. Such schemes are more complex to construct and solve than the explicit schemes discussed above, but may be more efficient because they are unconditionally stable and thus allow for a larger time step.

As an example of an implicit scheme, the Crank–Nicholson solution of the advection equation (9.6) is considered. Evaluating this equation at the intermediate time $(n + 1/2)\cdot\Delta t$ gives

$$\left(\frac{\partial H}{\partial t}\right)_k^{n+1/2} = -U\left(\frac{\partial H}{\partial x}\right)_k^{n+1/2} =$$

$$= -U\left(\left(\frac{\partial H}{\partial x}\right)_k^{n+1} + \left(\frac{\partial H}{\partial x}\right)_k^n\right). \qquad (9.26)$$

The derivatives are now replaced by central differences, and all values of H at the new time $(n + 1)\cdot\Delta t$ collected on the left-hand side. This gives

$$-\frac{1}{4}\mu H_{k-1}^{n+1} + H_k^{n+1} + \frac{1}{4}\mu H_{k+1}^{n+1} = \frac{1}{4}\mu H_{k-1}^n + H_k^n - \frac{1}{4}\mu H_{k+1}^n. \qquad (9.27)$$

If k_{max} represents the total number of gridpoints along the x-axis, equation (9.27) represents a set of k_{max} linear equations that can be solved by matrix inversion. This is an expedient procedure because the coefficient-matrix is tridiagonal (that is, with nonzero elements only on the diagonal plus or minus one column). As a result, there is no need to formally invert the k_{max} by k_{max} coefficient-matrix (which contains mostly zeros), and the solution can be coded very concisely (Press et al., 1992). When modeling glaciers, the advection velocity is not constant but depends nonlinearly on the ice thickness. This means that a fully implicit scheme becomes rather unpractical, as it requires the solution of a set of nonlinear equations.

The short introduction presented above highlights some of the more common problems encountered when constructing numerical models to simulate the behavior of ice sheets. The advection equation was chosen as example because this type of equation (hyperbolic equation) requires more stringent numerical schemes than do other types of equations. In particular, the diffusion equation is an example of a parabolic equation for which stable schemes are readily constructed that do not possess the nasty qualities discussed above (or, if they do, the effects are less manifest). For this reason, the continuity equation derived in Section 5.2 (which is an advection equation) is often recast as a diffusion equation. This is possible because the deformation of ice depends on the *gradient* of the quantity to be determined, namely, ice thickness, or rather, surface elevation. The next section discusses this issue in more detail.

9.3   MODEL DRIVEN BY SHEAR STRESS ONLY

To illustrate the steps involved in constructing numerical ice-flow models, this section starts with discussing the simplest model, namely, isothermal one-dimensional lamellar flow in the x-direction. In that case, the rate factor in the flow law is taken constant, and expressions for the depth-integrated ice flux are substituted in the continuity equation to compute evolution of ice thickness.

Where drag at the glacier base provides the sole resistance to flow, the depth-averaged ice velocity can be calculated from the lamellar flow model and is given by (Section 4.2)

$$U = \frac{2AH}{n+2}\tau_{dx}^n + U_b,\tag{9.28}$$

where U_b represents the sliding velocity. In the following discussion, basal sliding is not included so as to simplify the equations and more clearly demonstrate the procedure involved. Provided that the sliding velocity can be related to basal drag or driving stress (for example, using equation (7.9), or (7.41) if deformation of subglacial till is important), basal sliding is readily incorporated into the numerical model discussed in this section.

Conservation of ice volume is expressed by the continuity equation (5.18)

$$\frac{\partial H}{\partial t} = -\frac{\partial(HU)}{\partial x} + M.\tag{9.29}$$

Substituting expression (9.28) for the ice velocity into this equation gives

$$\frac{\partial H}{\partial t} = -\frac{\partial}{\partial x}\left(\frac{2A}{n+2}H^2\,\tau_{dx}^n\right) + M.\tag{9.30}$$

The driving stress can be calculated from the ice thickness, H, and slope of the ice surface, $\partial h/\partial x$ (Section 3.2) as

$$\tau_{dx} = -\rho g H \frac{\partial h}{\partial x}. \qquad (9.31)$$

The continuity equation now becomes

$$\frac{\partial H}{\partial t} = + \frac{\partial}{\partial x}\left(\frac{2A}{n+2}(\rho g)^n H^{n+2} \left|\frac{\partial h}{\partial x}\right|^{n-1} \frac{\partial h}{\partial x} \right) + M, \qquad (9.32)$$

or

$$\frac{\partial H}{\partial t} = \frac{\partial}{\partial x}\left(D \frac{\partial h}{\partial x} \right) + M. \qquad (9.33)$$

This is a parabolic differential equation that may be interpreted as a diffusion equation for the ice thickness, H. The diffusivity

$$D = \frac{2A}{n+2}(\rho g)^n H^{n+2} \left|\frac{\partial h}{\partial x}\right|^{n-1}, \qquad (9.34)$$

increases strongly with ice thickness and surface slope, due to the nonlinear character of Glen's flow law.

To solve the continuity equation (9.33) on a numerical grid, the diffusivity is first calculated at the gridpoints. To avoid confusion between exponents and grid coordinates, the following notation is used

$$H_{p,q} = H(p \cdot \Delta x, q \cdot \Delta t), \qquad (9.35)$$

where p and q represent integer numbers. Using central differencing, the diffusivity is

$$D_{p,q} = C H_{p,q}^{n+2} \left(\frac{\left| h_{p+1,q} - h_{p-1,q} \right|}{2\Delta x} \right)^{n-1}, \qquad (9.36)$$

with

$$C = \frac{2A}{n+2}(\rho g)^n, \qquad (9.37)$$

containing all constants. Note that C may vary along the flowline if the rate factor, A, is not constant.

The next step is to calculate the ice flux at intermediate gridpoints, using interpolated values for the diffusivity

$$F_{p+1/2,q} = \frac{1}{2}(D_{p,q} + D_{p+1,q})\frac{h_{p+1,q} - h_{p,q}}{\Delta x}. \tag{9.38}$$

Note that $F = -HU$ represents minus the ice flux.

Using the forward scheme for the time derivative in the continuity equation (9.33), the new ice thickness can now be calculated from

$$H_{p,q+1} = H_{p,q} + \frac{\Delta t}{\Delta x}(F_{p+1/2,q} - F_{p-1/2,q}) + M\,\Delta t. \tag{9.39}$$

The time step that can be used is rather small. For a linear diffusion equation (that is, an equation like (9.33) but with constant diffusivity, D), the criterion for stability is (Smith, 1985)

$$\Delta t \le \frac{(\Delta x)^2}{4D}. \tag{9.40}$$

In first approximation, this criterion may be used to estimate the maximum allowable time step for the nonlinear equation (9.39). A typical value for D is about 10^7 m²/yr but may be (much) larger near the edge of the ice sheet or at the start of an integration when the ice sheet is far from equilibrium. Taking a moderate grid spacing of 10 km, the maximum time step is only about 2.5 yr.

To numerically integrate the continuity equation, boundary conditions need to be imposed at the ends of the model domain. Strictly speaking, the starting profile (which may be zero ice thickness everywhere) is also a boundary condition, applying to $t = 0$ yr. However, it is common usage to refer to this "boundary" condition as the initial condition. Which other boundary conditions are used depends on the particular ice sheet being modeled. Of course, one should be careful not to impose boundary conditions that are too restrictive, such as prescribing the ice thickness at both ends of the model domain, or the ice flux at one end and the ice thickness at the other end.

As an example, consider the simple geometry of a horizontal continent bounded on one side by a deep ocean. At this end, the ice thickness is set to zero because the ice sheet cannot advance into the deep ocean. The surface mass balance is prescribed as a function of distance only, and decreasing toward the south

$$M = 0.5 - 0.5 \cdot 10^{-6}\,x, \tag{9.41}$$

with M in m ice depth per year and x the distance (in m) from the northern edge. M becomes negative (representing ablation) at $x = 1000$ km. Because the surface mass balance becomes sufficiently negative for a large distance, no boundary condition

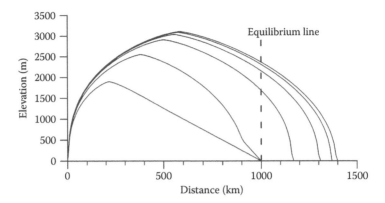

FIGURE 9.5 Results from a numerical integration of an ice sheet on a horizontal bed, bounded to the north (at 0 km) by a polar ocean. Profiles are shown for 1, 5, 10, 20, 30, 40, and 50 kyr of simulated time. The surface mass balance decreases linearly toward the south (in the direction of the x-axis).

needs to be prescribed at the lower end, other than that the ice thickness cannot become negative. Figure 9.5 shows the results of an integration starting with zero ice thickness everywhere. The horizontal grid spacing, Δx, equals 10 km and the time step, Δt, one year. At first, snow accumulates in areas at or above sea level where the surface mass balance is positive. After the ice cover becomes sufficiently thick, the ice starts to flow toward both margins, and in particular into the region with surface ablation. The southern edge of the ice sheet jumps from one gridpoint to the next, but this does not affect the results in a serious way. Changing the grid spacing leads to virtually the same ice-sheet profile as that shown in Figure 9.5. Apparently, if the ice-sheet edge is located in a region with negative surface mass balance, the margin position is well-defined and determined by large-scale ice sheet dynamics and consideration of mass conservation, rather than by the small-scale mechanics of the glacier snout.

The reason that the glacier snout advances is that the ice flux just beyond the edge is not zero, but determined by the diffusivity at the last gridpoint covered with ice. Thus, at the next gridpoint, the diffusion term in the continuity equation is nonzero, and positive in general, allowing a small layer of ice to form. This scheme works well where the bed is above sea level, but another treatment of the movement of the glacier edge is needed where the bed slopes downward below sea level. If the position of the glacier edge is defined by the flotation criterion (that is, H is set to zero when the ice becomes too thin to rest on the seafloor), the edge cannot advance. Using the continuity equation (9.33) to calculate the thickness change at the first gridpoint beyond the edge, the new ice thickness is a few meters at most, depending on the time step. In general, this is too small to overcome the flotation criterion where the bed is more than a few meters below sea level, and the thickness is set to zero. To allow for expansion of an ice sheet into a (shallow) sea, the formation of an ice shelf needs to be considered.

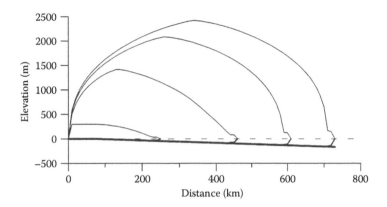

FIGURE 9.6 Results from a numerical integration of an ice sheet on a bed with constant slope, bounded to the north (at 0 km) by a polar ocean. Profiles are shown for 1, 5, 10, and 20 kyr of simulated time. The surface mass balance is taken constant.

The flow of ice shelves is governed by a balance between driving stress and resistance from gradients in longitudinal stress and lateral drag. As a result, modeling ice-shelf flow is essentially different from modeling lamellar flow, in which the driving stress is balanced entirely by drag at the glacier bed. It is possible to couple the two flow regimes to obtain a complete model for a grounded ice sheet discharging into an ice shelf to study the migration of the grounding line in more detail as discussed in Section 9.4. However, as a start, a simpler procedure can be adopted. Ice shelves react rather quickly to changes in environmental conditions (Van der Veen, 1986) and, in approximation, may be taken to be in equilibrium with their environment. In that case, the ice-shelf profile can be calculated using the equilibrium profiles derived in Section 5.5. Migration of the ice-sheet edge becomes possible if the ice shelf becomes sufficiently thick to run aground.

Figure 9.6 shows the growth of an ice sheet on a sloping bed. At the northern edge, the bed is just above sea level over a distance of 70 km, in order to initiate ice-sheet growth. The surface accumulation rate is constant, equal to 0.3 m/yr. The grounding line is assumed to be at the last gridpoint with grounded ice, and the thickness and speed at this point are used in the calculation of the ice-shelf profile (equation (5.77) for positive surface mass balance). As the grounded portion of the ice sheet builds up, the ice shelf also becomes thicker, until its thickness exceeds the flotation thickness and the ice-sheet edge advances one gridpoint.

It is important to test numerical models, whenever possible, against available data or against analytical solutions. Figure 9.7 shows the profile of an ice sheet on a horizontal bed with constant surface accumulation as calculated with the numerical model and according to the Vialov-profile (5.39). In the numerical model, the upstream boundary condition is $H(1) = H(2)$, so the ice divide is halfway in between the first two gridpoints. Prescribing the half-width at 505 km, the edge of the ice sheet is at x = 510 km. As the figure shows, the computed and analytical profiles are

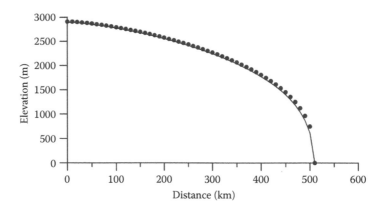

FIGURE 9.7 Comparison between the steady state ice-sheet profile calculated with the numerical model, and the analytical equilibrium profile (represented by the dots).

virtually the same. Of course, this is not a validation of the model per se. Rather, it shows that the numerical integration scheme is consistent with the analytical solution of the equations that the model tries to solve. Whether or not these equations actually apply depends on the particular situation being modeled. Time-dependent analytical solutions against which the model can be compared are discussed in Section 10.3.

The numerical model described above is readily expanded to include both horizontal directions. In that case, the diffusivity becomes

$$D = CH^{n+2}\left(\left(\frac{\partial h}{\partial x}\right)^2 + \left(\frac{\partial h}{\partial y}\right)^2\right)^{(n-1)/2}, \tag{9.42}$$

and the continuity equation is

$$\frac{\partial H}{\partial t} = +\frac{\partial}{\partial x}\left(D\frac{\partial h}{\partial x}\right) + \frac{\partial}{\partial y}\left(D\frac{\partial h}{\partial y}\right) + M. \tag{9.43}$$

These equations can be solved similarly as for the one-dimensional case (equation (9.39)), but to keep the solution stable, a relatively coarse horizontal grid has to be used (for example, Oerlemans, 1982). To allow for a finer grid size, most models employ somewhat different and more efficient solution schemes. For example, the Glimmer community ice sheet model includes three methods for solving this equation on a discrete numerical grid, namely, the alternating direction implicit method, a linearized semi-implicit method, and a nonlinear scheme (Rutt et al., 2009). These three methods are briefly described below.

To improve numerical stability, two staggered horizontal grids are used, with the same grid spacings, Δx and Δy, and offset by half a grid spacing (Figure 9.8). Following the notation of Rutt and others (2009), the grids are referred to as the (i, j) grid and the (r, s) grid. Ice thickness and ice flux are calculated on the (i, j) grid,

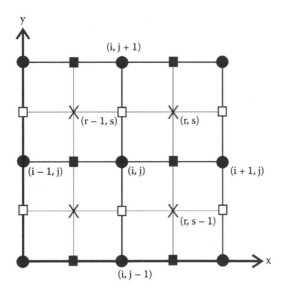

FIGURE 9.8 Grid geometry used in the Glimmer community ice sheet model. Ice thickness is known on the (i, j) grid (filled circles); the ice flux is calculated on the (i, j) grid at intermediate points: filled squares indicate where the flux in the x-direction is calculated, and open squares indicate where the flux in the y-direction is calculated. Diffusivities are calculated on the (r, s) grid, indicated by the X.

while diffusivities are calculated on the (r, s) grid. The diffusivity is proportional to the surface slope in both directions and these are estimated as

$$\left(\frac{\partial h}{\partial x}\right)_{r,s} = \frac{h_{i+1,j} - h_{i,j} + h_{i+1,j+1} - h_{i,j+1}}{2\,\Delta x}, \tag{9.44}$$

$$\left(\frac{\partial h}{\partial y}\right)_{r,s} = \frac{h_{i,j+1} - h_{i,j} + h_{i+1,j+1} - h_{i+1,j}}{2\,\Delta y}. \tag{9.45}$$

Ice thickness at gridpoint (r, s) is obtained by averaging values at the four surrounding gridpoints

$$H_{r,s} = \frac{1}{4}\left(H_{i,j} + H_{i+1,j} + H_{i+1,j+1} + H_{i,j+1}\right). \tag{9.46}$$

The ice flux in the x-direction is calculated at halfway points in the (i, j) grid (Figure 9.8; compare with equation (9.38)) as

$$F^{x}_{i+1/2,j} = \frac{1}{2}(D_{r,s} + D_{r,s-1})\frac{h_{i+1,j} - h_{i,j}}{\Delta x}. \tag{9.47}$$

Similarly, the ice flux in the other horizontal direction is

$$F^y_{i,j+1/2} = \frac{1}{2}(D_{r,s} + D_{r-1,s})\frac{h_{i,j+1} - h_{i,j}}{\Delta x}.$$ (9.48)

The first approach to discretize the continuity equation is the alternating direction implicit (or ADI) method, based on the Peaceman–Rachford formula (Mitchell and Griffiths, 1980, p. 59; c.f. Press et al., 1992, p. 861 ff). Although this scheme is not unconditionally stable because of the nonlinear dependence of the ice speed on the surface slope, it allows time steps that are an order of magnitude larger than can be used in the explicit approach, without a noticeable loss of accuracy (Huybrechts, 1992). The ADI method involves a two-step procedure. First, the ice thickness is calculated at the intermediate time, $t + 1/2\ \Delta t$, using an implicit scheme for the x-derivatives and the old values for the y-derivatives. Thus, the continuity equation is discretized as

$$2\frac{H^{t+1/2}_{i,j} - H^t_{i,j}}{\Delta t} = \frac{F^{x,t+1/2}_{i+1/2,j} - F^{x,t+1/2}_{i-1/2,j}}{\Delta x} + \frac{F^{y,t}_{i,j+1/2} - F^{y,t}_{i,j-1/2}}{\Delta y} + M_{i,j}.$$ (9.49)

Collecting all terms involving $t + 1/2$ on the left-hand side gives a set of tridiagonal equations for each grid row (y = constant) that can be solved for each row separately. Next, the implicit direction is reversed: the intermediate values are used to estimate the x-derivatives, while the y-derivatives are approximated by an implicit scheme. That is, the continuity equation is approximated as

$$2\frac{H^{t+1}_{i,j} - H^{t+1/2}_{i,j}}{\Delta t} = \frac{F^{x,t+1/2}_{i+1/2,j} - F^{x,t+1/2}_{i-1/2,j}}{\Delta x} + \frac{F^{y,t+1}_{i,j+1/2} - F^{y,t+1}_{i,j-1/2}}{\Delta y} + M_{i,j}.$$ (9.50)

Again, the resulting set of equations is solved for each grid column (x = constant) separately. A fuller description of this method can be found in Huybrechts (1992) or Rutt and others (2009).

The second numerical solution scheme implemented in the Glimmer model is a linearized semi-implicit method using the Crank–Nicholson scheme. Analogous to equation (9.26), the continuity equation is discretized as

$$\frac{H^{t+1}_{i,j} - H^t_{i,j}}{\Delta t} = \frac{F^{x,t+1}_{i+1/2,j} - F^{x,t+1}_{i-1/2,j}}{2\Delta x} + \frac{F^{y,t+1}_{i,j+1/2} - F^{y,t+1}_{i,j-1/2}}{2\Delta y}$$

$$+ \frac{F^{x,t}_{i+1/2,j} - F^{x,t}_{i-1/2,j}}{2\Delta x} + \frac{F^{y,t}_{i,j+1/2} - F^{y,t}_{i,j-1/2}}{2\Delta y} + M_{i,j}.$$ (9.51)

The ice flux at time $t + 1$ is calculated using diffusivities from the current time step and surface elevations at $t + 1$. That is

$$F^{x,t+1}_{i+1/2,j} = \frac{1}{2}(D^t_{r,s} + D^t_{r,s})\frac{h^{t+1}_{i+1,j} - h^{t+1}_{i,j}}{\Delta x}.$$ (9.52)

Equation (9.51) can be rearranged by collecting all the terms that involve thickness and surface elevation at time $t + 1$ on the left-hand side and moving the remaining terms to the right-hand side. This yields a set of $i_{max} \times j_{max}$ equations that can be solved using an iterative matrix solver. The lengthy equations can be found in Rutt and others (2009).

The third scheme to solve the continuity equation involves Picard iteration. In essence, the discrete continuity equation (9.51) is solved multiple times with diffusivities recalculated at each iteration step using the geometry from the previous iteration. This procedure is repeated until for all gridpoints the difference in ice thickness between successive iterations is smaller than some prescribed threshold value (Rutt et al., 2009).

9.4 FLOWBAND MODEL

The flowline model discussed in Section 9.3 is based on the assumption that the driving stress is balanced by drag at the glacier base (lamellar flow). This is not a realistic model for marine terminating outlet glaciers where gradients in longitudinal stress and lateral drag may be important in the balance of forces. Several full-stress flowband models have been developed, but these are based on the assumption of plane flow and do not incorporate lateral drag (for example, Hindmarsh, 2004; Price et al., 2007; Aschwanden and Blatter, 2009). Van der Veen and Whillans (1996) describe a model for an active ice stream with resistance to flow partitioned between basal and lateral drags. That model was extended to include gradients in longitudinal stress by Nick et al. (2009, 2010, 2012) and is described in this section.

The continuity equation (5.18) expresses conservation of ice volume along a flowband of constant width. To allow for diverging or converging flow, the width along the flowband is allowed to vary. Denoting the cross-sectional area by S, the continuity equation becomes

$$\frac{\partial S}{\partial t} = -\frac{\partial F}{\partial x} + 2MW, \tag{9.53}$$

where F represents the ice flux through the cross-section and W the half-width measured at the glacier surface. Neglecting effects from sloping side walls, the ice flux is given by $F = 2\,H\,U\,W$, with U the depth-averaged horizontal ice velocity averaged over the width of the glacier. Van der Veen and Whillans (1996) allow the width to change over time as the ice stream margins migrate inward or outward, but here the width is considered fixed in time. The continuity equation is then

$$\frac{\partial H}{\partial t} = -\frac{1}{2W}\frac{\partial F}{\partial x} + M. \tag{9.54}$$

The next step is to derive an equation from which the velocity, U, can be estimated. This requires consideration of force balance.

Force balance is expressed by equation (3.22). Most outlet glaciers are moving rapidly as a result of basal sliding, and the resistive stresses, R_{xx} and R_{yy}, may be considered constant with depth, and the balance equation becomes

$$\tau_{dx} = \tau_{bx} - \frac{\partial}{\partial x}(H R_{xx}) - \frac{\partial}{\partial y}(H R_{xy}). \tag{9.55}$$

The driving stress is calculated from the geometry of the glacier as

$$\tau_{dx} = -\rho g H \frac{\partial h}{\partial x}, \tag{9.56}$$

where ρ represents the ice density and h the surface elevation, taken to be constant in the transverse direction. The resistive stresses on the right-hand side of equation (9.55) can each be written in terms of the horizontal velocity, U.

The contribution from internal deformation to ice flow is ignored and all flow assumed to be associated with basal sliding. Adopting the modified Weertman sliding relation, basal drag is related to the sliding velocity as equation (7.25)

$$\tau_{bx} = \mu B_s N^q U^p, \tag{9.57}$$

where B_s is a sliding constant and μ a bed friction parameter that may be related to bed roughness and amount of basal water. Following Nick et al. (2010), the effective pressure at the bed, N, is set equal to the height above buoyancy (equation (7.57)), implying an easy connection between the subglacial drainage system and the open ocean. The sliding relation (9.57) then becomes

$$\tau_{bx} = \mu B_s \left(H - \frac{\rho_w}{\rho} D \right)^q U^p, \tag{9.58}$$

with ρ_w the density of water, and D the depth of the bed below sea level.

The stretching stress, R_{xx}, is linked to the stretching rate, $\dot{\varepsilon}_{xx}$, through the flow law (3.48). Neglecting the contribution of strain rates other than the stretching rate to the effective strain rate, the longitudinal stress is

$$R_{xx} = 2B\dot{\varepsilon}_{xx}^{1/n} = 2B \left(\frac{\partial U}{\partial x} \right)^{1/n}, \tag{9.59}$$

and

$$\frac{\partial}{\partial x}(H R_{xx}) = 2B \frac{\partial}{\partial x} \left[H \left(\frac{\partial U}{\partial x} \right)^{1/n} \right]. \tag{9.60}$$

The viscosity parameter, B, is taken constant along the flowline. Following the model described in Section 4.4, the assumption is made that resistance to flow from lateral drag is equally important across the full width of the glacier, and the lateral shear stress, R_{xy}, varies linearly across the glacier. Resistance to flow associated with lateral drag is then

$$F_s = \frac{H \tau_s}{W},$$
(9.61)

where τ_s represents the shear stress at the lateral margins. The corresponding transverse velocity profile is

$$U(y) = \left[1 - \left(\frac{y}{W} \right)^{n+1} \right] \frac{2W}{n+1} \left(\frac{\tau_s}{B} \right)^n,$$
(9.62)

with $y = 0$ at the center line. Averaging this expression over the width of the glacier gives

$$U = \frac{2W}{n+2} \left(\frac{\tau_s}{B} \right)^n.$$
(9.63)

Resistance to flow from lateral drag can now be written as

$$F_s = \frac{BH}{W} \left(\frac{(n+2)U}{2W} \right)^{1/n}.$$
(9.64)

Force balance can now be expressed in terms of the velocity, U, and its along-flow gradient, $\partial U / \partial x$. Substituting equations (9.56), (9.58), (9.60), and (9.64) in the balance equation (9.55) yields

$$-\rho g H \frac{\partial h}{\partial x} = \mu B_s \left(H - \frac{\rho_w}{\rho} D \right)^q U^p - 2B \frac{\partial}{\partial x} \left[H \left(\frac{\partial U}{\partial x} \right)^{1/n} \right] + \frac{BH}{W} \left(\frac{(n+2)U}{2W} \right)^{1/n}.$$
(9.65)

Inspection of this equation shows that at each location the velocity depends not only on the local geometry but also on neighboring velocities through the term involving gradients in longitudinal stress (the second term on the right-hand side). This means that the continuity equation cannot be written as a diffusion equation, and each time step the velocity has to be found by numerical iteration.

Nick et al. (2010) solve the force-balance equation (9.65) using the Newton iteration method (Press et al., 1992, Section 9.4). This method, also referred to as the method of tangents, is illustrated in Figure 9.9 and consists of extending the tangent line until it crosses zero to find the next guess. The balance equation is formally written as $G(U) = 0$, with derivative $G'(U) = \partial G(U)/\partial U$. Denoting successive guesses for the velocity by the subscript k, this iteration scheme is

$$U_k = U_{k-1} - \frac{G(U_{k-1})}{G'(U_{k-1})}.$$
(9.66)

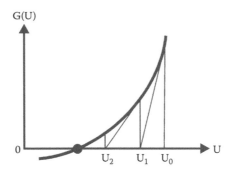

FIGURE 9.9 Illustrating the Newton iteration method.

The correct solution is reached when $G(U)$ is sufficiently small at all gridpoints. The iteration is somewhat complicated because the balance function, $G(U)$, depends not only on the velocity at the gridpoint under consideration but also on the values at the two neighboring gridpoints. Convergence usually occurs within a few iteration steps.

The numerical model employs a variable grid with the first gridpoint fixed at the upstream boundary, taken to be the ice divide. Ice thickness is evaluated at gridpoints, while velocity and the ice flux are calculated at intermediate gridpoints to keep the solution scheme stable. The last gridpoint is at the calving terminus, and the grid spacing is adjusted each time step as the terminus advances or retreats. Because of the high spatial resolution (a few hundred meters), grounding-line motion was found to be independent of the grid spacing. In the case where a floating ice tongue forms, a smooth transition in basal resistance from the grounded ice to the floating part is achieved because basal drag is inversely proportional to the effective basal pressure; this term decreases to near zero toward the grounding line as the glacier thins and approaches flotation.

Two boundary conditions are required for the numerical model. The upglacier boundary corresponds to the ice divide, and the surface slope and ice velocity are set to zero. The boundary condition at the calving terminus follows from the requirement that averaged over the ice thickness, the full stress, $\sigma_{xx}(z) = R_{xx} - \rho g(h - z)$, must be zero. The depth-averaged full stress is

$$\bar{\sigma}_{xx} = \frac{1}{H} \int_{-D}^{h} \sigma_{xx}(z)\,dz = R_{xx} - \frac{1}{2}\rho g H. \qquad (9.67)$$

This stress is partly balanced by water pressure where the bed is below sea level. Averaged over the ice thickness, the water pressure is

$$\bar{\sigma}_{w} = \frac{1}{H} \int_{-D}^{0} \rho_{w} g z\,dz = -\frac{1}{2}\rho_{w} g \frac{D^2}{H}. \qquad (9.68)$$

From these two expressions, and setting $\overline{\sigma}_{xx}$ equal to zero, the resistive stretching stress at the terminus is found to be given by

$$R_{xx}(L) = \frac{1}{2}\rho g\left(H - \frac{\rho_w}{\rho}\frac{D^2}{H}\right). \tag{9.69}$$

For a floating terminus, this expression is equivalent to the solution derived in Section 4.5 for a free-floating ice shelf spreading in the along-flow direction only. Sea ice, or a mélange of icebergs and bergy bits present in the proglacial fjord, may exert a (small) back stress, σ_b, on the calving terminus (Amundson et al., 2010), which would lower $R_{xx}(L)$. Including this back stress, the stretching rate at the terminus is

$$\frac{\partial U}{\partial x} = A\left[\frac{\rho g}{4}\left(H - \frac{\rho_w}{\rho}\frac{D^2}{H} - \frac{\sigma_b}{\rho g}\right)\right]^n. \tag{9.70}$$

In addition to the two velocity boundary conditions at both ends of the model domain, a calving criterion is needed to constrain the position of the calving front. Any of the calving relations discussed in Section 8.5 can be implemented. For the water-depth model, in which calving rate is proportional to the water depth at the terminus, at each time step the new terminus position is determined by the difference between the forward ice velocity and the calving rate causing the terminus to retreat. For the height-above-buoyancy calving model, the terminus position is at the location where the local ice thickness equals the minimum thickness required for the terminus to maintain the specified height above buoyancy. Similarly, for the crevasse-depth models, the position of the terminus is determined by the location where surface crevasses penetrate to sea level or where surface and bottom crevasses penetrate the full ice thickness.

The primary value of the flowband model is its ability to investigate how different external forcings may affect flow and stability of calving glaciers. All physical processes are included in a more or less realistic way: basal sliding, lateral drag, gradients in longitudinal stress, and iceberg calving. Internal deformation and proglacial sedimentation can be readily included if needed (Nick et al., 2009). Vieli and Nick (2011) and Nick et al. (2012) discuss series of model experiments aimed at identifying processes that may lead to large changes on Greenland outlet glaciers. Some of the results obtained with this flow-band model are discussed in Section 8.5.

9.5   CALCULATING THE TEMPERATURE FIELD

As discussed in Section 2.2, the rate of deformation of glacier ice depends strongly on the ice temperature, and glacier flow should be treated as a thermomechanically coupled problem. This means that the velocity and temperature must be determined simultaneously. The first attempt to do this is described by Jenssen (1977), who used a three-dimensional model to numerically solve the continuity and thermodynamic equations. Applying the numerical scheme to the Greenland Ice Sheet led to numerical instabilities, however, forcing time integrations to be discontinued after 1000 years of simulated evolution. To a large extent, this shortcoming may be attributed to the coarse grid used by Jenssen (1977). Available computer capacity restricted the size of the grid to 12 gridpoints in the east–west direction (spacing of 100 km), 12 gridpoints in the north–south direction (spacing of 200 km), and 10 gridpoints in the vertical (spacing varying from 5 to 300 m). In spite of the obvious disappointing performance of his model, the attempt by Jenssen (1977) can be considered the first successful model in which the ice flow is coupled to the temperature in a truly dynamic fashion. Later models are essentially the same as the Jenssen model, but more successful because of increased computer resources (allowing a finer grid) and perhaps more sophisticated numerical techniques than those used by Jenssen (1977).

The thermomechanical model consists of two parts, namely, solving the continuity equation for the evolution of the ice-sheet geometry, and solving the thermodynamic equation to obtain the ice temperature needed to calculate the rate factor in the continuity equation. In the most elaborate models (Jenssen, 1977; Huybrechts and Oerlemans, 1988; Rutt et al., 2009), the temperature is calculated on a two- or three-dimensional grid covering one or two horizontal directions and the vertical. Here, the two-dimensional (x, z) model is discussed to illustrate the principles. Because the ice-sheet geometry varies along the flowline, it is convenient to use the dimensionless vertical coordinate, s, introduced in Section 4.1, with s = 0 at the ice surface and s = 1 at the bed.

Conservation of heat is expressed by the thermodynamic equation (6.17). Horizontal diffusion of heat is assumed to be small compared with vertical diffusion, and internal melting (the last line in equation (6.17)) is neglected. Finally, lamellar flow is considered, with vertical shearing the only strain rate contributing to strain heating. The temperature equation then reduces to

$$\frac{\partial T}{\partial t} = -u\frac{\partial T}{\partial x} - w\frac{\partial T}{\partial z} + K\frac{\partial^2 T}{\partial z^2} + \frac{2}{\rho C_p}\dot{\varepsilon}_{xz}\,\tau_{xz}. \tag{9.71}$$

This equation is valid in the (x, z) coordinate system. To obtain its equivalent for the (x, s) coordinate system, the transformation formulas derived in Section 4.1 are used to write derivatives in the (x, z) system in terms of (x, s) derivatives. However, the model domain may change as the ice sheet evolves, which means that at a certain gridpoint, the level s does not always correspond to the same physical depth. In other words, the temperature T(s) may change as the ice thickness changes, even if the glacier was otherwise in thermal equilibrium. To account for this, the time derivative also needs to be rewritten. Similar to the derivation leading to equation (4.6), the following correspondence is found

$$\left(\frac{\partial T}{\partial t}\right)_z = \left(\frac{\partial T}{\partial t}\right)_s + \frac{1}{H}\left(\frac{\partial h}{\partial t} - s\frac{\partial H}{\partial t}\right)\left(\frac{\partial T}{\partial s}\right)_x. \tag{9.72}$$

Rewriting all derivatives, equation (9.71) becomes, in the (x, s) coordinate system

$$
\left(\frac{\partial T}{\partial t}\right)_s = -\frac{1}{H}\left(\frac{\partial h}{\partial t} - s\frac{\partial H}{\partial t}\right)\left(\frac{\partial T}{\partial s}\right)_x - u(s)\left(\frac{\partial T}{\partial x}\right)_s - \frac{u(s)}{H}\left(\frac{\partial h}{\partial x} - s\frac{\partial H}{\partial x}\right) +
$$

$$
+ \frac{w(s)}{H}\left(\frac{\partial T}{\partial s}\right)_x + \frac{K}{H^2}\left(\frac{\partial^2 T}{\partial s^2}\right)_x + \frac{2}{\rho C_p}\dot{\varepsilon}_{xz}\,\tau_{xz}.
$$

$$(9.73)$$

The next step is to calculate the velocity field and the strain heating at depth. For lamellar flow, the driving stress is balanced by drag at the glacier base, and the vertical shear strain rate becomes (c.f. Section 4.2)

$$
\dot{\varepsilon}_{xz} = A s \tau_{dx}^3,
$$

$$(9.74)$$

using n = 3 for the exponent in the flow law. Integrating from the base (s = 1) to some level s in the ice gives the velocity at that level

$$
u(s) = U_b + 2H\tau_{dx}^3 \int_s^1 A\,\bar{s}^3\,d\bar{s},
$$

$$(9.75)$$

where $U_b = u(1)$ represents the sliding velocity. The vertical velocity follows from the incompressibility condition (1.39). In the (x, s) system, this gives

$$
\left(\frac{\partial w}{\partial s}\right)_x = H\left(\frac{\partial u}{\partial x}\right)_s + \left(\frac{\partial h}{\partial x} - s\frac{\partial H}{\partial x}\right)\left(\frac{\partial u}{\partial s}\right)_x.
$$

$$(9.76)$$

At the surface, the vertical velocity is (taken positive when upward)

$$
w(0) = \frac{\partial h}{\partial t} + u(0)\frac{\partial h}{\partial x} - M.
$$

$$(9.77)$$

Integrating equation (9.76) from the surface to some depth s gives the vertical velocity

$$
w(s) = \frac{\partial h}{\partial t} + u(s)\left(\frac{\partial h}{\partial x} - s\frac{\partial H}{\partial x}\right) + \int_0^s \left(H\frac{\partial u}{\partial x} + u\frac{\partial H}{\partial x}\right)d\bar{s}.
$$

$$(9.78)$$

Finally, strain heating at level s is

$$
Q(s) = \frac{2\tau_{dx}^4}{\rho C_p} A s^4.
$$

$$(9.79)$$

With these expressions, the change in temperature can be calculated. The horizontal velocity is determined first, using the old temperatures to calculate the rate factor at depth. Next, the vertical velocity and strain heating are calculated. Substitution into the thermodynamic equation (9.73) gives $\partial T/\partial t$, allowing the new temperature to be estimated for time $t + \Delta t$.

The most efficient numerical scheme for solving the thermodynamic equation is a semi-implicit scheme. To illustrate how such a scheme is constructed, equation (9.73) is written as

$$\frac{\partial T}{\partial t} = a\frac{\partial^2 T}{\partial s^2} + b\frac{\partial T}{\partial s} + c\frac{\partial T}{\partial x} + d, \tag{9.80}$$

in which the coefficients, a, b, c, and d, are depth dependent and calculated using the old temperature. The horizontal advection term is also estimated from the old temperatures. The finite difference form of equation (9.80) is obtained analogous to the Crank–Nicholson approximation (9.27) discussed in Section 9.2. The result is

$$\left(-\frac{a}{2(ds)^2} + \frac{b}{4\,ds}\right)T_{p,r-1,q+1} + \left(\frac{1}{\Delta t} + \frac{a}{(ds)^2}\right)T_{p,r,q+1} +$$

$$+ \left(-\frac{a}{2(ds)^2} - \frac{b}{4\,ds}\right)T_{p,r+1,q+1} =$$

$$= \left(-\frac{a}{2(ds)^2} - \frac{b}{4\,ds}\right)T_{p,r-1,q} + \left(\frac{1}{\Delta t} - \frac{a}{(ds)^2}\right)T_{p,r,q} + \tag{9.81}$$

$$+ \left(\frac{a}{2(ds)^2} + \frac{b}{4\,ds}\right)T_{p,r+1,q} + \frac{c}{\Delta x}\left(T_{p,r,q} - T_{p-1,r,q}\right) + d.$$

In this expression, the subscripts p and r refer to the horizontal and vertical directions, respectively, and the subscript q to the time steps. Note that the horizontal advection term is approximated by upstream differencing to improve stability. At every horizontal gridpoint, $p\cdot\Delta x$, this set of tridiagonal equations is solved for the new temperatures at depth levels $r\cdot ds$ to obtain the new temperature at time $(q + 1)\cdot\Delta t$. The boundary conditions are the surface temperature,

$$T(0) = T_{air} - \gamma_a h, \tag{9.82}$$

where T_{air} represents the air temperature at sea level and γ_a the atmospheric lapse rate, and the temperature gradient at the glacier base,

$$\left(\frac{\partial T}{\partial s}\right)_{s=1} = +\frac{\gamma_g}{H} + \frac{\tau_{dx}\,U_b}{k\,H}. \tag{9.83}$$

In this last expression, γ_g represents the geothermal heat flux and k the thermal conductivity of ice. The first term on the right-hand side of equation (9.83) represents inflow of geothermal heat and the second term generation of heat at the glacier bed by basal sliding. Note that while the temperature near the base should decrease from the bed upward, the temperature gradient $\partial T/\partial s(1)$ is positive because the dimensionless vertical coordinate, s, increases toward the bed.

The evolution of the ice sheet is described by the continuity equation. The solution scheme is similar to that described in the previous section. The only difference is that the depth-averaged rate factor used in equation (9.34) for the diffusivity needs to be replaced by the vertical-mean value based on the calculated temperatures. This leads to the following expression for the diffusivity:

$$D = 2(\rho g)^3 H^5 \left|\frac{\partial h}{\partial x}\right|^2 \int_0^1 \int_s^1 A \bar{s}^3 \, d\bar{s}. \tag{9.84}$$

The Glimmer community ice-sheet model uses an iterative scheme to solve the temperature equation. For each column under consideration, the horizontal advection terms are estimated using temperatures at adjacent columns from the previous iteration. The remaining unknown temperatures in the column are discretized using the time-implicit Crank–Nicholson method to form a tridiagonal matrix equation. This scheme usually converges within a few iteration steps (Rutt et al., 2009).

9.6   GEODYNAMICS

The weight of ice sheets exerts a pressure on the underlying bedrock, which in response will deflect downward to restore equilibrium of forces. The deflection can be considerable under the central parts of ice sheets. For example, the Antarctic Ice Sheet is in places more than 4500 m thick. The equilibrium adjustment of the bed is roughly one-third of the ice thickness, so that the downward depression is about 1500 m. Generally, the mass balance at the glacier surface depends strongly on the elevation of the surface, and adjustment of the bedrock may be expected to have a large effect on the evolution of ice sheets. It is therefore important to include geodynamics in a numerical model that is used to simulate ice-sheet evolution over longer periods of time.

In most models, adjustments of the bed are calculated using a two-layer model for the earth deformation. The outer shell, or lithosphere, consists of rocks that are relatively cool and rigid and that do not deform significantly. The thickness of the lithosphere is about 100 km underneath ocean basins and about 200 km beneath continents. This is small compared with the radius of the earth, so that the lithosphere may be considered a thin elastic plate. Its lower boundary is often defined by the 1600 K isotherm, because rock beneath this isotherm is sufficiently hot for solid-state creep to occur. This lower layer of warm and deforming rock is called the asthenosphere. On geological time scales, the asthenosphere may be considered as a layer of viscous material.

The simplest model for bedrock adjustment is based on the principle of local isostatic equilibrium, illustrated in Figure 9.10. The lithosphere is assumed to be in

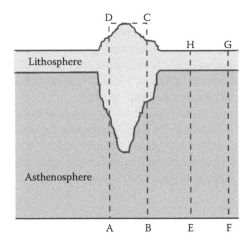

FIGURE 9.10 Illustration of isostatic equilibrium. The lithosphere under a surface load (mountain) bends downward such that the total mass in a column remains constant. (Reproduced from Oerlemans, J., and C. J. van der Veen, *Ice Sheets and Climate*, Reidel Publishing Co., Dordrecht, 1984. With permission from Springer Verlag.)

floating equilibrium with the underlying heavier substrate in the asthenosphere, and the total mass of a vertical column is constant. Thus, when a mountain or ice sheet is present, the lithosphere (made up of relatively light rock) will bend downward, replacing part of the heavier material in the asthenosphere, such that the total mass of column ABCD equals that of column EFGH. The deflection, w (taken positive when downward) follows immediately from the thickness of the ice sheet, H, and the ratio of ice density, ρ, to that of the substratum, ρ_m

$$w = \frac{\rho}{\rho_m} H \approx$$
$$\approx \frac{1}{3} H. \tag{9.85}$$

This relation is often used in analytical models that consider the equilibrium profile of ice sheets. However, the deflection is not instantaneous. Because of the viscous flow in the asthenosphere, isostatic equilibrium will be reached after thousands of years. This lagged response introduces an important feedback mechanism. As a start, the delay can be included in a numerical model by calculating bedrock sinking from (Oerlemans, 1980)

$$\frac{\partial b}{\partial t} = -\frac{1}{T_b}\left(\frac{1}{3}H + b - b_o\right), \tag{9.86}$$

where b represents the elevation of the bed (negative below sea level) and b_o the unperturbed bed elevation in the absence of ice. T_b is a time scale indicating how fast the system returns to isostatic equilibrium ($b = b_o - H/3$). The value of this time scale strongly influences the behavior of the model ice sheet (Oerlemans, 1980), so it seems worthwhile to use a more realistic model for bedrock adjustment that does not require prescribing the time scale for adjustment. This requires a model for the deflection of the lithosphere that takes into account the elastic properties of this layer, and a model that describes the viscous flow in the asthenosphere.

Deflection of the lithosphere is similar to the bending of a rigid elastic plate subject to a vertical load, q, and can be calculated from the general equation (c.f. Turcotte and Schubert, 2002, ch. 3)

$$D\nabla^4 w = q - (\rho_m - \rho)gw, \tag{9.87}$$

in which w represents the deflection, again taken positive when downward, and D the flexural rigidity of the lithosphere. The second term on the right-hand side represents the upward buoyancy force that arises from the displacement of heavier substratum. This force tends to restore the undisturbed situation ($w = 0$). Equation (9.87) is a linear equation, which means that the *total* deflection of the lithosphere under a surface load can be found by considering the sum of (many) point loads. In a numerical ice-sheet model, the ice thickness is known at discrete gridpoints, and each of these thicknesses may be considered a point load. Calculating the deflection caused by each of these point loads and adding the results gives the total deflection.

Consider first the one-dimensional case in which the ice-sheet profile is computed along a flowline. Equation (9.87) then reduces to

$$D\frac{d^4w}{dx^4} + (\rho_m - \rho)gw = q(x). \tag{9.88}$$

For a point load, Q_i, applied at $x = 0$, the solution is (Turcotte and Schubert, 2002, p. 125)

$$w_i(x) = \frac{Q_i}{8D}\alpha^3 e^{-\tilde{x}}(\cos\tilde{x} + \sin\tilde{x}), \tag{9.89}$$

where

$$\tilde{x} = \frac{|x|}{\alpha},$$

$$\alpha = \left(\frac{4D}{\rho_m g}\right)^{1/4}. \tag{9.90}$$

This solution is shown in Figure 9.11. An interesting and important feature is the presence of a forebulge at a distance of approximately 3α from the load. This forebulge results from the rigidity of the lithosphere. Observations suggest that the value of α is about 180 km (Walcott, 1970). With $\rho_m = 3300$ kg/m^3, this gives $D = 85 \cdot 10^{24}$ Nm for the flexural rigidity of the lithosphere. The value of D does not greatly influence the actual deflection, but the position of the forebulge is affected by this value.

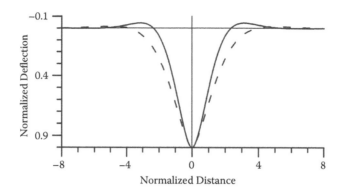

FIGURE 9.11   Deflection of the lithosphere under a point load applied at x = 0. The full curve represents the one-dimensional solution (9.89) and the dashed curve the two-dimensional solution (9.95). The deflection is normalized with the deflection directly under the point load, and the horizontal distance is normalized with the flexural length, α, and radius of relative stiffness, L_b, respectively.

If the gridpoints are denoted by a subscript i, the load at a gridpoint with grounded ice is

$$Q_i = \rho g H \Delta x, \tag{9.91}$$

in which Δx represents the grid spacing. If no grounded ice is present and the bed is below sea level, the weight of water (with density ρ_w) not present in the undisturbed situation needs to be taken into account. Denoting the actual bed elevation by b, and the undisturbed elevation by b_o, this water load is

$$Q_i = \rho_w g (b - b_o). \tag{9.92}$$

For each point load, the corresponding deflection can be calculated from the solution (9.89). The total deflection due to the presence of the entire ice sheet is then

$$w(x) = \sum_{i=1}^{I} w_i(x). \tag{9.93}$$

In Figure 9.12, the lithospheric deflection under the weight of an ice sheet is shown. Also shown in this figure is the deflection calculated from equation (9.85) for local isostatic adjustment. Differences between the two curves are generally small, except near the margins of the ice sheet. Because the topography near the ice-sheet edge may critically influence the position of the edge, especially in the case of a marine ice sheet grounded below sea level, it seems important to consider the elastic properties of the lithosphere when calculating bedrock adjustment.

For two-dimensional models, in which ice flow in both horizontal directions is considered, flexure of the lithosphere can be calculated in a similar manner as for

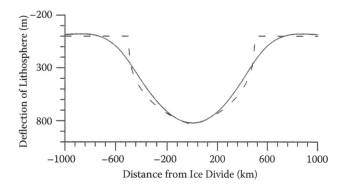

FIGURE 9.12 Deflection of the lithosphere due to the presence of an ice sheet, for local isostatic equilibrium (dashed curve) and taking into account the rigidity of the lithosphere (solid curve). The dashed curve mirrors the ice-sheet profile divided by a factor of 3. (Reproduced from Oerlemans, J., and C. J. van der Veen, *Ice Sheets and Climate*, Reidel Publishing Co., Dordrecht, 1984. With permission from Springer Verlag.)

the one-dimensional case discussed above. The response of the lithosphere to a point load is axisymmetric, so that equation (9.87) may be written as

$$D\left(\frac{\partial^2 w}{\partial r^2} + \frac{1}{r}\frac{\partial w}{\partial r}\right)^2 + (\rho_m - \rho)gw = q(r), \tag{9.94}$$

in which r represents the radial distance to the point load. For a point load, Q_i, applied at r = 0, the solution is (Brotchie and Silvester, 1969)

$$w_i(\tilde{r}) = \frac{Q_i L_b^2}{2\pi D}\,\text{kei}(\tilde{r}), \tag{9.95}$$

in which

$$\tilde{r} = \frac{r}{L_b},$$

$$L_b = \left(\frac{D}{\rho_m g}\right)^{1/4}. \tag{9.96}$$

In this expression, L_b represents the radius of relative stiffness. For the flexural rigidity estimated above, this length scale is about 130 km. The function $\text{kei}(\tilde{r})$ denotes the Kelvin function of zero order. This function is tabulated in mathematical handbooks such as Abramowitz and Stegun (1965), who also give polynomial approximations. The dashed line in Figure 9.11 shows the solution (9.95) as a function of normalized distance form the point load; this solution is symmetric around the vertical at the origin (r = 0). The total lithospheric deflection under an ice sheet is again found by summation of all deflection from point loads at the grid nodes.

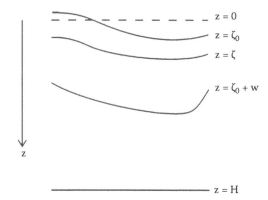

FIGURE 9.13 Geometry used in the derivation of the equation for the rate of bedrock adjustment. Note that the vertical coordinate, z, is positive downward. (Reproduced from Oerlemans, J., and C. J. van der Veen, *Ice Sheets and Climate,* Reidel Publishing Co., Dordrecht, 1984. With permission from Springer Verlag.)

The flexure of the lithosphere controls the equilibrium bedrock adjustment. The rate at which this equilibrium is achieved is controlled by the viscous properties of the asthenosphere. Flow in this layer is induced by pressures exerted by the lithosphere. Similar to the model for viscous flow of subglacial till, described in Section 7.3, the equation of motion in the asthenosphere is (compare equation (7.44) with $\theta = 0$)

$$\frac{\partial^2 u}{\partial z^2} = \frac{1}{\eta}\frac{\partial P}{\partial x}, \tag{9.97}$$

in which u represents the horizontal component of velocity in the asthenosphere and η the viscosity. The geometry is shown in Figure 9.13. As reference level, $z = 0$ is chosen; $z = \zeta_0$ denotes the unperturbed upper boundary of the asthenosphere and $z = \zeta$ the actual upper boundary. The lower boundary is at $z = H$, although this is a rather artificial boundary needed in the following derivation. Note that z is taken positive in the downward direction.

Flow in the asthenosphere will start when the lithosphere deflects downward and will cease when full adjustment to the ice load is achieved. Because deflection of the lithosphere displaces the upper boundary of the asthenosphere, the pressure, P, driving the flow may be written as

$$P = \rho_m g(\zeta_0 - \zeta + w). \tag{9.98}$$

Equation (9.97) can now be integrated twice with respect to z to yield the velocity profile

$$u(z) = \frac{g}{2\eta}(z^2 - 2Hz)\frac{\partial}{\partial x}(\zeta_0 - \zeta + w). \tag{9.99}$$

In deriving this expression, the following boundary conditions are used:

$$u = 0 \quad \text{at} \quad z = \zeta,$$

$$\frac{\partial u}{\partial z} = 0 \quad \text{at} \quad z = H. \tag{9.100}$$

The second boundary condition states that the asthenosphere does not exert a stress on the underlying material.

Considering a vertical column extending from $z = \zeta$ to $z = H$, conservation of mass requires

$$\frac{\partial \zeta}{\partial t} = -\frac{\partial}{\partial x} \int_{\zeta}^{H} u \, dz. \tag{9.101}$$

The integral is approximated as

$$\int_{\zeta}^{H} u \, dz \approx \int_{0}^{H} u \, dz. \tag{9.102}$$

Substituting the velocity profile (9.99) and calculating the integral gives

$$\frac{\partial \zeta}{\partial t} = \frac{g H^3}{3\eta} \frac{\partial^2}{\partial x^2}(\zeta_0 - \zeta + w). \tag{9.103}$$

If the lower boundary $z = H$ remains fixed, this equation describes how the thickness of the asthenosphere evolves. This thickness determines directly the depression of the bedrock under the ice (since the thickness of the lithosphere is assumed constant), so the rate of change of bedrock elevation is given by

$$\frac{\partial b}{\partial t} = \frac{g H^3}{3\eta} \frac{\partial^2}{\partial x^2}(b_0 - b + w). \tag{9.104}$$

For the two-dimensional case, a similar equation that includes flow in the other horizontal direction can be used.

The rate of bedrock adjustment is governed by the diffusion equation (9.104). The characteristic time scale for adjustment is

$$T_b = \frac{3\eta L^2}{g H^3}, \tag{9.105}$$

in which L represents the length scale of the load. This expression shows that T_b increases with the size of the ice sheet above. For small ice sheets, adjustment occurs relatively rapidly, but for large ice sheets, adjustment may take up to tens of thousands of years. This means that the earth's crust is still adjusting after the Northern Hemisphere glacial ice sheets disappeared some 10,000 years ago.

The diffusivity of the asthenosphere is

$$D_a = \frac{gH^3}{3\eta}. \tag{9.106}$$

Neither the thickness, H, nor the viscosity, η, of the asthenosphere is known. However, D_a can be estimated from measurements of glacial uplift. Walcott (1970) calculates a range for D_a from 35 to 50 km²/yr, based on observations in North America and Scandinavia. The results of coupled ice-flow and geodynamics models are rather insensitive to the particular choice of D_a, and of the flexural rigidity of the lithosphere, D, as well. This is probably because the equations that describe geodynamics in the present model are linear. This linearity also suggests that results will be similar for the two-dimensional case.

After bedrock adjustment has reached equilibrium, the height of the forebulge is rather small, about 4% of the depression under the center of the ice sheet (Figure 9.12). However, while the ice sheet is forming, this height may be considerably larger. This is shown in Figure 9.14 for an ice sheet that, after reaching equilibrium, is about 4000 m thick and has a half-width of about 500 km. The forebulge reaches a maximum height of about 240 m and subsequently decreases in size to the equilibrium value of about 15 m. This transient behavior is caused by the diffusive character of the flow in the asthenosphere. While the ice sheet is building up, its weight causes the substratum to spread horizontally, away from directly under the ice sheet. Because the flow is highly viscous, it takes a long time before the surrounding substratum starts to move out as well. As a result, material in the asthenosphere will initially pile up near the edge of the ice sheet until local pressure gradients are sufficiently large to drive this excess substratum farther out. Similarly, when the ice sheet melts away in a short period of time, the bedrock in the vicinity of the edge will sink a few

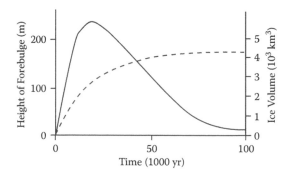

FIGURE 9.14 Growth and decay of the forebulge (solid line) when an ice sheet is building up to equilibrium. The dashed line gives the ice volume. (Reproduced from Oerlemans, J., and C. J. van der Veen, *Ice Sheets and Climate*, Reidel Publishing Co., Dordrecht, 1984. With permission from Springer Verlag.)

hundred meters, while the bedrock under the central part of the (former) ice sheet rises immediately.

The model for bedrock adjustment described above assumes that the lithosphere is continuous, with constant elastic properties. Stern and Ten Brink (1989) argue that this may not be appropriate for the Antarctic shield. They suggest that East and West Antarctica should be considered separately, with free edges at their common boundary (the Transantarctic Mountains). Applying this model results in smaller lithospheric deflection in the Transantarctic Mountains. Stern and Ten Brink (1989) also argue that the flexural rigidity may be much smaller for the Ross Embayment than the estimate given above. This would make bedrock adjustments more local in this region.

Other models have been proposed to describe adjustment of the bedrock under ice sheets (Lambeck and Nakiboglu, 1981; Lingle and Clark, 1985; Peltier, 1982, 1985, 1987). Peltier and Hyde (1987) argue that the time scale for adjustment in the viscous model (about 10,000 yr) is in sharp discord with geophysical constraints. Peltier (1985, 1988) suggests that the modeled transient forebulge may not be realistic, based on a comparison of modeled postglacial rebound with observed records of relative sea level. The problem is that the rheology of the asthenosphere cannot be measured directly, of course, but must be inferred from records of past sea-level stands (indicating postglacial rebound) and changes in ice volume. While the decrease in the total volume of ice stored on the earth during the last deglaciation can be estimated with reasonable accuracy, it is not clear which ice sheets contributed to the following rapid rise in sea level. In particular, the contribution from the West Antarctic Ice Sheet is, as of yet, disputed. Given the considerable uncertainty concerning the location and timing of the (un)loading of the lithosphere by the large ice sheets, inferring the viscous properties of the asthenosphere and deducing the most appropriate rheological model remains speculative. It is not immediately clear, however, that more advanced models for bedrock adjustment will significantly alter the modeled behavior of ice sheets.

9.7   ICE-SHELF MODELS

On floating ice shelves, basal drag is zero and the driving stress is balanced by gradients in longitudinal stress and by lateral drag. As a start, one-dimensional flow along the centerline of an ice shelf is considered, with the x-axis along the direction of flow. The balance of forces for this case is discussed in Section 4.5 where the following balance equation (4.68) is derived.

$$\frac{\partial}{\partial x}(H\,R_{xx}) = \frac{1}{2}\rho g\left(1 - \frac{\rho}{\rho_w}\right)\frac{\partial H^2}{\partial x} - \frac{H\tau_s}{W}. \qquad (9.107)$$

In this expression, ρ and ρ_w represent the density of ice and sea water, respectively. The second term on the right-hand side describes resistance to flow from lateral drag originating at the margins if the ice shelf is laterally confined. The assumption is made that the lateral shearing stress, R_{xy}, varies linearly across the width of the ice shelf, meaning that lateral drag supports the same fraction of driving stress across the full width. The shear stress at the margins, τ_s, is taken constant for now.

Integrating the balance equation (9.107) from the ice-shelf front (at $x = L$) to some distance $L - x$ upglacier gives

$$R_{xx} = \frac{1}{2}\rho g\left(1 - \frac{\rho}{\rho_w}\right)H - \frac{1}{H}\int_x^L \frac{H\tau_s}{W}\,d\bar{x}. \tag{9.108}$$

This expression allows the resistive stress, R_{xx}, to be calculated along the entire ice shelf. Invoking the constitutive relation (4.62), the stretching rate, and thus the along-flow gradient in ice velocity can then be calculated. If the ice velocity at the grounding line is known, the velocity on the ice shelf can be found from

$$U = \int_0^x \dot{\varepsilon}_{xx}\,d\bar{x} + U_o, \tag{9.109}$$

in which U_o represents the ice velocity at the grounding line.

The expressions given above show that the ice-shelf model is essentially different from the ice-sheet model discussed in Section 9.3. For lamellar flow, the driving stress is balanced by drag at the glacier bed, and the ice flux can be calculated from the local geometry (equation (9.28)). The ice-shelf model, on the other hand, requires two integrations along the entire ice shelf. The first integration, starting at the ice-shelf front, yields the resistive normal stress, R_{xx}, and the second integration, starting at the grounding line, gives the velocity, U.

With the velocities and stretching rate known along the entire ice shelf, the time evolution can be calculated from the continuity equation, written in the form similar to equation (5.67):

$$\frac{\partial H}{\partial t} = -U\frac{\partial H}{\partial x} - H\dot{\varepsilon}_{xx} + M. \tag{9.110}$$

Using a forward-in-time numerical scheme, this equation is readily solved. If the ice shelf front is advancing, a new grid needs to be defined at the start of each time step. Given the velocity at the last gridpoint, the new length of the ice shelf can be calculated and a new grid defined. The ice-shelf profile is then transferred from the old to the new grid, and thickness changes calculated at each of the new gridpoints. The thickness at the last gridpoint follows from conservation of mass, but with the restriction that this thickness cannot be larger than that of the gridpoint located just upstream. The time step must be chosen sufficiently small so that there is only one point on the new grid (namely, the last point) where at the beginning of the time step no ice is present. To prevent the ice shelf from becoming unrealistically large, either a critical minimum thickness can be specified or the maximum length of the ice shelf may be prescribed. Alternatively, a more realistic calving criterion (Section 8.5) can be prescribed.

Figure 9.15 shows an equilibrium profile calculated with the model described above. Also shown in this figure is the steady-state solution (5.77) derived in Section 5.5. The thickness and velocity at the grounding line are prescribed and kept constant

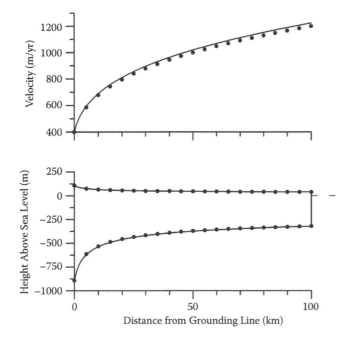

FIGURE 9.15   Equilibrium profile of an unconfined ice shelf as calculated with the numerical flowline model. Dots represent the analytical solution (5.77).

(at 1000 m and 400 m/yr, respectively), while the maximum length is set to 100 km. The velocity calculated with the time-evolving model is somewhat larger than that predicted by the analytical solution, which results in a slightly thicker ice shelf, although differences between the two profiles are undistinguishable in the lower panel of Figure 9.15. Because the velocity is determined by integrating the stretching rate along the centerline, the grid spacing needs to be small (1 km in the example shown here); if the spacing is too large, velocities are overestimated, especially near the grounding line where the stretching rate decreases rapidly in the direction of flow. By replacing the central difference approximation for the gradient in ice velocity with a more sophisticated numerical scheme, this effect could perhaps be avoided.

Including lateral drag has a large effect on the ice shelf. Figure 9.16 shows the steady-state profile of a 20 km wide ice shelf with a shear stress of 100 kPa acting at its lateral margins. The resistance offered by lateral drag allows the shelf to become considerably thicker, while the ice velocity is much smaller (so that the ice flux is the same as for the free-floating case).

The model described above was used by Van der Veen (1986) to investigate some properties of ice-shelf spreading and, in particular, the response of ice shelves to changes at the grounding line. A more elaborate version that includes calculation of the temperature field was applied by Lingle et al. (1991) to a flowband on the Ross Ice Shelf, to assess the response of this ice shelf to a CO_2-induced climatic warming. In both models, the ice speed is calculated from the stretching rate. However, this may not be entirely realistic for the Antarctic ice shelves. Measurements of velocity on the

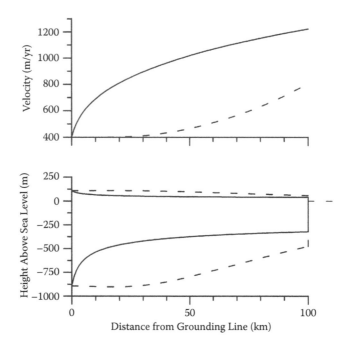

FIGURE 9.16   Effect of lateral drag on the equilibrium profile of an ice shelf. The solid profile is that of an unconfined ice shelf, while the dashed profile represents the equilibrium profile of an ice shelf with a total width of 20 km subject to a shear stress at the margins of 100 kPa.

Ross Ice Shelf suggest that the driving stress is mostly balanced by lateral drag, with gradients in longitudinal stress being small (Thomas and MacAyeal, 1982; Whillans and Van der Veen, 1993b). This suggests that the stretching rate may not be the best quantity to use for estimating the velocity of the ice shelf. Rather, the velocity should be calculated from lateral drag and balance of forces.

If the driving stress is fully supported by drag at the margins, the model developed in Section 5.6 can be used. The numerical scheme for solving the continuity equation for time evolution of the ice shelf is analogous to the scheme discussed in Section 9.3. Again, the ice flux is proportional to the local surface slope, allowing the continuity equation to be written as a diffusion equation, with diffusivity (from equation (5.90))

$$D = \frac{2AH}{n+2} W^{n+1} (\rho g)^n \left| \frac{\partial h}{\partial x} \right|^{n-1}. \tag{9.111}$$

An example of a steady-state profile calculated with this model is shown in Figure 9.17. The width of the ice shelf is 30 km. As boundary condition, the ice thickness as well as the ice velocity at the grounding line are prescribed, to make the model comparable to the ice-shelf model discussed previously. Because the ice velocity is linked to the driving stress, these boundary conditions implicitly prescribe the surface slope at

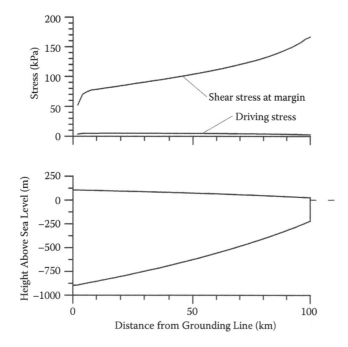

FIGURE 9.17   Equilibrium profile of an ice shelf on which the driving stress is balanced entirely by lateral drag. The total width of the ice shelf is 30 km.

the grounding line as well. This allows the diffusivity at the grounding line (needed to calculate the ice flux at the first intermediate gridpoint) to be calculated.

Contrary to the lamellar flow model described in Section 9.3, ice shelf models are not readily extended to include both horizontal dimensions. This is because resistance to flow from lateral drag or gradients in longitudinal stress is linked to spatial gradients in resistive stresses. This means that the balance equations must be solved by numerical iteration.

Vertical shear is small on ice shelves (Sanderson and Doake, 1979), and horizontal strain rates may be taken constant with depth. With basal drag zero under floating ice shelves, the force-balance equations derived in Section 3.1 become

$$\tau_{dx} = -\frac{\partial}{\partial x}(H R_{xx}) - \frac{\partial}{\partial y}(H R_{xy}), \qquad (9.112)$$

and

$$\tau_{dy} = -\frac{\partial}{\partial x}(H R_{xy}) - \frac{\partial}{\partial y}(H R_{yy}). \qquad (9.113)$$

Using the constitutive relation, the horizontal resistive stresses can be expressed in terms of strain rates (Section 3.3). The weight of an ice shelf is fully supported by the water below, so bridging effects may be neglected. Rewriting the balance equations in terms of velocity gradients gives

$$\frac{\tau_{dx}}{B} = -\frac{\partial}{\partial x}\left(H\dot{\varepsilon}_e^{1/n-1}\left(2\frac{\partial U}{\partial x} + \frac{\partial V}{\partial y}\right)\right) - \frac{\partial}{\partial y}\left(H\dot{\varepsilon}_e^{1/n-1}\frac{1}{2}\left(\frac{\partial U}{\partial y} + \frac{\partial V}{\partial x}\right)\right), \quad (9.114)$$

and

$$\frac{\tau_{dy}}{B} = -\frac{\partial}{\partial x}\left(H\dot{\varepsilon}_e^{1/n-1}\frac{1}{2}\left(\frac{\partial U}{\partial y} + \frac{\partial V}{\partial x}\right)\right) - \frac{\partial}{\partial y}\left(H\dot{\varepsilon}_e^{1/n-1}\left(\frac{\partial U}{\partial x} + 2\frac{\partial V}{\partial y}\right)\right). \quad (9.115)$$

The effective strain rate is given by

$$\dot{\varepsilon}_e^2 = \frac{1}{2}\left(\frac{\partial U}{\partial x}\right)^2 + \frac{1}{2}\left(\frac{\partial V}{\partial y}\right)^2 + \frac{1}{4}\left(\frac{\partial U}{\partial y} + \frac{\partial V}{\partial x}\right)^2. \quad (9.116)$$

In these expressions, U and V represent the components of velocity in the horizontal x- and y-directions, respectively.

Huybrechts (1992) uses the point relaxation technique to solve this set of coupled equations for the two components of velocity. Following his notation, the equations are formally written as

$$a_{i,j}\,U_{i,j} + c_{i,j} = 0, \quad (9.117)$$

and

$$b_{i,j}\,V_{i,j} + d_{i,j} = 0, \quad (9.118)$$

where the subscripts i and j refer to the gridpoint coordinates. The coefficients $a_{i,j}$, $b_{i,j}$, $c_{i,j}$, and $d_{i,j}$ are functions of the driving stress, effective strain rate, and velocities at neighboring gridpoints (c.f. Huybrechts, 1992, for the respective formulas). The neighboring velocities enter into these coefficients after approximating the gradients by central differences. The numerical solution starts with a guess for the two components of velocity and determining the local residuals

$$R\left(U_{i,j}^{(1)}\right) = a_{i,j}^{(1)}\,U_{i,j}^{(1)} + c_{i,j}^{(1)}, \quad (9.119)$$

$$R\left(V_{i,j}^{(1)}\right) = b_{i,j}^{(1)}\,V_{i,j}^{(1)} + d_{i,j}^{(1)}. \quad (9.120)$$

Here, the superscript refers to the iteration step. The next guess for the velocities is

$$U_{i,j}^{(2)} = U_{i,j}^{(1)} - \frac{R\left(U_{i,j}^{(1)}\right)}{a_{i,j}^{(1)}}, \quad (9.121)$$

$$V_{i,j}^{(2)} = V_{i,j}^{(1)} - \frac{R\left(V_{i,j}^{(1)}\right)}{b_{i,j}^{(1)}}. \quad (9.122)$$

If the coefficients were independent of the values at the surrounding gridpoints, the residuals would now be zero. However, because the new velocities affect calculations at the surrounding gridpoints, the procedure has to be repeated until the maximum residual on the grid is sufficiently small. According to Huybrechts (1992), convergence occurs rather slowly, and initially requires a few thousand iteration steps, starting with zero velocities everywhere.

This two-dimensional solution scheme for ice-shelf flow does not allow the continuity equation to be written as a diffusion equation. The ice flux is directly calculated from the ice velocities and ice thickness. This tends to make the solution unstable and may lead to (initially) small-amplitude oscillations in calculated ice thickness. To avoid these oscillations, Huybrechts (1992) applies spatial smoothing. The thickness at each gridpoint is set equal to the average thickness over a nine-grid square, assigning a 5% weight to the values at the eight surrounding gridpoints.

Determann (1991) applies a model similar to that of Huybrechts (1992) to simulate the flow of the Ronne-Filchner Ice Shelf in West Antarctica. The present-day ice thickness is prescribed, with the calving front held fixed at the current average position. At this front, the stretching rate is prescribed from solution (4.65) for an ice shelf spreading freely in both horizontal directions. Along the remaining ice-shelf margins, including around ice rises, the velocity parallel to the margin is set to zero except where the ice shelf is nourished by ice streams; for these ice streams, the width and mass flux across the grounding line are prescribed.

Special treatment is needed for ice rises and ice rumples where the ice shelf has run aground. These obstructions to flow affect the spreading rate and hence the velocity of the ice shelf. Ice rises can reach elevations of several hundred meters above the surrounding ice shelf and may be considered decoupled from the ice shelf. Zero velocity along their margins can be prescribed if the small amount of ice flowing outward into the surrounding shelf ice is ignored. Ice rumples, on the other hand, are more subtle features with surface elevations up to about 30 m higher than the surrounding ice. The most prominent rumples on the Ronne-Filchner Ice Shelf are the Doake Ice Rumples between Korff and Henry Ice Rises. While these rumples obstruct the flow due to local basal drag, the ice there forms an integral part of the ice shelf. Ideally, a full calculation of stresses around these obstacles should be included in the model, but this is not well feasible. Instead, Determann (1991) includes the effect of ice rumples in a parameterized way. Defining the height above buoyancy as

$$H_b = H - \frac{\rho_w}{\rho} D, \qquad (9.123)$$

in which D represents the water depth, a correction factor

$$r = 1 - \frac{H_b}{30}, \qquad H_b < 30\,\mathrm{m}, \qquad (9.124)$$

is applied to the two components of driving stress in the force-balance equations (9.114) and (9.115). If the height above buoyancy is greater than 30 m, the driving stress is zero and the obstruction is considered an ice rise with zero ice velocity. For $H_b = 0$, the correction factor is zero, corresponding to a free-spreading ice shelf.

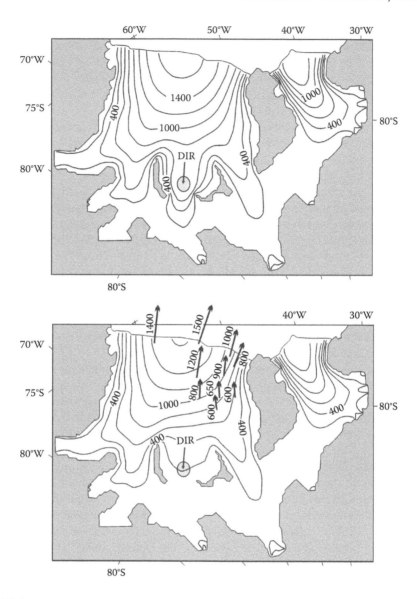

FIGURE 9.18   Contours of calculated ice velocity (interval: 200 m) on the Ronne–Filchner Ice Shelf. The upper panel is the result of a calculation that does not include the effect of the Doake Ice Rumples (DIR) on ice flow, while the lower panel is based on a calculation that accounts for these effects by lowering the local driving stress. Arrows in the lower panel represent observed velocities. (From Determann, J., *Ber. Polarforsch.*, 83, 1991.)

Figure 9.18 shows the effect of the ice rumples (indicated by DIR between the two ice rises) on the simulated flow of the ice shelf. If the obstructing effect of the rumples is not included, local speeds in excess of 800 m/yr are reached. Near the calving front of the Ronne Ice Shelf, velocities become greater than 1600 m/yr, compared with 1200 m/yr on the Filchner Ice Shelf. Including the slowing effect of the ice rumples eliminates the local velocity maximum between the two ice rises and also results in somewhat lower velocities downstream; the Filchner Ice Shelf is unaffected by these changes. The resulting velocity distribution agrees reasonably with measured speeds (indicated by the arrows in the panel on the right). Determann (1991) also conducts time-evolving calculations of the ice-shelf profile to investigate the sensitivity of the ice shelf to changes in surface accumulation and basal melting rates. As expected, interaction with the ocean through basal melting has the largest effect on the shelf thickness (as also concluded from the equilibrium profiles derived in Section 5.5; c.f. Figure 5.11).

Vieli et al. (2006) use observed velocities on the Larsen B Ice Shelf in the Antarctic Peninsula to constrain a two-dimensional ice-shelf model (equations (9.114)–(9.116)) and infer the spatial distribution of the viscosity parameter, B. They find strong weakening of ice in shear zones along the margins and suggest that these weak zones play a major role in controlling the flow of this ice shelf. Subsequent perturbation modeling experiments indicate that the acceleration of the Larsen B Ice Shelf prior to its collapse in 2002 may have resulted from further significant weakening of the already weak shear zones within the ice shelf (Vieli et al., 2007).

10 Dynamics of Glaciers and Ice Sheets

10.1 RESPONSE TO CHANGES IN SURFACE MASS BALANCE

How glaciers react to a (sudden) change in surface mass balance is often described by the kinematic wave theory. In 1885, L. de Marchi was the first to suggest the existence of such waves on glaciers (Marchi, 1911), and the theory was further developed by Finsterwalder (1907). Half a century later, Weertman (1958) and Nye (1958) reintroduced kinematic waves to glaciology and showed how this theory applies to glaciers. In a series of papers, Nye (1961; 1963a, b, c; 1965a, b) developed this theory further. The following discussion is based largely on these papers.

A kinematic wave is described by the following equation:

$$\frac{\partial q}{\partial t} + c\frac{\partial q}{\partial x} = b, \tag{10.1}$$

in which q represents a volume flux, c the speed of the wave, and b a local production term. The term *kinematic wave* was introduced by Lighthill and Whitham (1955a, b), although equation (10.1) was derived some hundred years earlier to describe flood movement traveling downstream on long rivers (Boussinesq, 1877; Seddon, 1900). Another application of the theory is traffic flow on long, crowded roads (Lighthill and Whitham, 1955b). Kinematic waves are essentially different from dynamic waves such as waves on the ocean surface. Newton's equations of motion do not play a role in kinematic waves (other than, perhaps, by linking the wave speed to the volume flux), as they do for dynamic waves that exist because of the inertia terms in the equations of motion. At any location, the kinematic wave possesses only one wave velocity whereas dynamic waves are characterized by at least two velocities traveling in both directions relative to the medium in which the wave occurs. Glacier flow is sufficiently slow that inertia or acceleration terms may be neglected (which reduces Newton's second law to equilibrium of forces), so that dynamic waves do not occur on glaciers.

Kinematic waves owe their existence to conservation of volume or mass (in traffic flow, this would be the number of cars), when a relation exists between the discharge, concentration, and position. The starting point is the continuity equation (5.18) expressing conservation of ice mass, or volume, if the density is taken constant. To keep the analysis simple, one-dimensional flow along the x-axis is considered. The width of the glacier is taken constant, and the bed is horizontal. This means that the terminus of the glacier is on land, and calving tidewater glaciers are excluded from the analysis. For an equilibrium glacier to exist, ablation near the terminus must be sufficiently large to compensate for accumulation in the upper regions of the glacier.

The continuity equation becomes

$$\frac{\partial H}{\partial t} + \frac{\partial (HU)}{\partial x} = M. \tag{10.2}$$

Assuming lamellar flow, the depth-averaged glacier speed is (Section 4.2)

$$U = -A_o H^{n+1} \left| \frac{\partial h}{\partial x} \right|^{n-1} \frac{\partial h}{\partial x}, \tag{10.3}$$

where

$$A_o = \frac{2A}{n+2} (\rho g)^n. \tag{10.4}$$

On a horizontal bed, the slope of the ice surface is equal to the thickness gradient. To simplify the notation,

$$\alpha = -\frac{\partial H}{\partial x}, \tag{10.5}$$

is used in the following analysis.

A reference or datum state is now defined, for example, the steady state of the glacier before the change in surface mass balance, with ice thickness H_o and ice flux $Q_o = H_o U_o$ (rather than considering the ice velocity, U, it is more convenient to consider the ice flux, $Q = HU$). The change in snowfall or ablation results in perturbations from this steady state. The perturbation in ice thickness, H_1 and ice flux, Q_1, are assumed to be small compared with the reference thickness and flux. This assumption allows the perturbation flux to be linearized using a series expansion

$$Q_1 = \left(\frac{\partial Q}{\partial H} \right)_o H_1 + \left(\frac{\partial Q}{\partial \alpha} \right)_o \alpha_1 =$$

$$= C_o H_1 + D_o \alpha_1, \tag{10.6}$$

where

$$C_o = (n+2) A_o H_o^{n+1} \alpha_o^n =$$

$$= (n+2) U_o, \tag{10.7}$$

and

$$D_o = n A_o H_o^{n+2} \alpha_o^{n-1} =$$

$$= \frac{n Q_o}{\alpha_o}. \tag{10.8}$$

These expressions are readily derived by writing $Q = HU$ and substituting equation (10.3) for the ice velocity.

The physical meaning of C_0 and D_0 can be understood by differentiating expression (10.6) for the perturbation flux with respect to time, giving

$$\frac{\partial Q_1}{\partial t} = C_0 \frac{\partial H_1}{\partial t} - D_0 \frac{\partial^2 H_1}{\partial x \partial t}. \tag{10.9}$$

The perturbation thickness can be eliminated from this equation by considering the continuity equation. The reference state satisfies continuity. Substituting $H = H_0 + H_1$ and $Q = Q_0 + Q_1$ for the total ice thickness and ice flux, respectively, into the continuity equation (10.2) and subtracting continuity for the reference state gives

$$\frac{\partial H_1}{\partial t} + \frac{\partial Q_1}{\partial x} = M_1, \tag{10.10}$$

as the continuity equation for the perturbation. In this expression, M_1 denotes the perturbation in surface accumulation. Using this equation to eliminate H_1 from equation (10.9) gives the following expression for the rate of change of Q_1

$$\frac{\partial Q_1}{\partial t} = C_0 M_1 - C_0 \frac{\partial Q_1}{\partial x} + D_0 \frac{\partial^2 Q_1}{\partial x^2} + D_0 \frac{\partial M_1}{\partial x}. \tag{10.11}$$

Comparing this expression with equation (10.1) shows that the perturbation is described by a kinematic wave equation. The first term on the right-hand side describes local production. The second term represents downstream advection of the kinematic wave, while the third term represents horizontal diffusion of the perturbation, with D_0 the diffusivity. The speed at which the wave travels downglacier is C_0; from equation (10.7) it follows that the wave travels at about five times the reference ice velocity.

If the perturbation flux is eliminated from equation (10.10) by substituting (10.6), the following equation for H_1 is obtained:

$$\frac{\partial H_1}{\partial t} = -H_1 \frac{\partial C_0}{\partial x} - \left(C_0 - \frac{\partial D_0}{\partial x} \right) \frac{\partial H_1}{\partial x} + D_0 \frac{\partial^2 H_1}{\partial x^2} + M_1. \tag{10.12}$$

This equation can be solved for the perturbation thickness if C_0 and D_0 are known functions of distance along the flowline. As noted above, C_0 is proportional to the ice velocity, U_0. The ice velocity increases over the accumulation area and is often largest in the middle regions of the glacier; in the ablation area, the velocity decreases, becoming small near the head and snout. A similar variation in C_0 may be expected. Similarly, D_0 may be expected to be of a form similar to the reference mass flux, that is, with a maximum in the middle part of the glacier. Based on these observations, Nye (1963b) adopts simple polynomial forms for these functions, namely,

$$C_0 = \frac{x}{T_c} \left(1 - \frac{x}{\ell} \right), \tag{10.13}$$

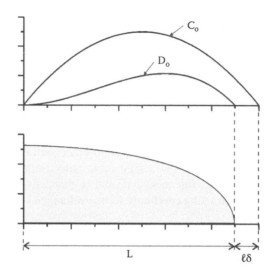

FIGURE 10.1   Along-flow variation in kinematic wave velocity, C_o, and diffusivity, D_o, as given by equations (10.37) and (10.38). For clarity $\delta = 0.1$ instead of the much smaller actual value ~0.006.

and

$$D_o = \frac{E x^2}{T_c}\left(1 - \delta - \frac{x}{\ell}\right). \tag{10.14}$$

The length of the reference glacier is

$$L = \ell(1 - \delta). \tag{10.15}$$

The small quantity, $\delta \sim 0.006$, is introduced to prevent C_o from being zero at the snout of the glacier (Figure 10.1). The parameter T_c is a time scale, estimated by Nye (1963b) to be about 6 yr. The dimensionless parameter, E, is about 0.9 according to Nye (1963b). Substituting expressions (10.13) and (10.14) into equation (10.12) for the perturbation thickness, gives

$$\frac{\partial H_1}{\partial t} = \left(-\frac{1}{T_c} + \frac{x}{2\,T_c\,\ell}\right) H_1 - \left[\frac{x}{T_c}\left(1 - \frac{x}{\ell}\right) - \frac{E x}{2\,T_c}(1 - \delta) + \frac{E x^2}{3\,T_c\,\ell}\right]\frac{\partial H_1}{\partial x} +$$

$$+ \frac{E x^2}{T_c}\left(1 - \delta - \frac{x}{\ell}\right)\frac{\partial^2 H_1}{\partial x^2} + M_1 \tag{10.16}$$

if the perturbation in accumulation is constant along the glacier.

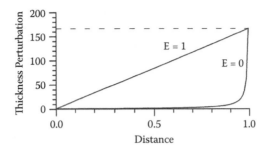

FIGURE 10.2   Change in ice thickness after a uniform stepwise increase in surface accumulation. The thickness perturbation is normalized by $M_1 T_c$, and the distance along the glacier by ℓ. For both solutions, $\delta = 0.006$. The solution for $E = 0$ corresponds to the case in which diffusion along the glacier is neglected.

An analytical solution can be found for a step-change in surface accumulation (or ablation, if M_1 is negative) that is constant along the glacier. For $E = 1$, the resulting change in ice thickness is

$$H_1(x) = M_1 T_c \left(1 + \frac{x}{\delta \ell} \right), \tag{10.17}$$

while for $E = 0$, the solution becomes

$$H_1(x) = M_1 T_c \left(1 - \frac{x}{\ell} \right)^{-1}. \tag{10.18}$$

For other values of E, the solution has to be determined numerically. For all values of E, the change in thickness at the upglacier boundary ($x = 0$) is equal to the perturbation in surface mass balance multiplied by the time scale T_c, while the change at the terminus equals $M_1 T_c / \delta$. A realistic value of E is about 1, and the thickness perturbation increases linearly with distance downglacier. If diffusion is neglected ($E = 0$), the thickness change is almost constant along most of the glacier and increases very rapidly near the terminus. In Figure 10.2 these two solutions are shown.

Adopting the value $E = 1$, the time-dependent solution can be found by first noting that the steady-state solution (10.17) is linear in x. Following Nye (1963b), the time-dependent solution is written as

$$H_1(x,t) = \sum_{k=0}^{K} b_k(t) \, x^k, \tag{10.19}$$

where the coefficients b_k are functions of time only. Substituting in equation (10.16) and equating the coefficients of like powers of x to zero (the equation must be satisfied for each value of x, which means that each coefficient must be zero) gives the solution

$$H_1(x,t) = M_1 T_c - M_1 T_c e^{-t/T_c} + M_1 T_c \left(\frac{1}{\delta} + \frac{2e^{-t/T_c}}{1-2\delta} - \frac{e^{-2\delta t/T_c}}{\delta(1-2\delta)} \right) \frac{x}{\ell}. \quad (10.20)$$

This expression shows that there are two time scales, namely, T_c and $T_c/2\delta$. The first time scale is associated with the speed at which the kinematic wave travels downglacier, while the second, much longer time scale is related to diffusion. With $\delta = 0.006$ and $T_c = 6$ yr, the time scale for diffusion is about 500 yr. It is worthwhile to consider the time scale for adjustment in more detail, and in particular the long diffusion time scale. This requires a closer look at the terminus of the glacier.

For the geometry under consideration, with the snout of the glacier on land and subject to ablation, the ice flux at the terminus approaches zero, but the velocity should remain finite. This means that the velocity must become independent of the ice thickness (which approaches zero at the terminus), so that the ice flux becomes proportional to the ice thickness, rather than proportional to H^4. Denoting the speed at the terminus by U_e, the flux at the terminus is $Q_o(L) = H(L) U_e$, and

$$C_o(L) = U_e, \quad (10.21)$$

instead of equation (10.7). From parameterization (10.13) for C_o, its value at the terminus is

$$C_o(L) = \frac{\delta \ell}{T_c}, \quad (10.22)$$

if terms with δ^2 may be neglected. Combining these two expressions gives the time scale for diffusion

$$T_d = \frac{1}{2} \frac{\ell}{U_e}. \quad (10.23)$$

Since δ is small, ℓ may be replaced by the actual length of the glacier, L, and this equation can be written as

$$T_d = f \frac{L}{U_e}, \quad (10.24)$$

with $f = 0.5$ according to Nye's model.

The problem with the time scale predicted by equation (10.24) is that it depends critically on what happens near the snout of the glacier. Also, equation (10.24) suggests that small glaciers have response times of several hundred years, which does not seem to be supported by records of climate and glacier fluctuations. Jóhannesson

et al. (1989a) argue that the long time scale predicted by Nye's theory is the consequence of the assumed dynamics near the snout and the overestimation of the change
in volume of the glacier after a sudden change in surface mass balance. Jóhannesson
et al. (1989b) discuss this issue in more detail and show that the predicted changes
in volume and speed at the terminus are highly sensitive to the choice of C_0 and D_0
along the glacier.

Van de Wal and Oerlemans (1995) use a numerical flow-band model to study
propagation of kinematic waves on glaciers. According to their results, these waves
travel at a speed of about 6 to 8 times the ice surface velocity. This compares favorably with the linear theory outlined above, which predicts a wave velocity of about
five times the ice surface speed. Van de Wal and Oerlemans (1995) attribute the difference to the increasing gradient in ice velocity in the ablation zone. Using a realistic geometry (the Hintereisferner in Austria), the model predicts an increase of about
10% in the local ice velocity, depending on the position along the glacier and on
the amplitude of the kinematic wave. Because of this moderate increase, kinematic
waves are difficult to detect from measurements of ice velocity alone. Van de Wal
and Oerlemans (1995) argue that synchronous changes in ice velocity on a glacier are
not an indication of kinematic waves and that many of the observations of what have
been classified as kinematic waves may in reality be manifestations of variations in
basal sliding. In order to correctly identify kinematic waves on glaciers, time series
of both ice velocity and surface elevation are needed because, in a kinematic wave,
these two quantities are coupled.

The polynomial forms (10.13) and (10.14) proposed by Nye (1963b) apply to
land-terminating glaciers with zero ice velocity at the terminus. For Greenland
and Antarctic outlet glaciers this is not a realistic assumption. When considering
flowlines extending from the interior to the grounding line, glacier speed is often
observed to increase steadily. To accommodate these observations in the context of
the kinematic wave theory, Van der Veen (2001) adopted an empirical approach.
Using measurements of glacier speed and surface slope on Petermann Glacier in
northwest Greenland, empirical parameterizations for the kinematic wave speed,
C_0, and diffusivity, D_0, can be found through regression. Substituting these relations
into the kinematic wave equation (10.12), glacier response to changes in surface
mass balance can be evaluated by numerically integrating this equation forward in
time. Figure 10.3 shows the calculated response of Petermann Glacier to a sudden
and uniform increase in surface accumulation of 0.2 m/yr. To allow for comparison
with Nye's solution, constant width of the glacier is assumed in this calculation.
Initially, the glacier responds by thickening at a rate equal to the increase in surface
accumulation. At the grounding line, the rate of thickness change decreases rapidly
and a new equilibrium is established after a few hundred years. At the ice divide,
500 km inland from the grounding line, adjustment takes considerably longer and a
new equilibrium is not established until after several thousands of years. The greatest thickening occurs at the ice divide. This result is opposite of the solution derived
by Nye (1963b), which shows the greatest change in ice thickness at the terminus
and no change at the divide (Figure 10.2). The reason for this contrasting response
is that in the parameterization of glacier speed adopted by Nye, no mass is permit-
ted to leave the glacier. The empirical velocity profile adopted by Van der Veen

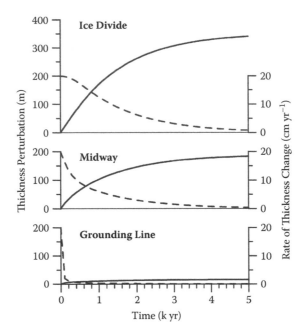

FIGURE 10.3 Ice sheet response to a uniform in surface accumulation (M = 0.2 m/yr). Constant width is assumed. The solid curves (scale on the left) show the thickness perturbation, and the dashed curves (scale on the right) show the rate of thickness change at three points along the 500 km long glacier. (From Van der Veen, C. J., *J. Geoph. Res.,* 106, 34047–34058, 2001. Copyright by the American Geophysical Union.)

(2001) allows part of the increased surface accumulation to be evacuated across the grounding line.

10.2   RESPONSE TO GROUNDING LINE THINNING

The kinematic wave theory discussed in Section 10.1 can also be applied to evaluate glacier response to imposed thinning at the grounding line. As shown by Alley and Whillans (1984), grounding-line thinning will be attenuated upglacier primarily through diffusion. Taking the diffusivity, D_o, constant along the glacier (as suggested by observations on Petermann Glacier; Van der Veen, 2001) and assuming constant flow-band width, the kinematic wave equation (10.12) simplifies to

$$\frac{\partial H}{\partial t} = D_o \frac{\partial^2 H}{\partial x^2}. \tag{10.25}$$

An analytical solution can be found following the similarity approach outlined by Turcotte and Schubert (2002, Section 4-15) for instantaneous heating or cooling of a semi-infinite half space. To simplify the analysis, the position of the grounding line is kept fixed at x = 0. This simplification is not as restrictive as it may appear to be.

Depending on the bed topography, thinning at the grounding line may result in retreat over some distance. In that case, the solution derived in this section applies to the part of the glacier upstream of the new grounding line position. The imposed thinning is then the change in ice thickness at this location.

To find a solution to the diffusion equation (10.25), thickness perturbations are normalized as

$$\hat{H}(x,t) = \frac{H(x,t)}{\Delta H(0)}, \tag{10.26}$$

where $\Delta H(0)$ represents the prescribed instantaneous thickness change at the grounding line at $t = 0$. This transformation does not change the form of the diffusion equation but simplifies the boundary conditions to

$$\hat{H}(x,0) = 0,$$

$$\hat{H}(0,t) = 1, \tag{10.27}$$

$$\hat{H}(\infty,t) = 0.$$

The diffusivity has dimension of $(\text{length})^2/\text{time}$; thus the quantity $\sqrt{D_o t}$ has the dimension of length. This length scale may be considered a characteristic diffusion length and is used to define a nondimensional similarity variable as

$$\eta = \frac{x}{2\sqrt{D_o t}}. \tag{10.28}$$

The factor 2 is introduced to simplify the solution. Using formal similarity theory, it can be shown that the solution, $\hat{H}(x,t)$, is a function of η only.

To rewrite the diffusion equation in terms of derivatives with respect to η, the chain rule for differentiation has to be invoked. Considering first the term on the left-hand side of equation (10.25), this gives

$$\frac{\partial \hat{H}}{\partial t} = \frac{d\hat{H}}{d\eta}\frac{\partial \eta}{\partial t} = \frac{d\hat{H}}{d\eta}\left(-\frac{1}{2}\frac{\eta}{t}\right). \tag{10.29}$$

Similarly,

$$\frac{\partial \hat{H}}{\partial x} = \frac{d\hat{H}}{d\eta}\frac{\partial \eta}{\partial x} = \frac{d\hat{H}}{d\eta}\left(\frac{1}{2\sqrt{D_o t}}\right), \tag{10.30}$$

and the second derivative is

$$\frac{\partial^2 \hat{H}}{\partial x^2} = \frac{\partial}{\partial x}\left[\frac{d\hat{H}}{d\eta}\left(\frac{1}{2\sqrt{D_o t}}\right)\right] = \frac{\partial}{\partial \eta}\left[\frac{d\hat{H}}{d\eta}\left(\frac{1}{2\sqrt{D_o t}}\right)\right]\frac{\partial \eta}{\partial x} = \frac{1}{4}\frac{1}{D_o t}\frac{d^2 \hat{H}}{d\eta^2}. \tag{10.31}$$

Substituting these results in equation (10.25) gives

$$-\frac{1}{2}\frac{\eta}{t}\frac{d\hat{H}}{d\eta} = \frac{1}{4}\frac{1}{D_o t}\frac{d^2\hat{H}}{d\eta^2}D_o,$$ (10.32)

or

$$-\eta\frac{d\hat{H}}{d\eta} = \frac{1}{2}\frac{d^2\hat{H}}{d\eta^2}.$$ (10.33)

Note that introducing the similarity variable, η, has reduced the partial differential equation (10.25) to the ordinary differential equation (10.33) and that the three boundary conditions (10.27) are replaced by the following two:

$$\hat{H}(\infty) = 0,$$

$$\hat{H}(0) = 1.$$ (10.34)

To integrate equation (10.33), define

$$\phi = \frac{d\hat{H}}{d\eta}.$$ (10.35)

Substituting in equation (10.33) gives

$$-\eta\phi = \frac{1}{2}\frac{d\phi}{d\eta},$$ (10.36)

or

$$-\eta d\eta = \frac{1}{2}\frac{d\phi}{\phi}.$$ (10.37)

Integration yields

$$-\eta^2 = \ln\phi - \ln C_1,$$ (10.38)

where C_1 is an integration constant. This equation can be rewritten as

$$\phi = C_1 e^{-\eta^2}.$$ (10.39)

Recalling the definition (10.35) of ϕ, one more integration gives the nondimensional thickness perturbation

$$\hat{H}(\eta) = C_1 \int_0^{\eta} e^{-\bar{\eta}^2} d\bar{\eta} + C_2.$$ (10.40)

At the grounding line, $\eta = 0$ and $\hat{H}(0) = 1$, giving $C_2 = 1$. Using

$$\int_0^\infty e^{-\bar{\eta}^2} \, d\bar{\eta} = \frac{\sqrt{\pi}}{2}, \tag{10.41}$$

applying the first boundary condition in (10.34), the other integration constant is found to be equal to $C_2 = -2/\sqrt{\pi}$.

Finally, the solution can be written as

$$\hat{H}(\eta) = 1 - \frac{2}{\sqrt{\pi}} \int_0^\eta e^{-\bar{\eta}^2} \, d\bar{\eta} = 1 - \mathrm{erf}(\eta) = \mathrm{erfc}(\eta). \tag{10.42}$$

In this equation, $\mathrm{erf}(\eta)$ is the error function and $\mathrm{erfc}(\eta)$ the complimentary error function. Values for these functions can be calculated using standard numerical routines (for example, Press et al., 1992, p. 213).

The solution (10.42) allows thickness perturbations to be calculated for any combination of distance from the grounding line and time after thinning. Figure 10.4 shows how the wave of adjustment travels upglacier following sudden thinning at $x = 0$ and $t = 0$. Figures 10.5 and 10.6 illustrate how the time scale for adjustment increases

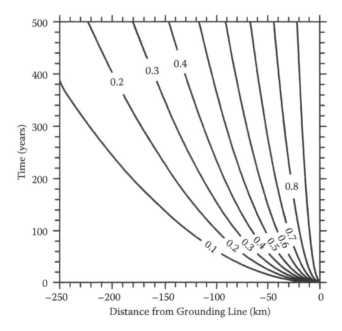

FIGURE 10.4 Transient glacier response to a sudden thinning at the grounding line at $t = 0$. Contours represent the thickness perturbation normalized by the imposed change at the grounding line. (From Van der Veen, C. J., *J. Geoph. Res.*, 106, 34047–34058, 2001. Copyright by the American Geophysical Union.)

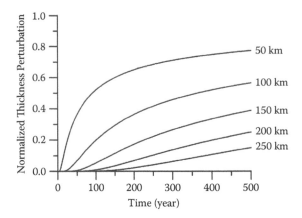

FIGURE 10.5 Normalized thickness perturbations at selected distance upstream of the grounding line after sudden thinning at the grounding line at t = 0. (From Van der Veen, C. J., *J. Geoph. Res.*, 106, 34047–34058, 2001. Copyright by the American Geophysical Union.)

rapidly in the upglacier direction. This is the direct consequence of the assumption that diffusion is the primary controlling process of glacier adjustment. For a diffusion equation such as equation (10.25), a time L^2/D_0 is required for thickness perturbations to propagate upglacier over a distance L. This time scale is proportional to the square of the distance from the grounding line, and the speed at which the wave propagates inland decreases rapidly toward the interior. For Petermann Glacier, $D_0 \approx 30 \times 10^6$ m²/yr, giving a wave propagation speed of ~30 km/yr at the grounding line. At a distance of 100 km inland, this speed has decreased to ~300 m/yr (Van der Veen, 2001).

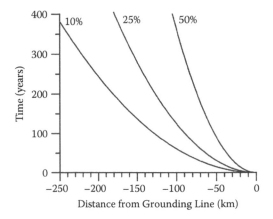

FIGURE 10.6 Time required for partial adjustment (labels indicate the amount of adjustment as a percentage of the equilibrium response) following sudden thinning at the grounding line at t = 0. (From Van der Veen, C. J., *J. Geoph. Res.*, 106, 34047–34058, 2001. Copyright by the American Geophysical Union.)

In the kinematic wave theory, response of an outlet glacier is described as a perturbation to a reference state. Possible feedback mechanisms that may lead to continued grounding-line retreat are not included in this theory. Feedback mechanisms that have been suggested as making marine terminating outlet glaciers unstable are discussed later in this chapter.

10.3   TIME-DEPENDENT SIMILARITY SOLUTIONS

In the previous section, similarity theory is applied to find analytical solutions describing how a sudden perturbation at the grounding line diffuses upglacier. The same methods can be used to find time-dependent exact analytical similarity solutions for the evolving profile of ice sheets. The first of such solutions applied to ice sheets are provided by Halfar (1981) for the case of two-dimensional flow along a flowline, and by Halfar (1983) for an axisymmetric ice cap. That solution was used by Nye (2000) to investigate the polar caps of Mars. Both the Halfar and Nye solutions apply to ice caps with zero surface accumulation. Bueler et al. (2005) expanded on these previous studies by allowing for nonzero accumulation. The motivation for these studies was primarily to provide a set of exact analytical time-dependent solutions that can serve as a tool for evaluating numerical ice-flow models. Bueler et al. (2005) note that "unfortunately the Halfar solution does not appear in the standard [glaciological] textbooks" (p. 291). The aim of this section is to remedy this deficiency.

Following Halfar (1983) and Nye (2000), an axisymmetric ice sheet on a horizontal bed and subject to zero surface accumulation is considered. In that case, the volume of the ice sheet is constant, and as the ice spreads out laterally, the thickness at the center decreases. The following derivation, based on Nye (2000), is somewhat lengthy, and readers not interested in the mathematical details can skip to the results.

The continuity equation for an axisymmetric ice sheet on a flat bed is derived in Section 5.4 and reads (with $M = 0$)

$$\frac{\partial H}{\partial t} = \frac{1}{r}\frac{\partial}{\partial r}\left[r\,A_o\,H^{n+2}\left|\frac{\partial H}{\partial r}\right|^{n-1}\frac{\partial H}{\partial r}\right], \tag{10.43}$$

where H represents the ice thickness and r the radial distance from the center. The radial ice flux, $Q(r, t)$, is given by

$$Q(r,t) = -A_o\,H^{n+2}\left|\frac{\partial H}{\partial r}\right|^{n-1}\frac{\partial H}{\partial r}, \tag{10.44}$$

and the continuity equation can also be written as

$$r\frac{\partial H}{\partial t} + \frac{\partial}{\partial r}(r\,Q) = 0. \tag{10.45}$$

Nye (2000) allows for depression of the initially flat bed by the weight of the ice (assuming equilibrium local isostatic adjustment), but for simplicity that effect is not included here.

As in Section 10.2 the ice thickness is nondimensionalized. For the case under consideration, an appropriate thickness scale is the thickness at the ice divide, $H_o(t)$. Then

$$\hat{H}(r,t) = \frac{H(r,t)}{H_o(t)}. \tag{10.46}$$

Similarly, if $R(t)$ denotes the radius of the ice sheet, a scaled radial coordinate is defined as

$$\hat{r} = \frac{r}{R(t)}. \tag{10.47}$$

The assumption is now made that the scaled ice thickness is a function of $\hat{r}$ only, independent of time, t, and $\hat{H}(r,t) = \hat{H}(r)$. This means that the scaled shape of the ice sheet keeps the same analytical form. However, because both the thickness at the divide and the radius of the ice sheet are allowed to change over time, the aspect ratio of the ice sheet, H/R, will change.

With these definitions, the first term on the left-hand side of the continuity equation (10.45) becomes

$$r\frac{\partial H}{\partial t} = r\frac{\partial}{\partial t}(H_o\,\hat{H}) = r\hat{H}\frac{dH_o}{dt} + rH_o\frac{d\hat{H}}{d\hat{r}}\frac{d\hat{r}}{dt}. \tag{10.48}$$

The term involving the change in scaled radius with time can be eliminated by considering that the volume of the ice sheet must remain constant (zero surface mass balance). This volume is proportional to $H_o R^2$ and thus

$$\frac{d}{dt}(H_o\,R^2) = R^2\frac{dH_o}{dt} + H_o\,2R\frac{dR}{dt} = 0, \tag{10.49}$$

from which it follows that

$$\frac{dR}{dt} = -\frac{R}{2H_o}\frac{dH_o}{dt}. \tag{10.50}$$

This gives

$$\frac{d\hat{r}}{dt} = \frac{d}{dt}\left(\frac{r}{R}\right) = -\frac{r}{R^2}\frac{dR}{dt} =$$

$$= \frac{r}{2RH_o}\frac{dH_o}{dt} = \frac{\hat{r}}{2H_o}\frac{dH_o}{dt}. \tag{10.51}$$

Substituting this expression in equation (10.48) gives

$$r\frac{\partial H}{\partial t} = r\hat{H}\frac{dH_o}{dt} + r\frac{d\hat{H}}{d\hat{r}}\frac{\hat{r}}{2}\frac{dH_o}{dt} =$$

$$= R\hat{r}\frac{dH_o}{dt}\left[\hat{H} + \frac{\hat{r}}{2}\frac{d\hat{H}}{d\hat{r}}\right] = \frac{R}{2}\frac{dH_o}{dt}\frac{d}{d\hat{r}}\left(\hat{r}^2\hat{H}\right). \tag{10.52}$$

The flux term in the continuity equation (10.45) can be rewritten as

$$\frac{\partial}{\partial r}(Qr) = \frac{d}{d\hat{r}}(QR\hat{r})\frac{d\hat{r}}{dr} = \frac{d}{d\hat{r}}(Q\hat{r}). \tag{10.53}$$

The continuity equation can now be written in terms of the normalized radial distance as

$$\frac{R}{2}\frac{dH_o}{dt}\frac{d}{d\hat{r}}(\hat{r}^2\hat{H}) + \frac{d}{d\hat{r}}(Q\hat{r}) = 0. \tag{10.54}$$

Integrating this equation with respect to $\hat{r}$ at constant t gives

$$\frac{R}{2}\frac{dH_o}{dt}(\hat{r}^2\hat{H}) + (Q\hat{r}) = G(t), \tag{10.55}$$

where G(t) is a function of time, t, only. The radial ice flux is then

$$Q = \frac{G}{\hat{r}} - \frac{1}{2}R\hat{r}\hat{H}\frac{dH_o}{dt}. \tag{10.56}$$

At the ice divide, $\hat{r} = 0$, and $Q = 0$ at all times. This boundary condition can only be satisfied if $G(t) = 0$.

For lamellar flow, the ice flux is also given by equation (10.44), which can be written in terms of normalized variables as

$$Q = A_o H^{n+2}\left(-\frac{\partial H}{\partial r}\right)^n = A_o\hat{H}^{n+2}H_o^{n+2}\left[\frac{d}{d\hat{r}}\left(-\hat{H}H_o\right)\frac{d\hat{r}}{dr}\right]^n =$$

$$= A_o\hat{H}^{n+2}H_o^{n+2}H_o^n\left[-\frac{d\hat{H}}{d\hat{r}}\frac{1}{R}\right]^n = A_o\hat{H}^{n+2}H_o^{2n+2}\frac{1}{R^n}\left[-\frac{d\hat{H}}{d\hat{r}}\right]^n. \tag{10.57}$$

Equating both expressions for the ice flux gives

$$-\frac{1}{2}R\hat{r}\hat{H}\frac{dH_o}{dt} = A_o\hat{H}^{n+2}H_o^{2n+2}\frac{1}{R^n}\left[-\frac{d\hat{H}}{d\hat{r}}\right]^n, \tag{10.58}$$

and, after rearranging,

$$2 A_o \frac{\hat{H}^{n+1}}{\hat{r}} \left[-\frac{d\hat{H}}{d\hat{r}} \right]^n = -\frac{R^{n+1}}{H_o^{2n+2}} \frac{dH_o}{dt}.$$

(10.59)

Now recall that the assumption is made that the scaled thickness, $\hat{H}$, is a function of the scaled radial distance, $\hat{r}$, only and independent of time. This means that both sides of equation (10.59) must be constant, say, equal to K. The thickness distribution can then be found by integrating the left-hand side with respect to $\hat{r}$, and the time dependence is obtained by integrating the right-hand side with respect to t.

Consider first the left-hand side of equation (10.59):

$$2 A_o \frac{\hat{H}^{n+1}}{\hat{r}} \left[-\frac{d\hat{H}}{d\hat{r}} \right]^n = K,$$

(10.60)

or

$$-\hat{H}^{1+1/n} d\hat{H} = \left(\frac{K}{2 A_o} \right)^{1/n} \hat{r}^{1/n} d\hat{r}.$$

(10.61)

This equation can be integrated to give

$$-\frac{n}{2n+1} \hat{H}^{2+1/n} = \left(\frac{K}{2 A_o} \right)^{1/n} \frac{n}{n+1} \hat{r}^{1+1/n} + C.$$

(10.62)

By definition, at the ice divide ($\hat{r} = 0$), the scaled ice thickness must equal unity, and $\hat{H}(0,t) = 0$, giving $C = -n/(2n + 1)$ for the integration constant. The second boundary condition is that the ice thickness is zero at the ice margin ($\hat{r} = 1$), and

$$K = 2 A_o \left(\frac{n+1}{2n+1} \right)^n.$$

(10.63)

The scaled thickness distribution can now be written as (Halfar, 1983; Nye, 2000)

$$\hat{H}^{2+1/n} + \hat{r}^{1+1/n} = 1.$$

(10.64)

Note the similarity between this profile and the steady-state profile derived in Section 5.4.

To find an expression for the time-dependent divide thickness, $H_o(t)$, the right-hand side of equation (10.59) is considered. At $t = 0$, the divide thickness equals $\tilde{H}_o$ and the ice sheet radius is $\tilde{R}$. The requirement that the ice volume remains constant then implies

$$H_o(t)R^2(t) = \tilde{H}_o \tilde{R}^2,$$

(10.65)

which can be used to eliminate R(t) from the right-hand side of equation (10.59). This gives

$$H_o^{-5(n+1)/2} \frac{dH_o}{dt} = -\left(\tilde{R}^2 \tilde{H}_o\right)^{-(n+1)/2} K. \tag{10.66}$$

Noting that the right-hand side is constant, integration yields

$$-\frac{2}{5n+3} H_o^{-(5n+3)/2} = -(\tilde{R}^2 \tilde{H}_o)^{-(n+1)/2} K t + C. \tag{10.67}$$

The integration constant follows from the initial boundary condition that $H_o(t) = \tilde{H}_o$ and is given by

$$C = -\frac{2}{5n+3} \tilde{H}_o^{-(5n+3)/2}. \tag{10.68}$$

With K given by expression (10.63), the solution for the thickness at the ice divide can now be written as

$$H_o(t) = \tilde{H}_o \left(1 + \frac{t}{T}\right)^{-2/(5n+3)}, \tag{10.69}$$

with

$$T = \frac{1}{(5n+3) A_o} \left(\frac{2n+1}{n+1}\right)^n \frac{\tilde{R}^{n+1}}{\tilde{H}_o^{2n+1}}. \tag{10.70}$$

Finally, the change in radius follows from equation (10.65) and is given by

$$R(t) = \tilde{R} \left(1 + \frac{t}{T}\right)^{1/(5n+3)}. \tag{10.71}$$

To summarize the preceding derivation, equation (10.64) gives the normalized ice-sheet profile as a function of normalized radial distance. Ice thickness is scaled with the thickness at the ice divide, and distance is scaled with the radius of the ice sheet. As the ice spreads out, the divide thickness decreases according to equation (10.69) and the radius increases as described by equation (10.71).

Figure 10.7 shows how the divide thickness decreases with time, and Figure 10.8 shows how the radius increases (using $n = 3$ for the flow law exponent). The greatest rates of change occur at the start ($t = 0$); as time progresses, the rate of adjustment decreases. The reason for this is that the aspect ratio (height-to-width ratio) diminishes according to (from equations (10.69) and (10.71))

$$\frac{H_o(t)}{R(t)} = \frac{\tilde{H}_o}{\tilde{R}} \left(1 + \frac{t}{T}\right)^{-3/(5n+3)}. \tag{10.72}$$

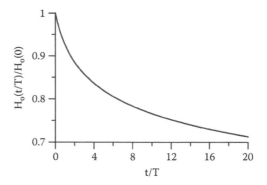

FIGURE 10.7   Decrease in ice thickness at the divide for the case of zero surface accumulation.

Thus, the slope of the ice surface decreases over time, resulting in slower ice speeds. Corresponding ice-sheet profiles are shown in Figure 10.9. Note that in this figure, the ice thickness is normalized by $\tilde{H}_o$ and the radial distance by $\tilde{R}$, instead of using equations (10.46) and (10.47). The reason for this is that these equations define the normalized thickness to vary from 1 at the divide to 0 at the margin, while the normalized radius increases from 0 to 1. Thus, the shape of the normalized profile (10.64) does not change over time.

The time scale for adjustment, T, is given by equation (10.70), with the constant, A_o, related to the rate factor, A, in Glen's flow law as (Section 5.3)

$$A_o = \frac{2A}{n+2}(\rho g)^n. \qquad (10.73)$$

Thus, the time scale depends on the temperature of the ice and on the initial divide thickness and radius. For ice close to the melting temperature, $A \approx 10^{-7}\ \mathrm{kPa}^{-3}\ \mathrm{yr}^{-1}$. Taking $\tilde{H}_o = 2$ km and $\tilde{R} = 500$ km for the initial geometry (at $t = 0$), the time scale is about 5000 years.

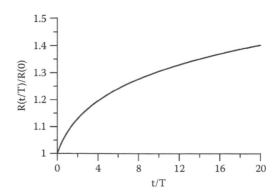

FIGURE 10.8   Increase in ice-sheet radius for the case of zero surface accumulation.

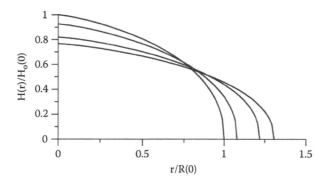

FIGURE 10.9 Surface profiles of an axisymmetric ice sheet for the case of zero surface accumulation, at times $t = 0$, $t = T$, $t = 5T$, and $t = 10T$. Note that the thickness is nondimensionalized by the divide thickness at $t = 0$, and the radial distance is nondimensionalized by the ice-sheet radius at $t = 0$.

As noted at the beginning of this section, Bueler et al. (2005) present similarity solutions corresponding to nonzero surface accumulation. Accumulation, M, is taken proportional to ice thickness and inversely proportional to time, such that

$$M = \frac{\lambda}{t}H, \tag{10.74}$$

with λ a constant parameter. Omitting the derivation here, the resulting ice-sheet profile is given by

$$H_\lambda(r,t) = \tilde{H}_0\left(\frac{t}{T}\right)^{-\alpha}\left[1 - \left(\left(\frac{t}{T}\right)^{-\beta}\frac{r}{\tilde{R}}\right)^{n/(2n+1)}\right]^{n/(2n+1)}, \tag{10.75}$$

where

$$\alpha = \frac{2 - (n+1)\lambda}{5n + 3},$$

$$\beta = \frac{1 + (2n+1)\lambda}{5n + 3}. \tag{10.76}$$

The thickness at the divide changes with time as

$$H_0(t) = \tilde{H}_0\left(\frac{t}{T}\right)^{-\alpha}, \tag{10.77}$$

while the time evolution of the radius is given by

$$R(t) = \tilde{R}\left(\frac{t}{T}\right)^\beta, \tag{10.78}$$

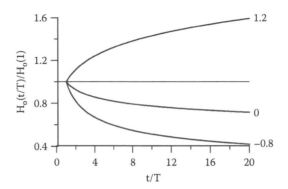

FIGURE 10.10 Change in ice thickness at the divide for three values of the accumulation parameter (indicated by the labels). $\lambda = 1.2$ corresponds to height-dependent accumulation, $\lambda = 0$ corresponds to zero surface accumulation, and $\lambda = -0.8$ corresponds to height-dependent ablation. In these solutions, $t = T$ corresponds to the initial profile.

(Bueler et al., 2005). In these solutions, the time, t, is defined somewhat different from the earlier solutions, and the initial time corresponds to $t = T$, as opposed to $t = 0$ as in the Halfar–Nye solution.

Figure 10.10 shows the change in ice thickness at the divide for three values of the accumulation parameter. The curve corresponding to $\lambda = 0$ is equivalent to the solution shown in Figure 10.7. The other two curves represent solutions for positive accumulation and for ablation. The corresponding change in radius is shown in Figure 10.11, while Figure 10.12 shows corresponding ice-sheet profiles at select times.

Bueler et al. (2005) propose a suite of tests for numerical ice-sheet models based on these exact, time-dependent analytical solutions. They proceed to test one numerical model and find that their explicit finite-difference scheme recovers the time-dependent

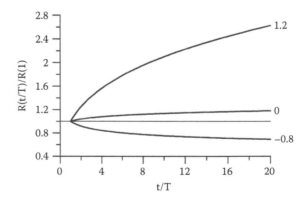

FIGURE 10.11 Change in ice-sheet radius for three values of the accumulation parameter (indicated by the labels). $\lambda = 1.2$ corresponds to height-dependent accumulation, $\lambda = 0$ corresponds to zero surface accumulation, and $\lambda = -0.8$ corresponds to height-dependent ablation. In these solutions, $t = T$ corresponds to the initial profile.

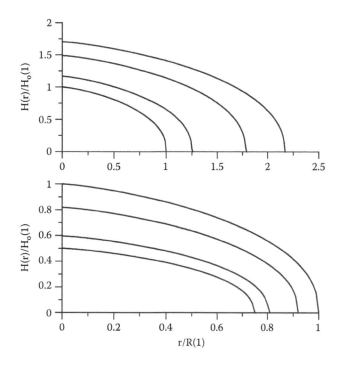

FIGURE 10.12   Surface profiles of an axisymmetric ice sheet for the case of height-dependent accumulation (top panel; $\lambda = 1.2$) and height-dependent ablation (bottom panel; $\lambda = -0.8$), at times $t = T$, $t = 2T$, $t = 6T$, and $t = 11T$. In these solutions, $t = T$ corresponds to the initial profile. The thickness is nondimensionalized by the divide thickness at $t = T$, and the radial distance is nondimensionalized by the ice-sheet radius at $t = T$.

analytical solutions. Rutt et al. (2009) compare the Glimmer community ice-sheet model against three of the time-dependent analytical solutions provided by Bueler et al. (2005) and find that, generally, model solutions converge toward the analytical solutions as the grid spacing and time step are decreased. Bueler et al. (2007) provide exact solutions to the thermomechanically coupled lamellar flow approximation in three spatial dimensions, but these are not similarity solutions. Sargent and Fastook (2010) construct manufactured analytical solutions for time-dependent and steady-state Stokes flow, both for two-dimensional flow along a flowline and for three-dimensional ice-sheet flow. Leng et al. (2012b) also derive solutions for three-dimensional, isothermal, nonlinear Stokes flow and use these to verify the parallel, high-order accurate, finite element, nonlinear Stokes ice-sheet model of Leng et al. (2012a). Results of this comparison show excellent agreement.

10.4   GLACIER SURGES

A glacier surge is a short period of comparatively fast motion following an extended period of normal flow. Meier and Post (1969) argue that the first and foremost characteristic that distinguishes surging glaciers is the fact that surges occur repeatedly.

This periodicity suggests that the longitudinal profile of a surging glacier is not stable. During the quiescent phase, which for a single glacier is fairly constant (of the order of 10 to 100 years), the ice speeds are generally small and the glacier thickens, most notably in the upper regions of the area affected by the surge. When the glacier reaches a certain thickness, a threshold may be reached and the ice speed increases by a factor of 10 or more and ice is rapidly drained from the upglacier area to the terminus. As a result, the thickness decreases rapidly in the upglacial regions, while near the terminus the thickness increases and the front may advance over several kilometers. This active phase, commonly referred to as the surge, is relatively short (months to a few years) compared with the quiescent phase. It is not clear what causes the surge to terminate, other than stating the obvious that conditions that allow the rapid flow cannot be maintained for prolonged periods of time, perhaps because the thickness of the glacier decreases rapidly during the surge.

One of the few glaciers that have been studied through several complete surge cycles is Medvezhiy Glacier located in the West Pamir Mountains in the Central Asian Republic of Tajikistan. The period of surges is 10 to 14 years, with surges having occurred in 1937, 1951, 1963, and 1973 (Dolgoushin and Osipova, 1975, 1978). Medvezhiy Glacier covers an area of about 25 km². Its large accumulation area is situated at an elevation of 4600 to 5500 m above sea level; the ablation area is nourished by a steep icefall and terminates at an elevation of 2840 m (during maximum advance) to 3000 m (during maximum retreat) above sea level. The surges are confined to the ice tongue below the ice fall. This tongue is about 8 km long and 100 to 200 m thick, and located in a narrow valley. The profiles before and after the 1963 surge are shown in the upper panel of Figure 10.13. Comparison of these profiles clearly shows that a considerable mass of ice was transferred from the upper part to the terminus, which is one of the characteristics of a glacier surge. In April 1963, the glacier snout started to advance, traveling a distance of 1600 m in a period of about two months. The increase in ice thickness in the lower part reached 150 m, while the surface in the upper part was lowered by up to 100 m. During the surge, the average ice speed was about 40 m/day, but a maximum of 100 m/day was reached during a two-week period apparently associated with the formation of a large horizontal fracture. After the surge, the terminal region became almost stagnant as the ice velocity decreased to a few cm per day. In the quiescent phase following the 1963 surge, the ice tongue consisted of two parts, namely, a zone of active ice, and the terminal inactive zone (Figure 10.13, middle panel). Because there was almost no flow of ice from the upper zone into the inactive zone, the mass balance of the inactive region was strongly negative and the thickness decreased steadily. In the active zone, the mass balance was positive due to inflow of ice from the accumulation area of the glacier. As this active lobe built up, it slowly expanded downglacier until a threshold was reached, and the next surge occurred in 1973 (Figure 10.13, lower panel).

A detailed record of changes in ice velocity during a surge is available for Variegated Glacier, Alaska. This glacier, with a length of ~20 km and located near Yakutat, Alaska, surged seven times during the 20th century with an average time between initiation of surges of 15 years (Lawson, 1997; Eisen et al., 2001). Eisen et al. (2001) found that a surge occurs when the cumulative mass balance reaches a threshold value and the time interval between successive surges depends on the

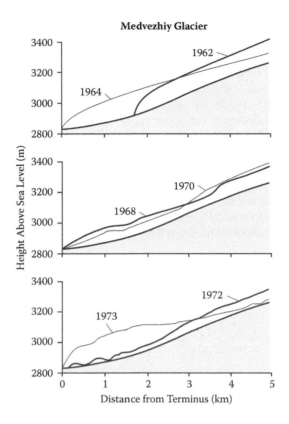

Medvezhiy Glacier

FIGURE 10.13 Longitudinal profiles of Medvezhiy Glacier, before and after the surges in 1963 and 1973. (After Dolgoushin, L. D., and G. B. Osipova, *Mater. Glyatsiologischeskikh Issled. Khronika. Obsuzhdeniya,* 32, 260–265, 1978.)

annual accumulation over that interval. The last recorded surge occurred in 1995 and was relatively small compared with earlier surges. This surge was initiated in the winter and terminated after two days of record high temperatures early in the following melt season (Eisen et al., 2005).

In 1973 a monitoring program was initiated to study the buildup of Variegated Glacier and the ensuing surge (Kamb et al., 1985; Raymond and Harrison, 1988). Between September 1973, and September 1981, the glacier thickened by as much as 60 m in the upper regions (about 60% of the total glacier length), while thinning occurred in the lower 40% of the glacier, up to about 40 m near the terminus. At the same time, the ice velocity increased in the upper region, while the terminus remained essentially stagnant. The boundary between the regions of net thickening and thinning slowly migrated downglacier at a speed of 0.3 km/yr. The location of maximum thickening rate appeared to propagate downglacier at a similar speed (Raymond and Harrison, 1988). The surge started in January 1982, with the velocity in the upper regions increasing from 2.6 to 9.2 m/day from late May to late June, and reaching a maximum of 10.4 m/day on June 26 (Kamb et al., 1985). Within a few hours after the occurrence of the peak, the velocity dropped to less than 5 m/day and steadily decreased to about 1 m/day by mid-August. During this slowing down, five major pulses of velocity occurred, during which the peak velocity reached 3–7 m/day. By the end of September 1982, it appeared that the surge had ended. This phase of the surge was restricted to the upper region of the glacier, and the lower parts remained unaffected by the increase and subsequent decrease in velocity. In October 1982, the velocity in the upper region started to increase again, leveling off at 5–7 m/day in January 1983. After a sudden drop in early February 1983, the velocity gradually increased, reaching a maximum of about 15 m/day in mid-June. During this second phase, the surge propagated into the lower regions of the glacier, and the ice velocity there rose from less than 0.2 m/day in early May to 40–60 m/day in June. The surge terminated on July 4, when the velocity along the entire length of the glacier dropped dramatically over the course of a few hours (Kamb et al., 1985). Figure 10.14 illustrates how the surge front propagated downglacier. Ahead of the

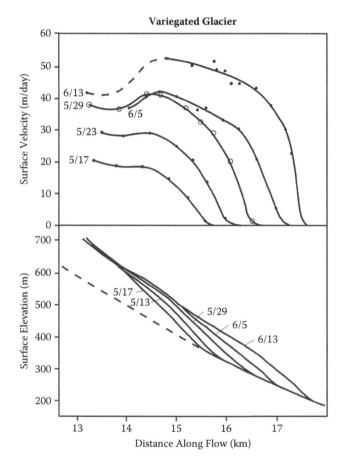

FIGURE 10.14 Propagation of the surge front on Variegated Glacier during May and June 1983. (From Kamb, B. et al., *Science,* 227, 469–479, 1985. Copyright 1985 by the American Association for the Advancement of Science.)

front, the ice was virtually stagnant, while the front advanced at an almost constant rate of 80 m/day. This rapid advance of the surge front, at a speed that exceeds the actual ice velocity, suggests that this advance is similar to a kinematic wave traveling along the glacier (Kamb et al., 1985).

It has long been recognized that surges are achieved by a switch in the basal conditions, allowing transition from slow to rapid flow. Initially, this transition was thought to be from a frozen glacier sole to a water-lubricated base, allowing rapid basal sliding. However, surges also occur in temperate glaciers such as Variegated Glacier, whose base remains well lubricated during the quiescent phase. This means that the no-sliding to sliding transition cannot explain why surges occur in these glaciers. Furthermore, a distinction needs to be made between glaciers resting on a hard and nondeformable bed and glaciers that rest on soft material. In view of the wide range of possible basal conditions, it is probably illusory to expect one theory of surges to apply to all glaciers. Nevertheless, a qualitative description can be given, based on the model presented by Fowler (1987b). This model is based on static external conditions, and in particular, changes in water input (seasonal or from individual rain events) are not included.

There are essentially two possibilities for the subglacial drainage system, namely, a system of Röthlisberger channels (Röthlisberger, 1972), or a linked-cavity system (Kamb, 1987). It could be argued that drainage in a water film at the bed represents a third possibility, but it is unlikely that such a film is of uniform thickness. Rather, one may expect the film to be thicker in some places, and almost pinched out in others. If so, there appears to be little difference between the water film and the linked-cavity system, and it may not be necessary to explicitly consider drainage through a film. The essential feature that allows transition from "slow" basal sliding to "rapid" basal sliding is the difference in water pressure in each of the drainage systems. Together with the assumption that the sliding speed is a function of this water pressure, it can be shown how the switch from one drainage system to the other leads to a surge.

In an R-channel, the effective pressure increases as the water flux through the channel increases (Röthlisberger, 1972; c.f. Section 7.5)

$$N_r \sim Q_w^{1/p}, \tag{10.79}$$

where $p \approx 12$. The effective pressure, N, is defined as the difference between the ice overburden pressure and the water pressure in the channel. Thus, the water pressure, P_w, decreases as the water flux, Q_w, increases. This means that water will tend to collect in the larger tunnels, and an arterial network of subglacial channels will develop with smaller channels feeding into larger tunnels. In a linked-cavity system, on the other hand, the water pressure increases as the water flux increases, and there is no tendency for water to collect in larger passageways. Provided that the sliding velocity is high and the water pressure not too large, a complicated network of meandering links that connect the cavities will remain stable (Kamb, 1987). Fowler (1987a) argues that for such a network to be able to discharge a reasonable amount of meltwater, some of the intercavity connections must enlarge to form small R-channels.

In that case, the effective pressure in the linked-cavity system is much smaller than that in an R-channel carrying the same water flux. That is

$$N_c = \delta N_r, \tag{10.80}$$

where N_r represents the effective pressure in an R-channel (equation (10.79)). The quantity δ depends on the degree of cavitation and the fraction of the bed that is free from cavities. For a typical glacier, δ may be 0.25 or smaller (Fowler, 1987b).

If the sliding velocity becomes sufficiently small, dissipational heat generated by the water flowing through a linked-cavity system is insufficient to maintain melting needed to counteract creep closure, and the drainage system may collapse into a tunnel system. On the other hand, if the sliding velocity becomes too large, a tunnel system will become unstable and a system of cavities connected by small channels will develop. Without going into the mathematical details, it can be shown (Fowler, 1987a) that the stability parameter, S, is a function of the ice velocity and the effective pressure

$$S \sim \frac{U}{N^n}, \tag{10.81}$$

where n = 3 represents the exponent in Glen's flow law for glacier ice. If S is smaller than some critical value, S_c, a tunnel system is stable and the effective basal pressure is large. But for $S > S_c$, the tunnel system collapses into a linked-cavity system, with (much) smaller effective pressure. This dependence is schematically illustrated in Figure 10.15, which shows the effective pressure, N, as a function of the stability parameter, S. The transition at $S = S_c$ may not be as sharp as shown in this figure, but for the present qualitative discussion, the picture shown in Figure 10.15 suffices. It may be noted that the implicit assumption is made here that the transition from tunnel to linked cavities occurs at the same critical value, S_c, as the transition from linked cavities to tunnels. This may not be true (Fowler, 1987a), but it is not clear whether introducing a hysteresis loop will significantly alter the qualitative aspects of the model.

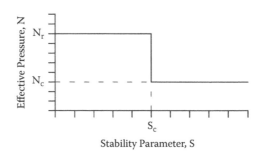

FIGURE 10.15   Schematic illustration of the dependency of the stability parameter, S, on the effective basal pressure, N. For small values of S, subglacial drainage is through tunnels, and the effective pressure is relatively large. For large values of S, drainage is through a linked-cavity system, with lower effective pressure.

Rather than considering how the effective pressure depends on the stability parameter, S, it is more instructive to consider how N depends on the sliding velocity, U. From the stability analysis summarized above, it follows that there are two critical velocities, namely,

$$U_1 \sim S_c N_r^n, \tag{10.82}$$

and

$$U_2 \sim S_c N_c^n. \tag{10.83}$$

For $U < U_1$ the tunnel system is stable and the effective pressure is equal to N_r, while for $U > U_2$, a linked-cavity system is stable and $N = N_c$. However, because the effective pressure in a tunnel is larger than that in a linked-cavity system, U_2 is smaller than U_1, and a range of velocities exists for which both drainage systems are possible. This is schematically illustrated in Figure 10.16, which shows the effective pressure as a function of the sliding velocity. To understand how this multivalued function leads to surging behavior, the sliding relation needs to be invoked.

While a universally applicable sliding relation has not yet been established, it is generally accepted that the sliding speed depends on basal drag and effective water pressure as in the generalized Weertman relation (c.f. Section 7.2)

$$U \sim \frac{\tau_b^m}{N^k}. \tag{10.84}$$

Using the functional relation between effective pressure and sliding speed shown in Figure 10.16, this sliding relation can be schematically represented by the curve shown in Figure 10.17. To adopt Fowler's (1987b) terminology, this curve is characterized by a "slow" branch and a "fast" branch.

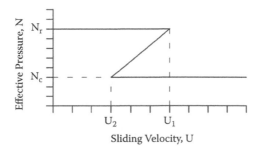

FIGURE 10.16   Schematic illustration between sliding velocity, U, and effective basal pressure, N. For sliding speeds smaller than U_1, a tunnel system with relatively large basal pressure is stable, while for $U > U_2$, the linked-cavity system with lower effective pressure is stable. Because $U2 < U_1$, there is a range of sliding velocities for which both drainage systems are possible.

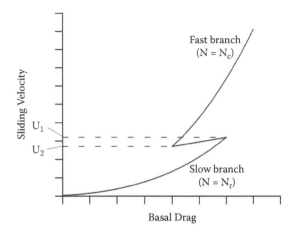

FIGURE 10.17   Schematic illustration of the relationship between sliding velocity and basal drag. The sliding relation is characterized by a slow and a fast branch. The coexistence of two possible sliding velocities for the same basal drag can lead to surge cycles.

Referring to Figure 10.18, a surge cycle can now be described. During the quiescent phase, the glacier builds up and its thickness increases, resulting in a larger driving stress. Making the reasonable assumption that the driving stress is primarily balanced by basal drag, the drag at the glacier bed gradually increases until a critical value is reached (point A in Figure 10.18). The sliding velocity has now become too large for subglacial tunnels to be maintained, and a linked-cavity system develops. As a result of the lower effective pressure, the ice velocity increases (point B). However, the (local) increase in mass discharge results in a loss of ice and a decrease in driving stress (and hence, basal drag). Thus, the sliding velocity decreases until sliding is too slow to maintain a linked-cavity system (point C). The cavities collapse and a tunnel system reforms, bringing the glacier in the slow-flow regime again (point D).

There are many factors that determine whether or not a particular glacier will exhibit surge-type behavior. First, surface accumulation must be large enough to allow the glacier to grow to the transition point A. If the steady-state size of the glacier is such that this point is never reached, the glacier will not switch to the fast-flow regime. Second, basal conditions must be such that the change in effective pressure (as described by the factor δ in equation (10.80)) is large enough. Generally, this requires a rough bed (Fowler, 1987b). Third, the difference between the two critical velocities, U_1 and U_2 (Figure 10.16), must be sufficiently large.

The model developed by Fowler (1987b) and outlined above applies only to glaciers whose bed is at the pressure melting temperature along most of the glacier length. This requirement does not apply to all surging glaciers. For example, Trapridge Glacier in the Yukon Territory, Canada, is a small (~4 km long) polythermal surge-type glacier resting on a bed of unconsolidated sediments (Clarke and Blake, 1991). The basal ice reaches the melting point only in the central area where the thickness is greatest. Near the terminus and the head of the glacier, the bed is

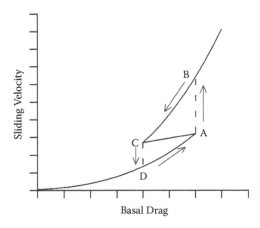

FIGURE 10.18 Schematic illustration of a complete surge cycle. See text for a description.

frozen and advance of the surge front appears to be associated with migration of the warm/cold boundary (Clarke, 1976; Clarke et al., 1984). Fowler et al. (2001) present a heuristic model for cyclic surging of such polythermal glaciers, with subglacial hydrology controlling both the flow of the glacier and deformation of subglacial till. The model produces oscillations (surges) if the permeability of unfrozen till is sufficiently small to impede water drainage and if the thickness of the till layer is small. Under those conditions, drainage of basal meltwater is restricted, resulting in higher water pressures and reduced effective basal pressure. At this stage, water is stored in "blisters," but eventually, discharge from these blisters becomes large enough to lower basal water pressure. Murray et al. (2003) argue that this model applies to the typical surge cycle of glaciers on Svalbard.

10.5   MARINE INSTABILITY

Many glaciologists have argued that, because of its marine nature, the West Antarctic Ice Sheet is inherently unstable and may respond drastically to (moderate) changes in climate (for example, Bentley, 1984; Hughes, 1973; Mercer, 1968, 1978; Thomas, 1979; Thomas and Bentley, 1978; Thomas et al., 1979; Weertman, 1974). This view is based on the idea that the large floating ice shelves that surround the grounded portion of this ice sheet regulate the discharge of inland ice. The grounding line tends to be unstable if the water depth is greater than some critical depth and the sea floor slopes downward toward the ice-sheet interior (as is the case in West Antarctica), and a reduction in size or complete removal of the ice shelves would initiate grounding-line retreat that may be irreversible and result in the collapse of the entire ice sheet.

Most of the West Antarctic Ice Sheet is grounded on a former sea floor, and currently the bed under the ice is well below sea level. A direct consequence of this is that the ice sheet is surrounded by floating ice shelves (indicated by dark shading in Figure 10.19). As the ice flows outward to the sea, its thickness decreases and, because the bed is below sea level, a point is reached where the weight of the ice becomes less than the upward buoyancy force of the sea water and the ice starts

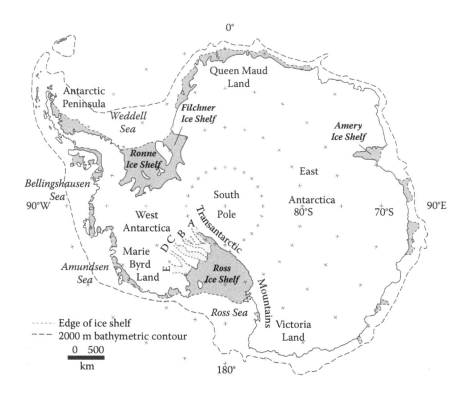

FIGURE 10.19 Location map of Antarctica. Dark shading represents the major ice shelves and light shading represents the five ice streams (labeled A–E; in 2001 these were renamed Mercer, Whillans, Kamb, Bindschadler, and MacAyeal Ice Streams, respectively) draining into the Ross Ice Shelf.

to float, forming an ice shelf. The boundary between grounded ice and floating ice shelves is called the *grounding line*. As shown in Figure 10.19, West Antarctica is almost completely surrounded by floating ice shelves, while smaller ice shelves are also found in East Antarctica. Ice shelves and floating ice tongues comprise about 57% of the Antarctic coast line. These ice shelves are nourished by ice streams and outlet glaciers, and by local accumulation. Mass loss occurs at the terminus where ice breaks off to form icebergs, and by basal melting.

Before discussing stability of the grounding line, it is instructive to have a closer look at the nature of the grounding zone and how different mapping methods have grounding-line positions that may differ by up to ~60 km (Brunt et al., 2010). Figure 10.20 shows a schematic view across the grounding zone. The line labeled G denotes the position of the actual grounding line, where the ice starts to float. This position can be determined accurately only through ground-based surveys such as radio-echo sounding or seismic methods (Brunt et al., 2010). Because this is impractical to do on a large scale, most reported grounding-line positions are based on the locations of one or more of the other surface features of the ground- ing zone. Vertical movements primarily associated with tides result in periodic

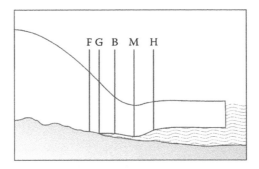

FIGURE 10.20   Schematic cross-section of the grounding zone showing the inland position of tidal flexure (F), the grounding line (G), the break in slope (B), the local minimum in surface elevation (M), and the hydrostatic point (H) where ice is in hydrostatic equilibrium. The grounding zone is defined as the region between points F and H. (From Brunt, K. M. et al., *Ann. Glaciol.*, 51(55), 71–79, 2010. Reprinted from the *Annals of Glaciology* with permission of the International Glaciological Society and the authors.)

vertical motions of the grounded ice. Point F corresponds to the landward limit of tidal flexure, and this position was used by Rignot (1998) to map the hinge line of Petermann Glacier in northwest Greenland. On the seaward end, the hydrostatic point H is the location where the ice shelf is in hydrostatic equilibrium with the ice shelf. The region between points F and H defines the grounding zone (Fricker et al., 2009; Brunt et al., 2010). In the grounding zone, there often is a local topographic minimum (point M in Figure 10.20) as well as one or more breaks in the surface slope (point B). Fricker et al. (2009, Table 1) discuss how different satellite sensors can be used to find different combinations of the features in the grounding zone. For comparison of grounding-line positions reported in different studies, it is thus important to make sure the same features are compared. Brunt et al. (2010) propose that the limit of tidal flexure (point F) is the most robust proxy for the location of the grounding line. On the other hand, for mass flux calculations into ice shelves, the location of point H is required because this represents the most upstream location where ice thickness can be estimated from the ice surface elevation and the flotation assumption (Brunt et al., 2010).

Mercer (1968, 1978) proposed that ice shelves are sensitive to moderate climate changes and cannot maintain their integrity once summer temperature or the annual mean temperature exceeds some threshold. Analysis of data on atmospheric warming and retreat of ice shelves surrounding the Antarctic Peninsula appears to confirm the existence of a thermal limit on ice-shelf viability (Vaughan and Doake, 1996). The observed pattern of ice-shelf retreat is consistent with a southward migration of the –5°C annual mean air temperature isotherm. Mercer (1968, 1978) further suggested that the peripheral ice shelves play a major buttressing role for discharge of inland ice, and their breakup would greatly enhance the discharge from the interior and could lead to the rapid collapse of the West Antarctic Ice Sheet.

The fringing ice shelves hypothesis summarized above is based on the observation that most of the ice shelves surrounding West Antarctica have formed in embayments and are locally in contact with the seabed. Resistance from the lateral walls and ice rises obstructs the flow of the ice shelves, and the spreading rate is significantly smaller than would be observed if the shelves were spreading freely. This effect is often described by the back pressure, defined as the downglacial integrated resistance to flow associated with lateral drag and resistance from ice rises (Section 4.5). For an ice shelf spreading in one direction only, the back pressure at any point on the ice shelf is calculated from

$$\sigma_b(x) = \frac{1}{H} \int_x^L \frac{H\tau_s}{W} \, d\bar{x}, \tag{10.85}$$

where τ_s represents the shear stress at the margins, W denotes the half width of the ice shelf and $x = L$ represents the position of the ice-shelf front (with the x-axis along the direction of flow). This back pressure reduces the spreading rate as compared with that of a free-spreading ice shelf. That is

$$\dot{\varepsilon}_{xx} = \theta \left(\frac{R_{xx}^{(0)} - \sigma_b}{B} \right)^n, \tag{10.86}$$

in which B and n represent the flow law parameters and θ accounts for transverse spreading (c.f. Section 4.5; for an ice shelf spreading in the x-direction only, $\theta = 2^{-n}$). In equation (10.86), $R_{xx}^{(0)}$ represents the stretching stress for a free-floating ice shelf on which the driving stress is balanced entirely by gradients in longitudinal stress. As derived in Section 4.5, this stress is given by

$$R_{xx}^{(0)} = \frac{\rho g}{2} \left(1 - \frac{\rho}{\rho_w} \right) H, \tag{10.87}$$

with ρ_w the density of sea water.

If the back pressure is zero and the ice shelf is spreading freely in one direction, the stretching rate is proportional to the nth power of the ice thickness

$$\dot{\varepsilon}_{xx} = \left[\frac{\rho g}{4B} \left(1 - \frac{\rho}{\rho_w} \right) \right]^n H^n. \tag{10.88}$$

This solution was derived first by Weertman (1957b) and represents the maximum spreading rate of an ice shelf of thickness H. The effect of lateral drag and ice rises is to reduce this spreading rate through the back pressure. Thus, if the back pressure decreases, due to thinning of the ice shelves (which reduces lateral drag and may result in the ungrounding of pinning points) or due to a shortening of the shelf (which reduces the length over which lateral drag acts), the creep rate for a given ice thickness increases.

How the creep rate affects thickness change can be seen by considering the continuity equation (Section 5.2). Considering flow in the x-direction only, this equation reads

$$\frac{\partial H}{\partial t} = -\frac{\partial (HU)}{\partial x} + M =$$

$$= -U\frac{\partial H}{\partial x} - H\frac{\partial U}{\partial x} + M. \tag{10.89}$$

Replacing the along-flow velocity gradient by the stretching rate (equation (10.86)) and using expression (10.87) for the stretching stress, gives

$$\frac{\partial H}{\partial t} = -U\frac{\partial H}{\partial x} - H\left(\frac{CH - \sigma_b}{2B}\right)^n + M, \tag{10.90}$$

with

$$C = \frac{\rho g}{2}\left(1 - \frac{\rho}{\rho_w}\right). \tag{10.91}$$

While this equation is derived from ice-shelf dynamics, it can also be applied to the grounding line, where the flow regime changes from ice-stream flow to ice-shelf spreading. Thus, the thickness at the grounding line is affected by three factors, namely, advection from upglacier (first term on the right-hand side), creep thinning (second term), and local accumulation or ablation, M. If these three terms sum to zero, the thickness at the grounding line does not change and the position of the grounding line is stationary. However, changing any of the three terms on the right-hand side of equation (10.90) results in a thickness change, which could initiate irreversible grounding-line retreat if feedback processes become important.

According to the second term on the right-hand side of equation (10.90), creep thinning at the grounding line depends strongly on the ice thickness, H. This dependency leads to an important feedback between creep thinning and seabed topography, as illustrated in Figure 10.21. Consider first the situation in which the seafloor slopes downward toward the exterior (left panel). If the grounding line advances, for example, because advection from upglacier increases, the thickness at the grounding line increases, leading to larger creep thinning, which slows the rate of advance. Similarly, if the grounding line retreats, creep thinning decreases, thus retarding the rate of retreat. So, for the geometry shown in the left panel, the interaction between creep thinning and ice thickness provides a negative feedback, slowing the rate of both advance and retreat. However, the situation shown in the panel on the right, with the seabed sloping downward toward the interior, is more representative of West Antarctica. Now the feedback is positive and an advance of the grounding line leads to a decrease in creep thinning, allowing the grounding line to advance further, while a retreat results in increased creep thinning and accelerated retreat.

The presence of a restricted ice shelf may have a stabilizing effect on grounding-line retreat or advance. At present, the calving fronts of the ice shelves are mostly

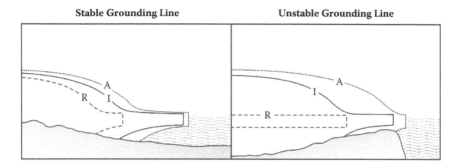

FIGURE 10.21 Interaction between grounding-line migration and basal topography. The panel on the left represents a stable configuration with the bed sloping downward from the ice-sheet interior. The panel on the right represents an unstable situation. Profiles labeled I represent the initial geometry, and profiles labeled A and R represent the geometry after advance and retreat, respectively. (From Thomas, R. H., *J. Glaciol.*, 24, 167–177, 1979. Reprinted from the *Journal of Glaciology* with permission of the International Glaciological Society and the author.)

at the end of the confining embayments, so an advance of the grounding line results in a shorter shelf length over which lateral drag acts. Thus, from equation (10.85), an advancing grounding line reduces the back pressure at the grounding line, which results in larger creep thinning, slowing the rate of advance. Similarly, retreat of the grounding line increases the back pressure, which slows the retreat. If the ice shelf becomes sufficiently large, the position of the grounding line may stabilize.

The model for studying grounding-line migration described above was developed by Thomas (1977) and applied to the Holocene retreat of the West Antarctic Ice Sheet by Thomas and Bentley (1978) and by Thomas (1985) to estimate the contribution of the polar ice sheets to rising sea level after a climatic warming. The main difficulty in these applications is to account for changes in advection from upglacier (the first term on the right-hand side of equation (10.90)) as the grounding line migrates. Thomas and Bentley (1978) calculate this term based on the assumption that the inland ice-sheet profile is parabolic, which allows the change in ice-sheet volume (and hence the discharge across the grounding line) to be calculated for a given change in grounding-line position. Thomas (1985) makes the assumption that a transition zone exists upstream from the grounding line. In this zone, the stretching rate is assumed constant (equal to the value calculated from equation (10.86)). With the velocity at the head of this transition zone known, the speed at the grounding line can be calculated from the stretching rate and the length of the transition zone. As the grounding line migrates, Thomas (1985) makes the additional assumption that both the length of the transition zone and the speed at the upglacial end of this zone remain constant. While the various assumptions that have been used affect the quantitative results, the qualitative behavior is the same, and these studies indicate that the West Antarctic Ice Sheet is unstable with respect to changes at the grounding line. In view of the model formulation, this result should come as no surprise.

The main objection against the instability model discussed above is the inconsistency in incorporating ice-shelf dynamics. Creep thinning is calculated from an expression based on ice-shelf dynamics, but the velocity at the grounding line or at the head of the transition zone is not affected by the presence of the ice shelf. Thus, an increase in stretching rate (due to a decrease in back pressure) leads to larger creep thinning but has no immediate effect on the ice speed. However, all strain rates affect this velocity through the effective strain rate, and an increase in stretching rate may be expected to result in an increase in speed at the grounding line as well. By not including this effect, the model becomes inherently unstable and prohibits a stable ice-sheet response to changes at the grounding line.

The inherent difficulty in modeling marine ice sheets appears to be how to merge the distinctly different flow regimes present in the system. On the grounded part in the interior, flow is dominated by vertical shear and can be described by the lamellar flow model. On floating ice shelves, basal drag is absent and the driving stress is balanced by gradients in longitudinal stress and by lateral drag. Coupling these flow regimes is a transition zone of unknown length. Despite some 40 years of theoretical and numerical modeling, debate continues about the nature of this flow transition and consequences for ice-sheet stability.

The first theoretical treatment of the stability of a grounding line is the study of Weertman (1974), who considered a two-dimensional ice sheet on a bed below sea level with initially constant slope. Bedrock adjustment to the ice load is included by assuming local isostatic equilibrium (equation (9.85)). By deriving an expression for the ice flux across the grounding line, steady-state grounding-line positions can be found by comparing the ice flux across the grounding line with integrated upglacier accumulation. The argument is that a small advance of the grounding line increases the size of the ice sheet and thus increases total accumulation over the grounded portion. If the downward slope of the bed at the grounding line is not large enough, the increase in total accumulation exceeds the increase in ice flux across the grounding line, and the ice sheet will continue to grow until the edge of the continental shelf is reached. Weertman's results show that an ice sheet that rests on a bed that was flat before isostatic adjustment and below sea level is inherently unstable. Depending on how deep the bed is below sea level, a small perturbation in grounding-line position (for example, resulting from a change in sea level) will cause continued retreat and the ice sheet will disappear entirely, or the grounding line will advance to the edge of the continental shelf. If the bed slopes away from the divide (as in the left panel of Figure 10.21), a stable ice sheet is possible.

An updated analysis of grounding-line stability can be found in Schoof (2007a, b; 2011), who applied a boundary layer theory to derive an expression for the ice flux across the grounding line. A more simple derivation of this flux is provided by Hindmarsh (2012, Appendix B). Weertman-type sliding is assumed with basal drag related to the sliding speed as in equation (7.9); that is

$$\tau_b = C_s |U|^{1/m-1} U. \tag{10.92}$$

Neglecting gradients in longitudinal stress and lateral drag, force balance reduces to a balance between driving stress and basal drag, and

$$C_s |U|^{1/m-1} U = -\rho g H \frac{\partial h}{\partial x}. \tag{10.93}$$

The second equation to consider is the continuity equation (10.90). Setting the back stress, σ_b, to zero, and assuming steady state, this equation becomes

$$U\frac{\partial H}{\partial x} + H\left(\frac{CH}{2B}\right)^n = M. \tag{10.94}$$

At the grounding line, the ice becomes afloat and the surface elevation is related to the ice thickness through hydrostatic equilibrium (equation (4.55)) and

$$h = \left(1 - \frac{\rho}{\rho_w}\right)H. \tag{10.95}$$

Using this relation to eliminate the thickness gradient from equations (10.93) and (10.94) gives the ice flux across the grounding line

$$Q = \left[\frac{2C^{n+1}}{C_s(2B)^n}\right]H^{(n+m+3)/(m+1)}. \tag{10.96}$$

More formal derivations, based on asymptotic expansions for the flow in boundary layers, are presented in a series of papers by C. Schoof. Flow of the ice sheet is described by the shallow ice approximation (or lamellar flow model), while interactions at the grounding line are represented through two boundary conditions that constrain the ice thickness at the grounding line (from flotation) and the ice flux across the grounding line as a function of local water depth (Schoof, 2007a, 2011, 2012). Applying a linear stability analysis, Schoof (2012) confirmed earlier inferences based on mass-balance arguments (Weertman, 1974; Schoof, 2007a, b) that the stability of equilibrium grounding-line positions is determined by the balance between discharge flux across the grounding line and total upglacier accumulation.

It could be argued that the sliding relation (10.93) adopted by Schoof is unrealistic because it does not account for the effect of basal pressure on the sliding speed. The more commonly used sliding relation (7.25) results in basal drag tending to zero toward the grounding line as the effective basal pressure approaches zero as the ice reaches floatation. Assuming that this lowered basal drag is compensated by lateral drag, a derivation similar to that above shows that the ice flux across the grounding line is proportional to H^{n+1} (Hindmarsh, 2012, equation (16)). Thus, for this case similar reasoning can be applied to show that grounding lines can be stable only where the bed slopes downward sufficiently steeply.

The crucial assumptions in these analytical models are that the peripheral ice shelf is unconfined and that gradients in longitudinal stress (termed *membrane stresses* by Hindmarsh, 2009) are unimportant in the balance of forces at the grounding line. This is valid when considering possible steady-state grounding-line positions and their stability for idealized geometries and, in particular, the effect of complete removal of a floating ice shelf on the stability of the grounded ice sheet. For the case where an ice shelf is maintained, buttressing by the shelf may act to stabilize the ice sheet (for example, Goldberg et al., 2009). The argument is analogous to the mass balance argument of Weertman (1974) and Schoof (2007a). Assuming the calving front of the ice shelf remains stationary, a small retreat of the grounding line would lengthen the ice shelf and increase back pressure at the grounding line, which would lower the stretching rate and thus mass loss across the grounding line.

Studies using numerical flowline models to investigate the stability of marine ice sheets have proven to be inconclusive. The earlier studies of Van der Veen (1985, 1987) suggested that marine ice sheets may be more stable than generally believed and that complete removal of a peripheral ice shelf does not lead to complete disintegration of the ice sheet. Similarly, the model studies of Nick et al. (2010, 2012) indicate that grounding-line dynamics are less sensitive to basal topography as previously suggested and stable grounding-line positions can be attained on a reverse bed slope with or without a floating ice tongue. On the other hand, using a full-Stokes numerical model, Durand et al. (2009a) reaffirmed the assertion that a marine ice sheet is unstable on an upsloping bed. To a large extent, these differences in model behavior may be attributed to how grounding-line dynamics are incorporated and to other numerical specifics.

Vieli and Payne (2005) investigated the ability of numerical flowline models to simulate grounding-line migration in response to changes in accumulation rate and sea level. An important finding is that models that employ a fixed numerical grid show a strong dependency on grid spacing and tend to predict irreversible changes in grounding-line position. In contrast, models that use a moving grid, in which the grounding line always coincides with a grid node, predict smaller changes in grounding-line position, and following a perturbation, a new stable equilibrium is reached. Moreover, the perturbation experiments are reversible, and migration of the grounding line is insensitive to the slope of the bed. A similar influence of grid size on grounding-line dynamics was found by Durand et al. (2009b) using a full Stokes numerical model. Docquier et al. (2011) concluded that for finite-difference models, the numerical implementation has an important effect on grounding-line migration. Models that employ a staggered grid (as in Section 9.3) tend to agree better with the analytical solutions of Schoof (2007a, b) than nonstaggered models, in which velocity and thickness are calculated at the same gridpoints.

The last few years have seen a proliferation of numerical models simulating flow and evolution of marine ice sheets. These models differ in their approximations of the flow of the grounded portion (for example, including gradients in longitudinal stress), boundary conditions at the grounding line (for example, prescribing the ice flux across the grounding line using the Schoof solution), numerics (for example, fixed or moving grid), and other factors. It falls outside the scope of this review to discuss each of these models. For an overview, the reader is referred to Pattyn et al. (2012), who present results of an intercomparison of 27 different flowline models. Validation was provided by comparing model predictions with the analytical solutions of Schoof (2007a, b). Models that are based on lamellar flow for the grounded part fail to reproduce the analytical solutions unless a parameterization of the ice flux across the grounding line is included. Because moving-grid models track movement of the grounding line explicitly, these types of models agreed best with the analytical solutions.

It is important to keep in mind that the marine instability hypothesis essentially consists of two parts, namely, an initial perturbation that initiates grounding-line retreat or advance, and the nonlinear dependency of stretching rate and ice flux across the grounding line on ice thickness, which may cause continued grounding-line migration until a new equilibrium position is reached. On time scales on the order of a decade or so (spanning the period of high spatial and temporal resolution observations on actual outlet glaciers), one favored perturbation is breakup of floating ice shelves or ice tongues and the release of back stress (as in the original model proposed by Mercer, 1968, 1978). Another possible grounding-line perturbation may be increased submarine melting associated with intrusion of relatively warm ocean water under the subshelf cavity (Holland et al., 2008; Motyka et al., 2011).

Observations of rapid changes and large negative mass balances of Greenland and Antarctic outlet glaciers have been well documented using various remote sensing platforms. However, using these observations to identify the physical processes initiating these changes has been hampered by the fact that, typically, multiple changes occur more or less simultaneously—at least at the time resolution at which observations are available—including glacier thinning, flow acceleration, retreat of the calving front, subaqueous melting at or near the grounding line, and increased surface ablation. Nevertheless, a growing consensus appears to be that observed changes in the dynamics of outlet glaciers can be attributed to perturbations at the ice-ocean interface.

Several studies have identified increases in longitudinal stress gradients caused by a large reduction or complete removal of the floating ice tongue as initiating large-scale changes in flow dynamics of bordering outlet glaciers. This view that glacier changes are initiated by an ice-marginal forcing appears to be supported by the observed accelerations of Jakobshavn Isbræ, and Helheim and Kangerdlugssuaq glaciers in Greenland (for example, Howat et al., 2007; Joughin et al., 2008b), Pine Island Glacier in West Antarctica (for example, Payne et al., 2004; Shepherd et al., 2004), and glaciers that used to drain into the Larsen B Ice Shelf prior to its collapse (for example, De Angelis, 2003; Scambos et al., 2004) and elsewhere in the Antarctic Peninsula (for example, Rignot et al., 2004). All of these glaciers underwent changes in the extent of their floating ice tongue, which was coincident with a sudden flow acceleration and glacier thinning. Numerical modeling studies appear to support the notion that release of back stress initiated glacier speedup and thinning (for example, Dupont and Alley, 2005), and further indicate that small changes in the balance of forces at the ice-ocean interface can give rise to velocity increases that are consistent with observations (Payne et al., 2004).

A difficulty with the back-stress model is that, generally, the reduction in back stress at the grounding line is, by itself, not sufficient to affect the stress balance on the grounded part in such a way that rapid accelerations extending several tens of km inland will result. To understand this, consider the force-balance term associated with gradients in longitudinal stress (the second term on the right-hand side of the balance equation (3.22)). Taking the longitudinal stress, R_{xx}, constant with depth, resistance to flow associated with gradients in this stress is given by

$$F_{lon} = -\frac{\partial(H R_{xx})}{\partial x}. \tag{10.97}$$

A decrease in back stress at the grounding line ($x = 0$) results in an increase in R_{xx}(0) and must be balanced by increased flow resistance on the grounded part because initially the glacier geometry, and thus the driving stress, does not change. Now assume that the increase in stretching stress at the grounding line is instantaneously transmitted over a distance L (the coupling length; Kamb and Echelmeyer, 1986a; Hindmarsh, 2009). Integrating equation (10.97) from $x = 0$ to $x = L$ gives

$$\int_0^L F_{lon}\, dx = H(0)R_{xx}(0) - H(L)R_{xx}(L). \tag{10.98}$$

Averaged over the distance L, resistance associated with gradients in longitudinal stress is then

$$\overline{F}_{lon} = H(0)R_{xx}(0) - H(L)R_{xx}(L). \tag{10.99}$$

The coupling length represents the distance over which changes in back stress are transmitted upglacier, and $R_{xx}(L)$ remains unchanged, initially. Ignoring for simplicity variations in thickness, the average decrease in $\overline{F}_{lon}$ is then

$$\Delta\overline{F}_{lon} = \frac{H(0)}{L}\Delta R_{xx}(0). \tag{10.100}$$

Now consider the situation in which the back stress at the grounding line is reduced by 100 kPa. For an ice thickness of 1000 m and using a coupling length of about seven ice thicknesses (Kamb and Echelmeyer, 1986a), this would give an average increase in resistive forces of ~15 kPa. Applying the coupling-length formula derived from scaling principles in Hindmarsh (2009, p. 1755) gives L = 4 km, and the required increase in basal and/or lateral drag is 25 kPa. In the case of Jakobshavn Isbræ, the driving stress ranges from 200 to 400 kPa, and such minor adjustments to resistive stresses appear to be inconsequential and cannot explain the observed doubling of glacier speed (Van der Veen et al., 2011). Note that the above estimate of changes in the role of resistance to flow associated with gradients in longitudinal stress is consistent with the assumption of Schoof (2007a, b) that these stresses are unimportant on the grounded part of a marine ice sheet.

It could be argued that the increase in stretching stress at the grounding line is balanced by a greater increase in resistive stresses, but over a much shorter distance than the coupling length. In that case, resistive stresses upstream of the grounding line would initially not be affected, yet observations on Jakobshavn Isbræ show that changes in glacier speed and stretching stress occurred almost instantaneously over the lower 30–40 km of the glacier. If these changes in speed resulted solely from changes in boundary conditions at the glacier's lower end, one would expect to see a kinematic wave of adjustment traveling upglacier. The force-balance analysis of Van der Veen et al. (2011) indicates that gradients in longitudinal stress remained small and unimportant during the speedup of Jakobshavn Isbræ. Moreover, during the acceleration, the driving stress did not increase significantly, as would be the case if the upglacier adjustment resulted from steeper slopes migrating inland. Instead, these authors surmise that the velocity changes resulted from weakening of the ice in lateral shear margins and perhaps basal properties. Similarly, the analysis of Joughin et al. (2012) suggests that while loss of buttressing following breakup of the glacier tongue may have contributed to the initial speedup, other processes must have been responsible for continued retreat and speedup of Jakobshavn Isbræ. These authors suggest that thinning-induced changes in the effective basal pressure dominate near-terminus glacier behavior.

Quantitative process studies such as those of Van der Veen et al. (2011) and Joughin et al. (2012) for Jakobshavn Isbræ demonstrate the difficulties faced when interpreting observations. Both weakening of the lateral shear margins (as proposed by Van der Veen et al., 2011) and changes in basal water pressure (Joughin et al., 2012) are plausible explanations for the observed speedup, but additional data to discern between these explanations are currently lacking. Further, the role of confining icebergs and bergy bits in the adjacent fjord on calving rate and the role of ocean melting at the grounding line have yet to be quantified. Consequently, predictions about future behavior of this particular glacier remain fraught with uncertainty.

The conclusions of Van der Veen et al. (2011) and Joughin et al. (2012) regarding possible processes most likely responsible for rapid changes observed on Jakobshavn Isbræ apply to that glacier only and cannot be generalized to other glaciers and to the question of stability of marine ice sheets in general. As pointed out by Truffer and Echelmeyer (2003), a continuum of streaming behavior exists, characterized by different dynamics flow regimes. Jakobshavn Isbræ represents one end member of the spectrum, with high driving stress supported by basal and lateral drag. On the other end of the spectrum are the Siple Coast ice streams moving over very weak beds and with the low, driving stress primarily supported by lateral drag. Stability of these different types of fast-moving glaciers may be entirely different, and concerted quantitative interpretations of data of rapid change pertaining to a range of glaciers, combined with numerical and theoretical modeling, are needed to address the question of stability of marine-terminating glaciers.

11 Interpreting Observations

11.1 INTRODUCTORY REMARKS

To describe the flow of a particular glacier, the appropriate model needs to be selected first. In some instances, the choice is obvious. For example, the lamellar flow model discussed in Section 4.2 is based on the assumption that the driving stress is balanced by drag at the glacier base. For floating ice shelves, basal drag is zero and the lamellar flow model does not apply. Most often, the choice is less clear. For example, many of the West Antarctic ice shelves have formed in embayments, and lateral drag may provide much, if not all, of the resistance to flow. It is not immediately obvious whether the ice-shelf model described in Section 4.5 applies best or whether the lateral-drag model of Section 4.4 should be used. On the Siple Coast ice streams, the concave surface profile has led some authors to propose that the flow along these drainage routes is primarily controlled by gradients in longitudinal stress and that other resistive stresses may be neglected when modeling these ice streams. Because the selected model determines to a large extent the predicted behavior of the glacier under consideration, it is important to include the major sources of flow resistance. To be on the safe side, one could use a complete model in which all stresses are calculated at depth. Increasingly, higher-order full-Stokes models are being applied to model glacier flow, but this may not always be a very practical solution, especially when trying to model the evolution of an ice sheet over an extended period of time. Therefore, one often seeks simpler models that realistically include the most important physical processes, treating the less important processes in a parameterized way. This means that the relative importance of the potential sources of resistance to flow needs to be known.

The essential difference between the models described in Chapter 4 is the source of flow resistance. In the lamellar flow model (Section 4.2), basal drag provides the sole resistance, while in the lateral drag model (Section 4.4), drag at the glacier sides is the controlling resistance. On free-floating ice shelves, only gradients in longitudinal stress oppose the driving stress (Section 4.5). Selecting any of these models to describe a certain glacier requires at least some data such as measured surface velocities and surface slopes to evaluate whether the model applies to the glacier being modeled. Too often, modelers develop models that are based mostly on limitations imposed by numerical or analytical techniques, with justification provided by "physical intuition." There is nothing wrong with developing hypothetical models, but claims that such models apply to actual glaciers or ice sheets when measurements indicate otherwise reveal more about the modeler than the model reveals about the glacier.

The force-budget technique discussed in Chapter 3 was developed to determine the mechanical controls on a glacier. The basic equations express balance of forces, stating that the driving stress is balanced by resistance to flow from gradients in longitudinal stress, lateral drag, and basal drag. The driving stress is estimated from glacier geometry, while the terms involving gradients of the resistive stresses are estimated from measured surface velocities. Basal drag is estimated from the requirement that the sum of all forces acting on a section of glacier must be zero. Taking the x-axis in the average direction of flow, force balance in this direction is

$$\tau_{dx} = \tau_{bx} - \frac{\partial}{\partial x}(H\bar{R}_{xx}) - \frac{\partial}{\partial y}(H\bar{R}_{xy}). \tag{11.1}$$

A similar equation applies to force balance in the second horizontal direction.

Equation (11.1) applies to the full ice thickness, and the resistive stresses in this equation are the depth-averaged stretching stress and lateral shear stress. It is possible to extend the force balance to calculate velocities, strain rates, and stresses at depth, as was done by Whillans et al. (1989) for Byrd Glacier in East Antarctica. The solution scheme is described in Van der Veen (1989; 1999b, Sections 3.4 and 3.5), but for most applications, it is sufficient to consider depth-averaged force balance. This so-called isothermal block-flow model has the great merit of being simple to carry out. Its drawback is that the viscous terms in the balance equations (the last two terms in equation (11.1)) tend to be overestimated because surface values of velocity and strain rates are applied to the entire glacier thickness. Thus, the inferred value for basal drag represents a limiting value. The other extreme value for basal drag can be found by setting the viscous terms equal to zero, that is, by equating basal drag to the driving stress. The actual value for basal drag may be expected to fall in between these two extremes.

Section 11.2 discusses application of the isothermal block-flow model to Byrd Glacier, updating the earlier results of Whillans et al. (1989). These calculations are best performed on regularly spaced data. To minimize errors introduced by measurement uncertainty and small-scale variations in velocity and surface elevation, the horizontal grid spacing should be on the order of one to two ice thicknesses so that derivatives are calculated over distances of two to four ice thicknesses. For Byrd Glacier, which is ~25 km wide, this gives enough spatial resolution to estimate lateral drag across the glacier. For narrow glaciers, this approach is not well suited. For example, the fast-moving part of Jakobshavn Isbræ in West Greenland is only 2 to 4 km wide. With a thickness of ~1 km, an appropriate grid spacing would be 1 to 2 km, which would give one or two gridpoints across the glacier. Clearly, this is insufficient to find a meaningful estimate for lateral drag. Decreasing the grid spacing would increase uncertainties in calculated force-balance terms. Therefore, for narrow glaciers, a two-step approach to evaluating force balance is proposed. First, the role of gradients in longitudinal stress in opposing the driving stress is estimated by considering flow along the central flowline as described in Section 11.3. Next, transects of velocity across the glacier are used to evaluate lateral drag at regular intervals along the glacier (Section 11.4). By combining the results from both calculations, basal drag can be estimated.

The force-balance technique is a diagnostic tool to investigate location and magnitude of resistance to flow on glaciers. By conducting calculations for successive time periods, changes in resistive forces can be evaluated, as was done by Van der Veen et al. (2011) for a 10-year period during which Jakobshavn Isbræ underwent large changes. While determining *how* partitioning of flow resistance is a necessary first step, it does not necessarily provide an answer to the question of *why* this partitioning changed.

11.2   LOCATING MECHANICAL CONTROLS

To illustrate the steps involved in the force-budget technique, application of the isothermal block-flow model to Byrd Glacier is considered. Byrd Glacier is one of the major outlet glaciers of the East Antarctic Ice Sheet, draining through the Transantarctic Mountains into the Ross Ice Shelf. Figure 11.1 shows the geometry of the lower reach of the glacier. The surface elevation is derived from ASTER (Advanced Spaceborne Thermal Emission and Reflection Radiometer) satellite imagery (Stearns, 2007), while the bed topography was determined from airborne radar sounding conducted in 2011–2012 by the Center for Remote Sensing of Ice Sheets at the University of Kansas (S. P. Gogineni, pers. comm., 2012). The lower panel shows the height above buoyancy, defined as

$$H_{ab} = H + \frac{\rho_w}{\rho} H_b, \tag{11.2}$$

with H the ice thickness (third panel), H_b the bed elevation (second panel; negative when below sea level), and ρ_w and ρ the density of sea water and ice, respectively. On the floating ice shelf, $H_{ab} = 0$, and the zero-contour in this map indicates the position of the grounding line where ice transitions from grounded to floating.

Original point measurements of surface and bed elevation were gridded to obtain values at regular gridpoints in the (x, y) coordinate system, with the x-axis horizontal and mainly downglacier, and the y-axis perpendicular and counterclockwise to the x-axis in map view. In the following, values at gridpoints denoted by two counters, namely, i = 1, 2, ..., I for the x-direction and j = 1, 2, ..., J for the y-direction, (i, j) = (1, 1) corresponds to the gridpoint in the lower left corner of the maps in Figure 11.1. The grid spacing is 2 km in both horizontal directions.

The first step is to calculate the two components of driving stress, given by

$$\tau_{dx} = -\rho g H \frac{\partial h}{\partial x}, \tag{11.3}$$

$$\tau_{dy} = -\rho g H \frac{\partial h}{\partial y}, \tag{11.4}$$

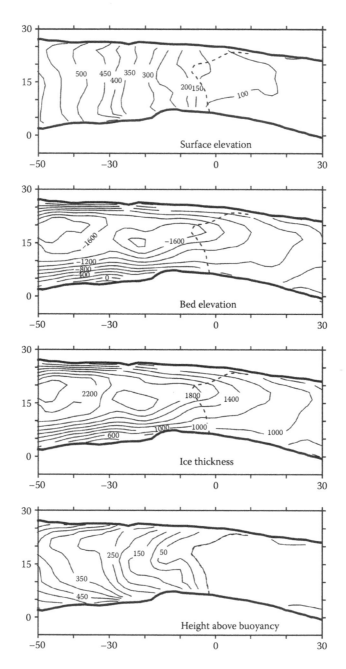

FIGURE 11.1   Surface elevation (contour interval: 50 m above sea level), bed elevation (contour interval: 200 m above sea level), ice thickness (contour interval: 200 m), and height above buoyancy (contour interval: 50 m) on Byrd Glacier. Ice flow is from left to right. The dashed line represents the grounding line as determined from the flotation criterion. Glacier data in this and the following figures have been transferred to a local coordinate system oriented such that ice flow is predominantly in the x-direction.

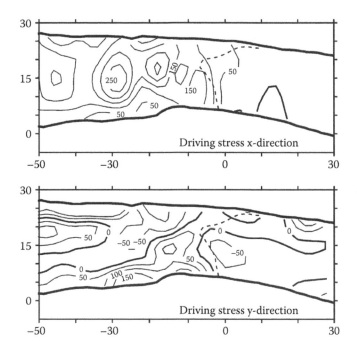

FIGURE 11.2   Two components of the driving stress on Byrd Glacier (contour interval: 50 kPa).

in which h represents the elevation of the ice surface. With values of H and h known at gridpoints (i, j), the two components of driving stress can be estimated at gridpoints using central differencing, and

$$\tau_{dx}(i,j) = -\rho g H(i,j) \frac{h(i+1,j) - h(i-1,j)}{2\Delta x}, \tag{11.5}$$

$$\tau_{dy}(i,j) = -\rho g H(i,j) \frac{h(i,j+1) - h(i,j-1)}{2\Delta y}, \tag{11.6}$$

where $\Delta x = \Delta y$ represents the spacing between neighboring gridpoints. Figure 11.2 shows values of the two components of driving stress. Values of the driving stress are small on the ice shelf and larger and more variable on the grounded portion. While the flow of the glacier is approximately in the x-direction, there is a significant component of driving stress in the transverse direction. This component is associated with longitudinal flow stripes on the glacier.

The next step is to calculate surface strain rates from measured surface velocities. These velocities are derived from feature tracking on two ASTER satellite images (12/5/2005–1/28/2007; Stearns, 2007). The two components of the velocity in the local coordinate system are shown in Figure 11.3. Greatest speeds are near the

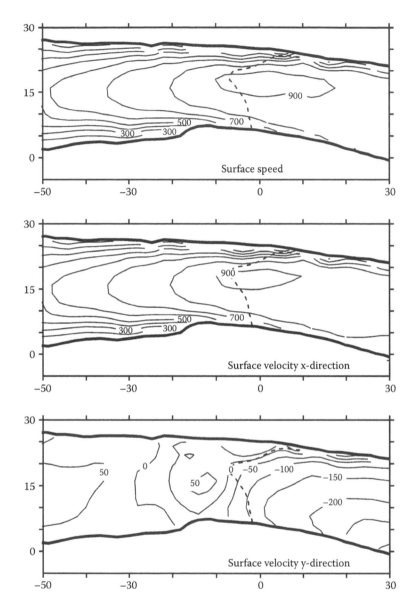

FIGURE 11.3 Measured surface velocity on Byrd Glacier (contour interval: 100 m/yr in the upper two panels, 50 m/yr in the lower panel).

grounding line and in the center of the glacier. The velocity decreases to zero at the lateral margins.

Because depth-averaged flow is considered, only three strain rate components need to be considered, namely, longitudinal stretching or compression, $\dot{\varepsilon}_{xx}$, transverse stretching or compression, $\dot{\varepsilon}_{yy}$, and shearing, $\dot{\varepsilon}_{xy}$. From the definitions given in Section 1.2, these strain rates are linked to velocity gradients. Denoting the velocity in the x-direction by U and that in the y-direction by V, applying finite differencing to equation (1.34) allows the strain rates to be estimated from the following expressions:

$$\dot{\varepsilon}_{xx}(i,j) = \frac{\partial U}{\partial x}(i,j) = \frac{U(i+1,j) - U(i-1,j)}{2\Delta x}, \tag{11.7}$$

$$\dot{\varepsilon}_{yy}(i,j) = \frac{\partial V}{\partial y}(i,j) = \frac{V(i,j+1) - V(i,j-1)}{2\Delta y}, \tag{11.8}$$

$$\dot{\varepsilon}_{xy}(i,j) = \frac{1}{2}\frac{\partial U}{\partial y}(i,j) + \frac{1}{2}\frac{\partial V}{\partial x}(i,j) = \frac{1}{2}\frac{U(i,j+1) - U(i,j-1)}{2\Delta x} + \frac{1}{2}\frac{V(i+1,j) - V(i-1,j)}{2\Delta x}. \tag{11.9}$$

Calculated strain rates are shown in Figure 11.4. The lower panel shows the effective strain rate, defined as

$$2\dot{\varepsilon}_e^2 = \dot{\varepsilon}_{xx}^2 + \dot{\varepsilon}_{yy}^2 + \dot{\varepsilon}_{zz}^2 + 2\left(\dot{\varepsilon}_{xy}^2 + \dot{\varepsilon}_{xz}^2 + \dot{\varepsilon}_{yz}^2\right). \tag{11.10}$$

For the present case, vertical shearing is neglected, while the vertical strain rate, $\dot{\varepsilon}_{zz}$, follows from the incompressibility condition. The effective strain rate then reduces to

$$\dot{\varepsilon}_e^2 = \dot{\varepsilon}_{xx}^2 + \dot{\varepsilon}_{yy}^2 + \dot{\varepsilon}_{xx}\dot{\varepsilon}_{yy} + \dot{\varepsilon}_{xy}^2. \tag{11.11}$$

Invoking the flow law (3.47) and (3.48), the resistive stresses are calculated from strain rates as

$$R_{xx} = B\dot{\varepsilon}_e^{1/n-1}(2\dot{\varepsilon}_{xx} + \dot{\varepsilon}_{yy}), \tag{11.12}$$

$$R_{yy} = B\dot{\varepsilon}_e^{1/n-1}(\dot{\varepsilon}_{xx} + 2\dot{\varepsilon}_{yy}), \tag{11.13}$$

$$R_{xy} = B\dot{\varepsilon}_e^{1/n-1}\dot{\varepsilon}_{xy}. \tag{11.14}$$

Applying these relations yields the surface values for the resistive stresses R_{xx} and R_{xy}, shown in Figure 11.5. Longitudinal tension (upper panel) is generally positive. The ice-shelf portion (beyond x = 0 km) is under about 100 kPa tension. In the grounded portion, this quantity is more variable and locally reaches 250 kPa. Side shear (lower panel) is largest near the lateral margins of the glacier, as expected. Where the glacier is grounded, side shear varies in a complex way across the glacier, but for the ice-shelf portion it decays within about 7 km of the margin and is near zero in the middle portion.

In the balance of forces, gradients in resistive stresses are important (see Section 3.2), and the next step is to use horizontal gradients in the resistive stresses shown in

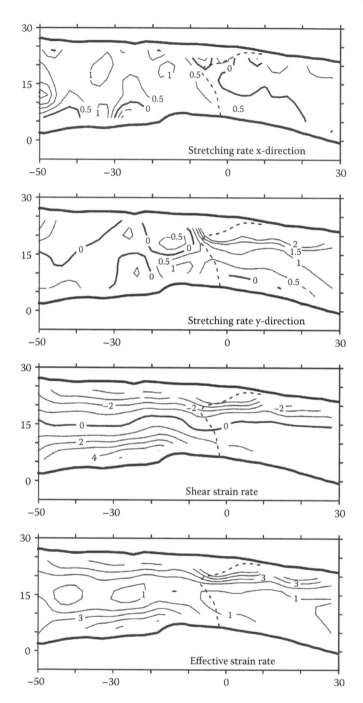

FIGURE 11.4   Derived surface strain rates on Byrd Glacier (contour interval: 0.5×10^{-3} yr^{-1} in the upper two panels, 1×10^{-3} yr^{-1} in the lower two panels).

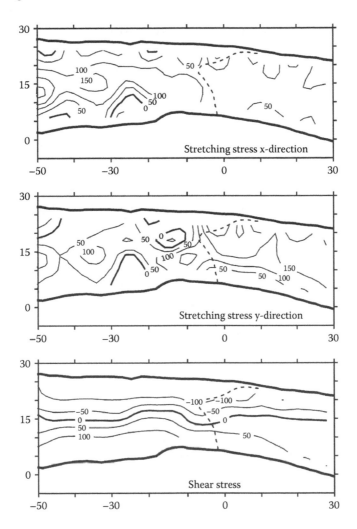

FIGURE 11.5   Resistive stresses on Byrd Glacier (contour interval: 50 kPa).

Figure 11.5 to estimate resistance to flow from longitudinal stress gradients and from lateral drag. First, gradients in longitudinal stress at gridpoints (i, j) are calculated from

$$\frac{\partial}{\partial x}(HR_{xx})(i, j) = \frac{H(i+1, j)R_{xx}(i+1, j) - H(i-1, j)R_{xx}(i-1, j)}{2\Delta x}. \quad (11.15)$$

Lateral drag is calculated from

$$\frac{\partial}{\partial y}(HR_{xy})(i, j) = \frac{H(i, j+1)R_{xy}(i, j+1) - H(i, j-1)R_{xy}(i, j-1)}{2\Delta y}. \quad (11.16)$$

Basal drag can now be calculated from the force-balance equation

$$\tau_{bx} = \tau_{dx} + \frac{\partial}{\partial x}(HR_{xx}) + \frac{\partial}{\partial y}(HR_{xy}). \tag{11.17}$$

A similar expression applies to the second horizontal direction. In Figure 11.6 the results of the force-budget calculation for the approximate direction of flow are shown. The driving stress is shown in the upper panel. The next two panels show the gradient in longitudinal stress and resistance to flow associated with side drag. Adding the three panels together yields the basal drag (last panel). Comparison of the top and bottom panels shows that fluctuations in driving stress are largely muted by the gradient terms, resulting in smaller variations in basal drag. Nevertheless, basal drag is far from uniform. As expected, it is zero under the floating ice-shelf portion. On the grounded part, basal drag is large and variable and appears to be concentrated at isolated spots ("sticky spots").

Because of the relatively simple geometry of Byrd Glacier, force-balance terms in the x-direction correspond mostly to those in the flow direction and can be interpreted as such. This need not be the case for glaciers where flow may be turning in the downglacier direction. For such glaciers, a final step is needed to complete the force-balance calculations, namely, rotation to a local flow-following coordinate system.

At each location, the direction of ice flow is given by (Section 1.1)

$$\phi(i, j) = atan\left(\frac{V(i, j)}{U(i, j)}\right), \tag{11.18}$$

with ϕ defined as the angle between the velocity direction and the x-axis (Figure 1.2). This angle is shown in Figure 11.7 and is small on the grounded part. On the ice shelf, the angle increases as the flow turns clockwise onto the Ross Ice Shelf. This flow direction defines a Cartesian coordinate system with the $\tilde{x}$-direction at each location aligned with the flow at that location, and the $\tilde{y}$-direction perpendicular to the flow direction. Generally, the orientation of the $(\tilde{x}, \tilde{y})$ coordinate system differs at each location on the glacier as the flow direction changes along the glacier. In the following expressions, these flow directions are indicated by subscripts l (along flow) and c (cross flow), respectively. Results for the coordinate transformation are shown only for the floating section of Byrd Glacier.

The driving stress is a vector and rotates according to equations (1.5) and (1.6). The component of driving stress in the (local) flow direction is then

$$\tau_{dl} = \tau_{dx} \cos\phi + \tau_{dy} \sin\phi, \tag{11.19}$$

while the cross-flow component is given by

$$\tau_{dc} = -\tau_{dx} \sin\phi + \tau_{dy} \cos\phi. \tag{11.20}$$

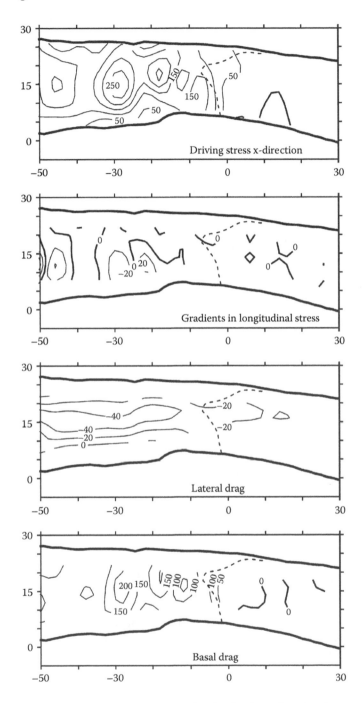

FIGURE 11.6 Force-balance terms on Byrd Glacier for the x-direction (contour interval: 50 kPa in top and bottom panels, 20 kPa in second and third panels).

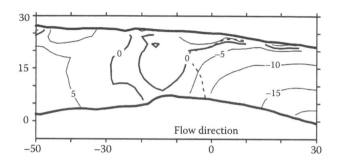

FIGURE 11.7   Angle of flow direction measured relative to the x-direction on Byrd Glacier (contour interval: 5 degrees).

The other terms in the force-balance equation transform similarly. The rotated driving stress is shown in the upper panels of Figure 11.8. For this particular example, there is not much difference between the rotated driving stress components and those in the (x, y) coordinate system because the rotation angle is relatively small.

The three surface strain rates form a 2 × 2 tensor and rotate to the flow-following coordinate system according to (Jaeger, 1969, p. 7; cf. Section 1.1)

$$\dot{\varepsilon}_{ll} = \dot{\varepsilon}_{xx} \cos^2\phi + \dot{\varepsilon}_{yy} \sin^2\phi + 2\dot{\varepsilon}_{xy} \sin\phi\cos\phi, \tag{11.21}$$

$$\dot{\varepsilon}_{cc} = \dot{\varepsilon}_{xx} \sin^2\phi + \dot{\varepsilon}_{yy} \cos^2\phi - 2\dot{\varepsilon}_{xy} \sin\phi\cos\phi, \tag{11.22}$$

$$\dot{\varepsilon}_{lc} = (\dot{\varepsilon}_{xx} - \dot{\varepsilon}_{yy})\sin\phi \cos\phi + \dot{\varepsilon}_{xy} (\cos^2\phi - \sin^2\phi). \tag{11.23}$$

The lower panels in Figure 11.8 compare two of the rotated strain rates to the corresponding strain rates in the (x, y) coordinate system. To find the resistive stresses in the flow-following coordinate system, equations similar to (11.21)–(11.23) are applied.

In addition to estimating the terms in the balance of forces, associated uncertainties should also be estimated. This can be done using the formulas given in Section 1.3 and is not discussed here for the Byrd Glacier example.

11.3   ESTIMATING THE ROLE OF GRADIENTS IN LONGITUDINAL STRESS

The longitudinal resistive stress, R_{xx}, is linked to the along-flow stretching rate or longitudinal gradient in the downstream component of ice velocity through the flow law (3.48). If the dynamic centerline of a glacier is considered, the only nonzero strain rates contributing to the effective strain rate are the along-flow stretching rate and the lateral spreading rate. Along-flow stretching is calculated from the gradient in ice velocity

$$\dot{\varepsilon}_{xx} = \frac{\partial U}{\partial x}, \tag{11.24}$$

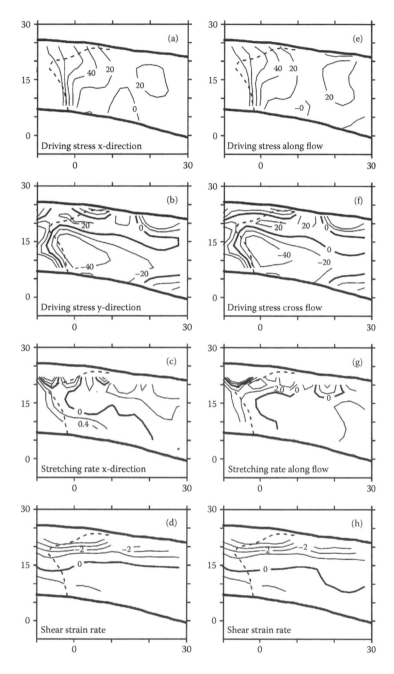

FIGURE 11.8   Two components of driving stress in the original (x, y) coordinate system (panels a and b) and in the flow-following $(\tilde{x}, \tilde{y})$ coordinate system (panels e and f) (contour interval: 20 kPa), and stretching rate and shear strain rate in the original (x, y) coordinate system (panels c and d) and in the flow-following $(\tilde{x}, \tilde{y})$ coordinate system (panels g and h) (contour interval: 0.2×10^{-3} yr^{-1} in panels c and f, and 1.0×10^{-3} yr^{-1} in panels d and h).

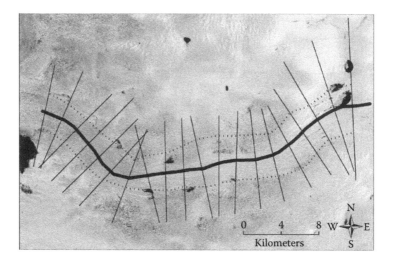

FIGURE 11.9   Map showing the central flowline of the lower 30 km of Jakobshavn Isbræ, and location of velocity transects used to estimate resistance to flow from lateral drag. The dotted lines mark the lateral shear margins. (From Van der Veen, C. J., J. C. Plummer, and L. A. Stearns, *J. Glaciol.*, 57, 770–782, 2011. Reprinted from the *Journal of Glaciology* with permission of the International Glaciological Society and the authors.)

with U the component of velocity in the direction of flow and the x-axis following the flow direction. As shown in Figure 11.9, this axis may be curvilinear as in the case of Jakobshavn Isbræ, West Greenland. Lateral spreading is estimated from the width of the glacier, making the assumption that the ice must remain in contact with the margins. That is,

$$\dot{\varepsilon}_{yy} = \frac{U}{W}\frac{\partial W}{\partial x}. \tag{11.25}$$

This term may be neglected where the glacier width, W, is approximately constant in the flow direction.

On fast-moving outlet glaciers, the contribution to ice discharge from internal deformation is small, and in good approximation, measured surface velocities may be used to estimate the depth-averaged strain rate and resistive stress. Of importance to the balance of forces is how the depth-integrated resistive stress varies along the flowline. That is, resistance to flow from gradients in longitudinal stress is given by

$$F_{lon} = \frac{\partial H R_{xx}}{\partial x}, \tag{11.26}$$

and this term must be compared to the driving stress calculated from the geometry of the glacier to evaluate the importance of longitudinal stress gradients to the flow

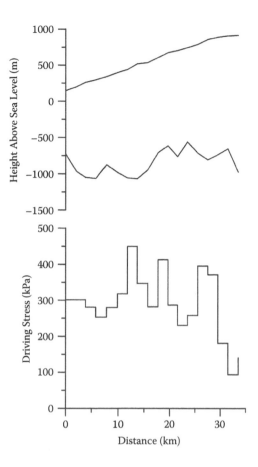

FIGURE 11.10 Surface and bed elevation (2005) along the central flowline of Jakobshavn Isbræ and calculated driving stress. (After Van der Veen, C. J., J. C. Plummer, and L. A. Stearns, *J. Glaciol.,* 57, 770–782, 2011.)

of the glacier. To illustrate the steps involved, results of a calculation for the lower 30 km of Jakobshavn Isbræ are discussed next.

Figure 11.9 shows the central flowline and the lateral boundaries of Jakobshavn Isbræ, as well as across-flow transects used to estimate lateral drag. Section 11.4 discusses how these boundaries were determined and how resistance to flow from lateral drag is estimated. For the present discussion, only the glacier geometry and velocity along the flowline need to be considered. This geometry is shown in the upper panel of Figure 11.10. Surface elevations derive from airborne laser altimetry conducted in 2005, while the bed topography was measured by airborne radar sounding conducted by the Center for Remote Sensing of Ice Sheets at the University of Kansas over multiple years (Gogineni et al., 2001; Van der Veen et al., 2011). From this geometry, the driving stress shown in the lower panel is calculated.

Equally important as calculating the value of the driving stress is to estimate the uncertainty or error in this quantity. This error is estimated using equation (1.40) for propagation of independent errors. Let

$$\alpha = -\frac{\partial h}{\partial x}, \tag{11.27}$$

denote the slope of the ice surface. The driving stress is then given by

$$\tau_{dx} = \rho g H \alpha. \tag{11.28}$$

Neglecting errors in the ice density, ρ, and gravitational constant, g, applying equation (1.40) gives

$$(\Delta \tau_{dx})^2 = \rho g \alpha (\Delta H)^2 + \rho g H (\Delta \alpha)^2, \tag{11.29}$$

or

$$\left(\frac{\Delta \tau_{dx}}{\tau_{dx}}\right)^2 = \left(\frac{\Delta H}{H}\right)^2 + \left(\frac{\Delta \alpha}{\alpha}\right)^2. \tag{11.30}$$

The surface slope is calculated as the difference between the surface elevations at two locations, x_1 and x_2, and the uncertainty is given by (compare with equation (1.44))

$$\Delta \alpha = \frac{\Delta h}{x_2 - x_1} \sqrt{2}. \tag{11.31}$$

For the present case, the uncertainty in surface elevation is estimated to be about 0.6 m. Surface slopes are calculated over a nominal distance of 4 km, giving an uncertainty in the slope of $\sim 0.21 \times 10^{-3}$. The uncertainty in ice thickness is ~ 10 m. With an average ice thickness of 1500 m and an average slope equal to 0.028, the resulting uncertainty in driving stress is

$$\left(\frac{\Delta \tau_{dx}}{\tau_{dx}}\right)^2 = \left(\frac{10}{1500}\right)^2 + \left(\frac{0.21 \times 10^{-3}}{0.028}\right)^2, \tag{11.32}$$

and

$$\frac{\Delta \tau_{dx}}{\tau_{dx}} = 10^{-2}. \tag{11.33}$$

Thus, for this particular example, the error in driving stress is rather small, being about 1% of the value of the driving stress. Because the relative error is so small, error bars are not shown in the lower panel of Figure 11.10.

Next, gradients in longitudinal stress are estimated. The input data for this calculation are surface velocities (Figure 11.11, upper panel) measured from satellite interferometric image pairs collected in October 2005, using standard speckle tracking techniques (Joughin et al., 2008b). Neglecting the contribution of shear strain rates

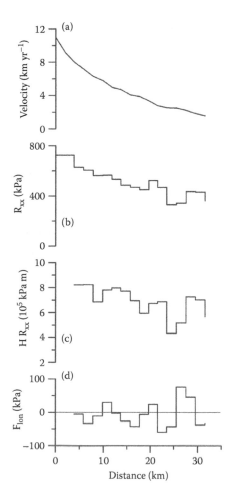

FIGURE 11.11 Steps in the calculation of resistance to flow from gradients in longitudinal stress. Panel a shows the measured surface velocity from which stretching rates (not shown) are calculated. Using the flow law, the corresponding resistive stress (panel b) is calculated. This stress is multiplied by the ice thickness (panel c) and the derivative of this quantity gives F_{lon} (panel d). (After Van der Veen, C. J., J. C. Plummer, and L. A. Stearns, *J. Glaciol.*, 57, 770–782, 2011.)

to the effective strain rate, the longitudinal resistive stress is linked to the stretching rate by invoking the flow law,

$$R_{xx} = 2B\dot{\varepsilon}_{xx}^{1/n},\qquad (11.34)$$

where n = 3 is the flow-law exponent and B the temperature-dependent viscosity parameter. Strain rates are estimated from measured surface velocities and taken to be constant with depth. The value $B = 400$ kPa yr$^{1/3}$ corresponding to an effective depth-averaged ice temperature of $-10°$C, is used here. The calculated stretching stress is shown in the second panel in Figure 11.11. The third panel shows gradients in longitudinal stress multiplied by the ice thickness, as called for in equation (11.25). Taking the derivative, resistance to flow associated with gradients in longitudinal stress is obtained. The result is shown in the lower panel in Figure 11.11 and suggests that this term in the balance of forces is small compared to the driving stress. There are spatial variations in F_{lon}, but it is not clear whether these have an important effect on the dynamics of the glacier. A linear regression of the curve shown in the third panel gives an average contribution to the large-scale force balance of ~10 kPa, compared with an average driving stress of ~250 kPa.

The uncertainty in the calculated force-balance term, F_{lon}, can be estimated using the expressions given in Section 1.3. For this example, the uncertainty is rather small.

11.4 ESTIMATING RESISTANCE FROM LATERAL DRAG

Resistance from lateral drag is linked to the transverse gradient in lateral shear (the second term on the right-hand side of the balance equation (11.1)). Thus, where transects of velocity across a glacier are available, lateral drag can be estimated. As an example, consider the velocity transect across Jakobshavn Isbræ shown in Figure 11.12. Following the model described in Section 4.4, the assumption is made that lateral shearing is the dominant strain rate, so that the lateral shear stress, R_{xy}, can be estimated from equation (4.47). That is

$$R_{xy} = B \left(\frac{1}{2} \frac{\partial U}{\partial y} \right)^{1/n}, \tag{11.35}$$

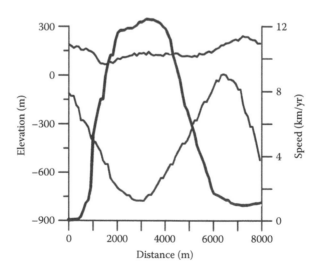

FIGURE 11.12 Surface and bed elevation (scale on the left) and surface velocity (bold curve; scale on the right) along a transect across Jakobshavn Isbræ. (From Van der Veen, C. J., J. C. Plummer, and L. A. Stearns, *J. Glaciol.,* 57, 770–782, 2011. Reprinted from the *Journal of Glaciology* with permission of the International Glaciological Society and the authors.)

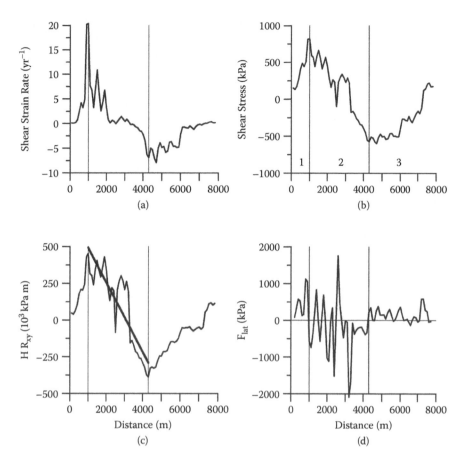

FIGURE 11.13 Steps in the calculation of resistance to flow from lateral drag. Panel a shows the measured surface velocity from which stretching rates (not shown) are calculated. Using the flow law, the corresponding resistive stress (panel b) is calculated. This stress is multiplied by the ice thickness (panel c) and the derivative of this quantity gives F_{lat} (panel d).

where U represents the discharge velocity perpendicular to the transect. This velocity is a function of the transverse distance, y. The corresponding shear strain rate and shear stress are shown in panels a and b of Figure 11.13. Recalling that resistance to flow from lateral drag is given by

$$F_{lat} = -\frac{\partial}{\partial y}(H R_{xy}),$$ (11.36)

three different regions can be identified along the transect (panel b): in the central part (region 2), the shear stress decreases in the y-direction and lateral drag opposes the driving stress, while in both outboard regions (1 and 3) the shear stress increases in the y-direction and lateral drag acts in the same way as does the driving stress. Thus, the slower-moving outboard ice is being dragged along by the fast-moving interior part of the glacier. As a result, basal drag must be greater than the local driving stress in the outboard regions (Whillans and Van der Veen, 1997, 2001; Van der Veen et al., 2007). The minimum and maximum values of lateral shear stress correspond to the two inflection points in the velocity curve shown in Figure 11.12 and define the boundaries of the ice stream (Van der Veen et al., 2011). The vertical lines in Figure 11.13 demarcate the lateral margins of the ice stream at this particular transect.

The curve of lateral shear stress, R_{xy}, shows many small-scale variations that may be associated with small-scale variations in driving stress, or locally important gradients in longitudinal stress. As a result of such variations, applying equation (11.36) to estimate lateral drag across the ice stream typically yields a noisy curve (Figure 11.13, panel d). The main reason why local variability and, to some extent, measurement uncertainties become important when estimating lateral drag is that spatial derivatives in the transverse direction are calculated over relatively short distances. According to equation (1.44), the error in calculated strain rate is inversely proportional to the distance over which the velocity gradient is calculated. When considering along-flow stretching, this distance is typically several tens of km. However, to calculate lateral drag, derivatives are obtained from velocity determinations that are spaced 1 km or less. Thus, for the same uncertainty in velocity, the error in shear strain rate is about 100 times the error in stretching rate. Transverse gradients need to be calculated over much shorter distances because the width of an ice stream or outlet glacier is comparatively small, up to about 30 km for West Antarctic ice streams and only a few km for Jakobshavn Isbræ, compared with a length of up to several hundreds of km. To minimize the uncertainty in calculated lateral drag, the width-averaged balance of forces is considered.

As derived in Section 4.4, the average lateral resistance acting on a section of glacier of unit width can be calculated, using the values of the shear stress, τ_s, at both margins:

$$F_{lat} = \frac{H_1 \tau_{s1} - H_2 \tau_{s2}}{W}. \tag{11.37}$$

In this expression, the subscripts 1 and 2 refer to the southern and northern margin, respectively, and W represents the distance between the two margins. The total width of the transect shown in Figure 11.12 is 3.3 km, while the shear stresses at the margins are 819 kPa and −570 kPa, with corresponding thicknesses of 553 and 683 m, respectively. Substituting these values into equation (11.37) gives $F_{lat} = 255$ kPa. In other words, averaged over the width of the ice stream, lateral drag supports about 255 kPa of driving stress. Using the formulas for error propagation discussed in Section 1.3, the estimated error in this value is 15 kPa.

Alternatively, F_{lat} can be estimated from linear regression of HR_{xy} against transverse distance, with the regression restricted to the region between the lateral shear margins. The slope of the regression line shown in Figure 11.13c corresponds to a width-averaged lateral drag of 237 kPa with an uncertainty of 23 kPa. Within error limits, both approaches yield the same result.

The relative importance of lateral drag in controlling the flow Jakobshavn Isbræ can now be determined for the entire flowline by comparing F_{lat} calculated for each transect shown in Figure 11.9 to the width-averaged driving stress. This comparison is shown in Figure 11.14 and indicates that over the lower 10 km or so, lateral drag

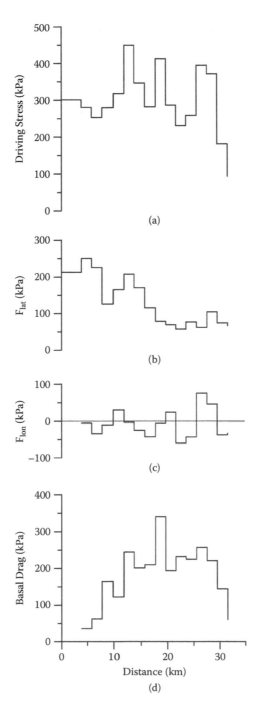

FIGURE 11.14   Force-balance terms along the central flowline of Jakobshavn Isbræ. (After Van der Veen, C. J., J. C. Plummer, and L. A. Stearns, *J. Glaciol.,* 57, 770–782, 2011.)

is the main source of flow resistance. As discussed in Section 11.3, gradients in longitudinal stress are small and contribute little to the balance of forces. Subtracting F_{lat} and F_{lon} from the driving stress yields basal drag (Figure 11.14, panel d). For the upstream part of the flowline, the driving stress is primarily balanced by drag at the glacier base. Toward the glacier front, basal resistance decreases.

Van der Veen et al. (2011) consider force balance on Jakobshavn Isbræ at three different times (1995, 2000, and 2005) to evaluate changes in location and magnitude of flow resistance that accompanied the doubling of ice discharge and the contemporaneous surface lowering. They propose the rapid changes on this major outlet glacier resulted from weakening of the lateral shear margins; but without additional observations, the available measurements of surface speed and geometry alone are insufficient to identify the actual physical processes responsible for the observed changes in partitioning of flow resistance.

References

Abramowitz, M., and I. A. Stegun (1965). *Handbook of mathematical functions*. Dover Publ. Inc., New York, 1046 pp.

Albrecht, T., and A. Levermann (2012). Fracture field for large-scale ice dynamics. *Journ. Glaciol.* 58, 165–176.

Alley, R. B. (1989a). Water-pressure coupling of sliding and bed deformation I. Water system. *Journ. Glaciol.* 35, 108–118.

Alley, R. B. (1989b). Water-pressure coupling of sliding and bed deformation: II. Velocity-depth profiles. *Journ. Glaciol.* 35, 119–129.

Alley, R. B. and I. M. Whillans (1984). Response of the East Antarctic Ice Sheet to sea-level rise. *Journ. Geoph. Res.* 89(C), 6487–6493.

Alley, R. B. et al. (1986). Deformation of till beneath ice stream B, West Antarctica. *Nature* 322, 57–59.

Alley, R. B. et al. (1987a). Till beneath ice stream B. 3. Till deformation: Evidence and implications. *Journ. Geoph. Res.* 92(B), 8921–8929.

Alley, R. B. et al. (1987b). Till beneath ice stream B. 4. A coupled ice-till flow model. *Journ. Geoph. Res.* 92(B), 8931–8940.

Ambach, W. (1968). The formation of crevasses in relation to the measured distribution of strain-rates and stresses. *Archive fur Meteorologie, Geophysik und Bioklimatologie* 17(A), 78–87.

Amundson, J. M. et al. (2010). Ice mélange dynamics and implications for terminus stability, Jakobshavn Isbræ, Greenland. *Journ. of Geoph. Res.* 115, F01005, doi: 10.1029/2009JF001405.

Anandakrishnan, S. et al. (2007). Discovery of till deposition at the grounding line of Whillans Ice Stream. *Science* 315, 1835–1838.

Anderson, M. G., and P. D. Bates (eds.) (2001). *Model validation: Perspectives in hydrological science*. New York: John Wiley & Sons, 500 pp.

Anderson, M. P., and W. W. Woessner (1992). *Applied groundwater modeling: Simulation of flow and advective transport*. San Diego: Academic Press, 381 pp.

Andreas, E. L. (1987). A theory for the scalar roughness and the scalar transfer coefficients over snow and sea ice. *Bound. Layer Meteorol.* 38, 159–184.

Arthern, R. J., and D. J. Wingham (1998). The natural fluctuations of firn densification and their effect on the geodetic determination of ice sheet mass balance. *Climatic Change* 40, 605–624.

Aschwanden, A., and H. Blatter (2009). Mathematical modeling and numerical simulation of polythermal glaciers. *Journal of Geophysical Research* 114, F01027, doi:10.1029/2008JF001028.

Ayer, A. J. (1953). *Language, truth, and logic*. New York: Dover, 160 pp.

Azuma, N., and A. Higashi (1985). Formation processes of ice fabric patterns in ice sheets. *Ann. Glaciol.* 6, 130–134.

Bader, H. (1954). Sorge's law of densification of snow on high polar glaciers. *Journ. Glaciol.* 2, 319–323.

Baker, R. W. (1981). Textural and crystal-fabric anisotropies and the flow of ice masses. *Science* 211, 1043–1044.

Baker, R. W. (1982). A flow equation for anisotropic ice. *Cold Reg. Sci. Techn.* 6, 141–148.

Balise, M. J. (1988). *The relation between surface and basal velocity variations in glaciers, with application to the mini-surges of Variegated Glacier*. PhD Thesis, University of Washington, Seattle, 205 pp.

Balise, M. J., and C. F. Raymond (1985). Transfer of basal sliding variations to the surface of a linearly viscous glacier. *Journ. Glaciol.* 31, 308–318.

Bassis, J. N., and C. C. Walker (2012). Upper and lower limits on the stability of calving glaciers from the yield strength envelope of ice. *Proc. Royal Soc Ser. A* 462, 913–932.

Bassis, J. N. et al. (2007). Seismicity and deformation associated with ice-shelf rift propagation. *Journ. Glaciol.* 53, 523–536.

Batchelor, G. K. (1967). *An introduction to fluid dynamics.* Cambridge, U.K.: Cambridge University Press, 615 pp.

Benn, D. I., N. R. J. Hulton, and R. H. Mottram (2007b). "Calving laws," "sliding laws," and the stability of tidewater glaciers. *Ann. Glaciol.* 46, 123–130.

Benn, D. I., C. R. Warren, and R. H. Mottram (2007a). Calving processes and the dynamics of calving glaciers. *Earth Sci. Rev.* 82, 143–179.

Benthem, J. P., and W. T. Koiter (1973). Asymptotic approximations to crack problems. In: *Mechanics of fracture, Vol. 1: Methods of analysis and solutions of crack problems* (ed. G. C. Sih). Leiden, Netherlands: Noordhoff, 131–178.

Bentley, C. R. (1984). Some aspects of the cryosphere and its role in climate change. In: *Climate processes and climate sensitivity* (ed. J. E. Hansen and T. Takahashi). Geophysical Monograph 29, American Geophysical Union, Washington, DC, 207–220.

Bentley, C. R., J. W. Clough, K. C. Jezek, and S. Shabtaie (1979). Ice-thickness patterns and the dynamics of the Ross Ice Shelf, Antarctica. *Journ. Glaciol.* 24, 287–294.

Bindschadler, R. (1983). The importance of pressurized subglacial water in separation and sliding at the glacier bed. *Journ. Glaciol.* 29, 3–19.

Bindschadler, R. A. (1984). Jakobshavns glacier drainage basin: A balance assessment. *Journ. Geoph. Res.* 89(C), 2066–2072.

Birchfield, G. E., J. Weertman, and A. T. Lunde (1981). A paleoclimate model of Northern Hemisphere ice sheets. *Quat. Res.* 15, 126–142.

Blankenship, D. D. et al. (1986). Seismic measurements reveal a saturated porous layer beneath an active Antarctic ice stream. *Nature* 322, 54–57.

Blankenship, D. D. et al. (1987). Till beneath ice stream B. 1. Properties derived from seismic travel times. *Journ. Geoph. Res.* 92(B), 8903–8911.

Boas, W., and E. Schmid (1931). Zur deutung der deformations texturen von metallen. *Techn. Physik.* 12, 71–75.

Bond, G. C., and R. Lotti (1995). Iceberg discharges into the North Atlantic on millennial time scales during the last glaciation. *Science* 267, 1005–1010.

Boulton, G. S. (1974). Processes and patterns of glacial erosion. In: *Glacial geomorphology* (ed. D. R. Coates). State University of New York, Binghamton, 41–87.

Boulton, G. S. (1986). A paradigm shift in glaciology? *Nature* 323, 18.

Boulton, G. S., K. E. Dobbie, and S. Zatsepin (2001). Sediment deformation beneath glaciers and its coupling to the subglacial hydraulic system. *Quat. Int.* 86, 3–28.

Boulton, G. S., and R. C. A. Hindmarsh (1987). Sediment deformation beneath glaciers: Rheology and geological consequences. *Journ. Geoph. Res.* 92(B), 9059–9082.

Boulton, G. S., and A. S. Jones (1979). Stability of temperate ice caps and ice sheets resting on beds of deformable sediment. *Journ. Glaciol.* 24, 29–43.

Boussinesq, J. (1877). Essai sur la théorie des eaux courantes. *Mém. prés. Acad. Sci. Paris* 23, 680 pp.

Braithwaite, R. J. (1995). Aerodynamic stability and turbulent sensible-heat flux over a melting ice surface, the Greenland ice sheet. *Journ. Glaciol.* 41, 562–571.

Brock, B. W., I. C. Willis, and M. J. Sharp (2000). Measurement and parameterization of albedo variations at Haut Glacier d'Arolla, Switzerland. *Journ. Glaciol.* 46, 675–688.

Brock, B. W., I. C. Willis, and M. J. Sharp (2006). Measurement and parameterization of aerodynamic roughness length variations at Haut Glacier d'Arolla, Switzerland. *Journ. Glaciol.* 52, 281–297.

Broek, D. (1986). *Elementary engineering fracture mechanics (4th ed.)*. Dordrecht: Kluwer Academic Publishers, 516 pp.

Brotchie, J. F., and R. Silvester (1969). On crustal flexure. *Journ. Geoph. Res.* 74, 5240–5252.

Brown, C. S., M. F. Meier, and A. Post (1982). *Calving speed of Alaska tidewater glaciers, with application to Columbia Glacier.* USGS Professional Paper. 1258-C, 13 pp.

Brown, R. A. (1991). *Fluid mechanics of the atmosphere.* Academic Press, San Diego, 489 pp.

Brunt, D. (1939). *Physical and dynamical meteorology.* Cambridge Univ. Press, Cambridge, 428 pp.

Brunt, K. M. et al. (2010). Mapping the grounding zone of the Ross Ice Shelf, Antarctica, using ICESat laser altimetry. *Ann. Glaciol.,* 51(55), 71–79.

Brutsaert, W. (1975). On a derivable formula for long-wave radiation from clear skies. *Water Resour. Res.* 11, 742–744.

Budd, W. F. (1970a). The longitudinal stress and strain-rate gradients in ice masses. *Journ. Glaciol.* 9, 19–27.

Budd, W. F. (1970b). Ice flow over bedrock perturbations. *Journ. Glaciol.* 9, 29–48.

Budd, W. F., and T. H. Jacka (1989). A review of ice rheology for ice-sheet modelling. *Cold Reg. Sci. Techn.* 16, 107–144.

Budd, W. F., and D. Jenssen (1987). Numerical modelling of the large-scale water flux under the West Antarctic Ice Sheet. In: *Dynamics of the West Antartcic Ice Sheet* (eds. C. J. van der Veen and J. Oerlemans). Reidel Publishing Co., Dordrecht, 293–320.

Budd, W. F., D. Jenssen, and I. N. Smith (1984). A three-dimensional time-dependent model of the West Antarctic Ice Sheet. *Ann. Glaciol.* 5, 29–36.

Budd, W. F., P. L. Keage, and N. A. Blundy (1979). Empirical studies of ice sliding. *Journ. Glaciol.* 23, 157–170.

Bueler, E., J. Brown, and C. Lingle (2007). Exact solutions to the thermomechanically coupled shallow-ice approximation: Effective tools for verification. *Journ. Glaciol.* 53, 499–516.

Bueler, E. et al. (2005). Exact solutions and verification of numerical models for isothermal ice sheets. *Journ. Glaciol.* 51, 291–306.

Busch, N. E. (1973). On the mechanics of atmospheric turbulence. In: *Workshop on micrometeorology* (ed. D. A. Haugen). *Amer. Meteor. Soc.* Boston, MA, 1–65.

Businger, J. A. (1973). Turbulent transfer in the atmospheric surface layer. In: *Workshop on micrometeorology* (ed. D. A. Haugen). *Amer. Meteor. Soc.* Boston, MA, 67–100.

Chorin, A. J., and J. E. Marsden (1992). *A mathematical introduction to fluid mechanics (3rd ed.).* New York: Springer-Verlag, 169 pp.

Clarke, G. K. C. (1976). Thermal regulation of glacier surging. *Journ. Glaciol.* 16, 231–250.

Clarke, G. K. C. (2005). Subglacial processes. *Annual Review Earth Planet. Sci.* 33, 247–276.

Clarke, G. K. C., and E. W. Blake (1991). Geometric and thermal evolution of a surge-type glacier in its quiescent stage: Trapridge Glacier, Yukon Territory, Canada, 1969–1989. *Journ. Glaciol.* 37, 158–169.

Clarke, G. K. C., S. G. Collins, and D. E. Thompson (1984). Flow, thermal structure, and subglacial conditions of a surge-type glacier. *Can. Journ. Earth Sci.* 21, 2909–2912.

Colbeck, S. C. (1973). *Theory of metamorphism of wet snow.* CRREL Research Report 313. Cold Regions Research and Engineering Laboratory, Hanover, NH, 11 pp.

Conway, H. et al. (1999). Past and future grounding-line retreat of the West Antarctic Ice Sheet. *Science* 286, 280–283.

Cook, S. et al. (2012). Testing the effect of water in crevasses on a physically based calving model. *Ann. Glaciol.* 53, 90–96.

Cuffey, K. M., and W. S. B. Paterson (2010). *The physics of glaciers* (4th ed.). Burlington, MA: Butterworth-Heinemann, 693 pp.

Dahl-Jensen, D. (1985). Determination of the flow properties at Dye 3, south Greenland, by bore-hole-tilting measurements and perturbation modelling. *Journ. Glaciol.* 31, 92–98.

Dahl-Jensen, D. (1989). Two-dimensional thermo-mechanical modelling of flow and depth-age profiles near the ice divide in central Greenland. *Ann. Glaciol.* 12, 31–36.

Dalrymple, P. C. (1966). A physical climatology of the Antarctic Plateau. In: *Studies in Antarctic meteorology* (ed. M. J. Rubin). American Geophysical Union, Washington, DC, Antarctic Research Series 9, 195–231.

Dalrymple, P. C., H. H. Lettau, and S. H. Wollaston (1966). South Pole micrometeorology program: Data analysis. In: *Studies in Antarctic meteorology* (ed. M. J. Rubin). American Geophysical Union, Washington, DC, Antarctic Research Series 9, 13–57.

Das, S. B. et al. (2008). Fracture propagation to the base of the Greenland Ice Sheet during supraglacial lake drainage. *Science* 320, 778–781.

De Angelis, H., and P. Skvarca (2003). Glacier surge after ice shelf collapse. *Science* 299, 1560–1562.

Denby, B. (1999). Second order modeling of turbulence in katabatic flows. *Bound. Layer Meteorol.* 92, 67–100.

Denton, G. H., and T. J. Hughes (eds.) (1981). *The last great ice sheets.* New York: John Wiley, 484 pp.

Determann, J. (1991). Das fliessen von schelfeisen: numerische simulationen mit der methode der finiten differenzen (The flow of ice shelves: Numerical simulations using the finite difference method). *Ber. Polarforsch.* 83, 82 pp.

Docquier, D., L. Perichon, and F. Pattyn (2011). Representing grounding line dynamics in numerical ice sheet models: Recent advances and outlook. *Surv. Geoph.* 32, 417–435.

Dolgoushin, L. D., and G. B. Osipova (1975). Glacier surges and the problem of their forecasting. *Int. Assoc. Hydr. Sci. Publ.* 104, 292–304.

Dolgoushin, L. D., and G. B. Osipova (1978). Balance of a surging glacier as the basis for forecasting its periodic advances. *Mater. Glyatsiologischeskikh Issled. Khronika. Obsuzhdeniya* 32, 260–265.

Dupont, T., and R. B. Alley (2005). Assessment of the importance of ice-shelf buttressing to ice sheet flow. *Geoph. Res. Lettrs.* 32, L04503, doi:10.1029/2004GL022024.

Durand, G. et al. (2009a). Marine ice sheet dynamics: Hysteresis and neutral equilibrium. *Journ. Geoph. Res.* 114, F03009, doi:10.1029/2008JF001170.

Durand, G. et al. (2009b). Full Stokes modeling of marine ice sheets: Influence of the grid size. *Ann. Glaciol.* 50(52), 109–114.

Durham, W. A., L. A. Stern, and S. H. Kirby (2001). Rheology of ice I at low stress and elevated confining pressure. *Journ. Geoph. Res.* 106, 11031–11042.

Duval, P. (1981). Creep and fabrics of polycrystalline ice under shear and compression. *Journ. Glaciol.* 27, 129–140.

Echelmeyer, K. A., and W. D. Harrison (1990). Jakobshavns Isbræ, West Greenland: Seasonal variations in velocity—or lack thereof. *Journ. Glaciol.* 36, 82–88.

Echelmeyer, K. A., and B. Kamb (1986). Stress-gradient coupling in glacier flow: II. Longitudinal averaging in the flow response to small perturbations in ice thickness and surface slope. *Journ. Glaciol.* 32, 285–298.

Eisen, O., W. D. Harrison, and C. F. Raymond (2001). The surges of Variegated Glacier, Alaska, USA, and their connection to climate and mass balance. *Journ. Glaciol.* 47, 351–358.

Eisen, O. et al. (2005). Variegated Glacier, Alaska, USA: A century of surges. *Journ. Glaciol.* 51, 399–406.

Engelder, T. (1993). *Stress regimes in the lithosphere.* Princeton, NJ: Princeton University Press, 457 pp.

Engelhardt, H., W. D. Harrison, and B. Kamb (1978). Basal sliding and conditions at the glacier bed as revealed by bore-hole photography. *Journ. Glaciol.* 20, 469–508.

Engelhardt, H. et al. (1990). Physical conditions at the base of a fast moving Antarctic ice stream. *Science* 248, 57–59.

Erdogan, F., and G. C. Sih (1963). On the crack extension in plates under plane loading and transverse shear. *Journ. Basic Eng.* 85, 519–527.

Ericksen, J. L., and R. S. Rivlin (1954). Large elastic deformations of homogeneous anisotropic materials. *Journ. Ration. Mech. Anal.* 3, 281–301.

Fetzer, J. H. (1988). Program verification: The very idea. *Comm. ACM* 31, 1048–1063.

Finsterwalder, S. (1907). Die theorie der gletscherschwankungen. *Zeitsch. Gletscherk. Eiszeitforschung und Geschichte des Klimas* 2, 81–103.

Fitzpatrick, M. F., R. E. Brandt, and S. G. Warren (2004). Transmission of solar radiation by clouds over snow and ice surfaces: A parameterization in terms of optical depth, solar zenith angle, and surface albedo. *Journ. Clim.*, 17, 266–275.

Flowers G. E. (2008). Subglacial modulation of the hydrograph from glacierized basins. *Hydrol. Proc.* 22, 3903–3918.

Forbes, J. D. (1859). *The theory of glaciers.* Edinburgh, UK: Adam and Charles Black, North Bridge, 278 pp.

Fowler, A. C. (1987a). Sliding with cavity formation. *Journ. Glaciol.* 33, 255–267.

Fowler, A. C. (1987b). A theory of glacier surges. *Journ. Geoph. Res.* 9(B), 9111–9120.

Fowler, A. C., T. Murray, and F. S. L. Ng (2001). Thermally controlled glacier surging. *Journ. Glaciol.* 47, 527–538.

Fricker, H. A. et al. (2009). Mapping the grounding zone of the Amery Ice Shelf, East Antarctica, using InSAR, MODIS, and ICESat. *Ant. Sci.* 21, 515–532.

Frolich, R. M., and C. S. M. Doake (1988). Relative importance of lateral and vertical shear on Rutford Ice Stream, Antarctica. *Ann. Glaciol.* 11, 19–22.

Frolich, R. M. et al. (1987). Force balance of Rutford Ice Stream, Antarctica. *Int. Assoc. Hydr. Sci. Publ.* 170, 323–331.

Fujita, S., M. Nakawo, and S. Mae (1987). Orientation of the 700-m Mizuho core and its strain history. *Proc. NIPR Symp. on Polar Meteo. and Glaciol.* 1, 122–131.

Funk, M., and H. Röthlisberger (1989). Forecasting the effects of a planned reservoir which will partially flood the tongue of Unteraargletscher in Switzerland. *Ann. Glaciol.* 13, 76–81.

Gagliardini, O., and T. Zwinger (2007). The ISMIP-HOM benchmark experiments performed using the finite-element code Elmer. *The Cryosphere* 2, 67–76.

Gagliardini, O. et al. (2007). Finite-element modeling of subglacial cavities and related friction law. *Journ. Geoph. Res.* 112, F02027, doi:10.1029/2006JF000576.

Gdoutos, E. E. (1993). *Fracture mechanics.* Dordrecht: Kluwer Acad. Publishers, 369 pp.

Geiger, R. (1966). *The climate near the ground.* Harvard, MA: Harvard University Press, 310 pp.

Glen, J. W. (1952). Experiments on the deformation of ice. *Journ. Glaciol.* 2, 111–114.

Glen, J. W. (1958). The flow law of ice: A discussion of the assumptions made in glacier theory, their experimental foundations, and consequences. *Int. Assoc. Hydr. Sci. Publ.* 47, 171–183.

Glen, J. W. (1975). *The mechanics of ice.* CRREL Monograph 11-C2b. Hanover, NH: Cold Regions Research and Engineering Laboratory, 47 pp.

Glen, J. W., and M. F. Perutz (1954). The growth and deformation of ice crystals. *Journ. Glaciol.* 2, 397–403.

Goddard, W. B. (1975). Description of a surface temperature equilibrium energy balance model with application to Arctic pack ice in early spring. In: *Climate of the Arctic* (eds. G. Weller and S. A. Bowling). Fairbanks, AK: Geoph. Inst. Univ. Alaska, 230–237.

Gogineni, S. et al. (2001). Coherent radar ice thickness measurements over the Greenland Ice sheet. *Journ. Geoph. Res.*, 106, 33,761–33,772.

Goldberg, D., D. M. Holland, and C. Schoof (2009). Grounding line movement and ice shelf buttressing in marine ice sheets. *Journ. Geoph. Res.* 114, F04026, doi:10.1029/2008JF001227.

Goldsby, D. L. (2006). Superplastic flow of ice relevant to glacier and ice-sheet mechanics. In: *Glacier science and environmental change* (ed. P. G. Knight). Malden, MA: Blackwell, 308–314.

Goldsby, D. L., and D. L. Kohlstedt (2001). Superplastic deformation of ice: Experimental observations. *Journ. Geoph. Res.* 106, 11,017–11,030.

Goldthwait, R. P. (1971). Introduction to till, today. In: *Till: a symposium.* (ed. R. P. Goldthwait). Columbus, OH: Ohio State University Press, 3–26.

Grainger, M. E., and H. Lister (1966). Wind speed, stability, and eddy viscosity over melting ice surfaces. *Journ. Glaciol.* 6, 101–127.

Green, A. E. (1964a). A continuum theory of anisotropic fluids. *Proc. Cambridge Philos. Soc.* 60, 123–128.

Green, A. E. (1964b). Anisotropic simple fluids. *Proc. Royal Soc. London, Ser. A* 279, 437–445.

Greuell, W. (1992). Hintereisferner, Austria: Mass-balance reconstruction and numerical modelling of the historical length variations. *Journ. Glaciol.* 38, 233–244.

Greuell, W., and C. Genthon (2003). Modelling land-ice surface mass balance. In: *Mass balance of the cryosphere: Observations and modeling of contemporary and future changes* (eds. J. L. Bamber and A. J. Payne). Cambridge, UK: Cambridge University Press, 117–168.

Greuell, W., and T. Konzelmann (1994). Numerical modelling of the energy balance of the Greenland Ice Sheet. Calculations for the ETH-Camp location (West Greenland, 1155 m a.s.l.). *Global Planet. Change* 9, 91–114.

Greuell, W., and J. Oerlemans (1986). Sensitivity studies with a mass balance model including temperature profile calculations inside the glacier. *Zeitsch. Gletscherk. Glazialgeol.* 22, 101–124.

Greuell, W., and J. Oerlemans (1989a). The evolution of the englacial temperature distribution in the superimposed ice zone of a polar ice cap during a summer season. In: *Glacier fluctuations and climatic change* (ed. J. Oerlemans). Dordrecht: Kluwer Acad. Publ., 289–303.

Greuell, W., and J. Oerlemans (1989b). Energy balance calculations on and near Hintereisferner, Austria, and an estimate of the effect of greenhouse warming on ablation. In: *Glacier fluctuations and climatic change* (ed. J. Oerlemans). Dordrecht: Kluwer Acad. Publ., 305–323.

Greve, R., and H. Blatter (2009). *Dynamics of ice sheets and glaciers.* Berlin: Springer Verlag, 287 pp.

Gudmundsson, G. H. (1997). Basal-flow characteristics of a linear medium sliding frictionless over small bedrock undulations. *Journ. Glaciol.* 43, 71–79.

Gudmundsson, G. H. (2003). Transmission of basal variability to a glacier surface. *Journ. Geoph. Res.* 108, 2253, doi: 10.1029/2002JB002107.

Gudmundsson, G. H. (2008). Analytical solutions for the surface response to small amplitude perturbations in boundary data in the shallow-ice-stream approximation. *The Cryosphere* 2, 77–93.

Halfar, P. (1981). On the dynamics of ice sheets. *Journ. Geoph. Res.* 86, 11065–11072.

Halfar, P. (1983). On the dynamics of ice sheets 2. *Journ. Geoph. Res.* 88, 6043–6051.

Hallet, B. (1981). Glacial abrasion and sliding: Their dependence on the debris concentration in basal ice. *Ann. Glaciol.* 2, 23–28.

Hambrey, M. J. (1976). Structure of the glacier Charles Rabots Bre, Norway. *Geol. Soc. Amer. Bull.* 87, 1629–1637.

Hambrey, M. J., and F. Müller (1978). Structures and ice deformation in the White Glacier, Axel Heiberg Island, Northwest Territories, Canada. *Journ. Glaciol.*, 20, 41–66.

Harbor, J. M. (1992). Application of a general sliding law to simulating flow in a glacier cross-section. *Journ. Glaciol.* 38, 182–190.

Harper, J. T. et al. (2010). Vertical extension of the subglacial drainage system into basal crevasses. *Nature*, 467, 579–582.

Heinrich, H. (1988). Origin and consequences of cyclic ice rafting in the northeast Atlantic Ocean during the past 130,000 years. *Quat. Res.* 29, 143–152.

Helsen, M. M. et al. (2008). Elevation changes in Antarctica, mainly determined by accumulation variability. *Science* 320, 1626–1629.

Hindmarsh, R. C. A. (2004). A numerical comparison of approximations to the Stokes equations used in ice sheet and glacier modeling. *Journ. of Geoph. Res.* 109, F01012, doi: 10.1029/2003JF000065.

Hindmarsh, R. C. A. (2009). The role of membrane-like stresses in determining the stability and sensitivity of the Antarctic Ice Sheet: Back pressure and grounding line motion. *Phil. Trans. Royal Soc. Ser. A* 364, 1733–1767.

Hindmarsh, R. C. A. (2012). An observationally validated theory of viscous flow dynamics at the ice-shelf calving front. *Journ. Glaciol.* 58, 375–387.

Hindmarsh, R. et al. (2006). Draping or overriding: The effect of horizontal stress gradients on internal layer architecture in ice sheets. *Journ. Geoph. Res.* 111, F02018, doi: 10.1029/2005JF000309.

Hobbs, P. V. (1974). *Ice physics.* Oxford: Clarendon Press, 837 pp.

Hogan, A. W., and A. J. Gow (1997). Occurrence of frequency thickness of annual snow accumulation layers at South Pole. *Journ. Geoph. Res.* 102(D), 14021–14027.

Holdsworth, G. (1968). Primary transverse crevasses. *Journ. Glaciol.*, 8, 107–129.

Holland, D. M. et al. (2008). Acceleration of Jakobshavn Isbræ triggered by warm subsurface ocean waters. *Nature Geosc.* 1, 659–664.

Holmgren, B. (1971). *Climate and energy exchange on a sub-polar ice cap in summer. (Parts A–F).* Uppsala, Sweden: Meteorol. Inst. Uppsala Univ. Medd., 362 pp.

Hooke, R. LeB. (1981). Flow law for polycrystalline ice in glaciers: Comparison of theoretical predictions, laboratory data, and field measurements. *Rev. Geoph. Space Phys.* 19, 664–672.

Hooke, R. LeB. (1989). Englacial and subglacial hydrology: A qualitative review. *Arctic Alpine Res.* 21, 221–233.

Hooke, R. LeB., and P. J. Hudleston (1980). Ice fabrics in a vertical flow plane, Barnes Ice Cap, Canada. *Journ. Glaciol.* 25, 195–214.

Hooke, R. LeB. et al. (1989). A 3 year record of seasonal variations in surface velocity, Storglaciären, Sweden. *Journ. Glaciol.* 35, 235–247.

Hopkins, W. (1844a). On the motion of glaciers. *Trans. Cambridge Philos. Soc.*, 8, 50–74.

Hopkins, W. (1844b). On the motion of glaciers. *Trans. Cambridge Philos. Soc.*, 8, 159–169.

Hopkins, W. (1862). On the theory of the motion of glaciers. *Philos. Trans. Royal Soc. London*, 152, 677–745.

Houghton, H. G. (1985). *Physical meteorology.* Cambridge, MA: MIT Press, 442 pp.

Howat, I. M., I. Joughin, and T. A. Scambos (2007). Rapid changes in ice discharge from Greenland outlet glaciers. *Science* 315, 1559–1561.

Howat, I. M. et al. (2008). Synchronous retreat and acceleration of southeast Greenland outlet glaciers 2000–2006: Ice dynamics and coupling to climate. *Journ. Glaciol.* 54, 646–660.

Hughes, T. J. (1973). Is the West Antarctic Ice Sheet disintegrating? *Journ. Geoph. Res.* 78, 7884–7910.

Hughes, T. J. (1985). The great Cenozoic Ice Sheet. *Palaeogeogr. Clim. Ecol.* 50, 9–43.

Hugi, J. (1830). *Naturhistorische Alpenreise.* Solothurn: Amiet-Lutiger, 378 pp.

Hulbe, C. L., and C. M. LeDoux (2011). MOA-derived structural feature map of the Ronne Ice Shelf. Boulder, CO: National Snow and Ice Data Center. Digital Media.

Hulbe, C. L., C. LeDoux, and K. Cruikshank (2010). Propagation of long fractures in the Ronne Ice Shelf, Antarctica, investigated using a numerical model for fracture propagation. *Journ. Glaciol.*, 56, 459–472.

Hutchinson, J. W. (1976). Bounds and self-consistent estimates for creep of polycrystalline materials. *Proc. Royal Soc. London A,* 348, 101–127.

Hutchinson, J. W. (1977). Creep and plasticity of hexagonal polycrystals as related to single crystal slip. *Metall. Trans. A*, 8, 1465–1469.

Hutchinson, J. W. (1979). Recent developments in non-linear fracture mechanics. In: *Proc. 7th Canadian Cong. Applied Mech.*, Sherbrooke, May 27–June 1, 24–36.

Hutter, K. (1981). The effect of longitudinal strain on the shear stress of an ice sheet: In defense of using stretched coordinates. *Journ. Glaciol.* 27, 39–56.

Hutter, K. (1983). *Theoretical glaciology: Material science of ice and the mechanics of glaciers and ice sheets.* Dordrecht: Reidel, 510 pp.

Hutter, K., F. Legerer, and U. Spring (1981). First-order stresses and deformations in glaciers and ice sheets. *Journ. Glaciol.* 27, 227–270.

Huybrechts, P. (1992). The Antarctic Ice Sheet and environmental change: A three-dimensional modelling study. *Ber. Polarforsch.* 99, 241 pp.

Huybrechts, P., and J. Oerlemans (1988). Evolution of the East Antarctic Ice Sheet: A numerical study of thermo-mechanical response patterns with changing climate. *Ann. Glaciol.* 11, 52–59.

Hvidberg, C. (1996). Steady-state thermomechanical modelling of ice flow near the centre of large ice sheets with the finite-element technique. *Ann. Glaciol.* 23, 116–123.

Iken, A. (1973). Schwankungen der oberflächen-geschwindigkeit des White Glacier, Axel Heiberg Island. *Zeitsch. Gletscherk. Glazialgeol.* 9, 207–219.

Iken, A. (1981). The effect of the subglacial water pressure on the sliding velocity of a glacier in an idealized numerical model. *Journ. Glaciol.* 27, 407–421.

Ingraffea, A. R. (1987). Theory of crack initiation and propagation in rock. In: *Fracture Mechanics of Rocks* (ed. B. K. Atkinson). San Diego: Academic Press, 71–110.

Iverson, N. R., and R. M. Iverson (2001). Distributed shear of subglacial till due to Coulomb slip. *Journ. Glaciol.* 47, 481–488.

Jacka, T. H., and W. F. Budd (1989). Isotropic and anisotropic flow relations for ice dynamics. *Ann. Glaciol.* 12, 81–84.

Jacka, T. H., and M. Maccagnan (1984). Ice crystallographic and strain rate changes with strain in compression and extension. *Cold Reg. Sci. Techn.* 8, 169–186.

Jacobel, R. W., and S. M. Hodge (1995). Radar internal layers from the Greenland summit. *Geoph. Res. Ltrs.* 22, 587–590.

Jaeger, J. C. (1969). *Elasticity, fracture, and flow.* London: Chapman & Hall, 268 pp.

Jaeger, J. C., and N. G. W. Cook (1976). *Fundamentals of rock mechanics.* New York, NY: Halsted Press, 585 pp.

Jenssen, D. (1977). A three-dimensional polar ice-sheet model. *Journ. Glaciol.* 18, 373–389.

Jezek, K. C. (1984). A modified theory of bottom crevasses used as a means for measuring the buttressing effect of ice shelves on inland ice sheets. *Journ. Geoph. Res.* 89, 1925–1931.

Jezek, K. C., C. R. Bentley, and J. W. Clough (1979). Electromagnetic sounding of bottom crevasses on the Ross Ice Shelf, Antarctica. *Journ. Glaciol.* 24, 321–330.

Jóhannesson, T., C. F. Raymond, and E. D. Waddington (1989a). A simple method for determining the response time of glaciers. In: *Glacier fluctuations and climatic change* (ed. J. Oerlemans). Dordrecht: Kluwer, 343–352.

Jóhannesson, T., C. F. Raymond, and E. D. Waddington (1989b). Time-scale for adjustment of glaciers to changes in mass balance. *Journ. Glaciol.* 35, 355–369.

Johnson, A. F. (1977). Creep characterization of transversely-isotropic metallic materials. *Journ. Mech. Phys. Solids* 25, 117–126.

Johnson, A. M. (1970). *Physical processes in geology.* San Francisco: Freeman, Cooper & Co., 577 pp.

Johnson, J. V., and J. W. Staiger (2007). Modeling long-term stability of the Ferrar Glacier, East Antarctica: Implications for interpreting cosmogenic nuclide inheritance. *Journ. Geoph. Res.* 112, F03S30, doi: 10.1029/2006JF000599.

Jonas, J. J., and F. Müller (1969). Deformation of ice under high internal shear stresses. *Canad. Journ. Earth Sci.* 6, 963–968.

Joos, G. (1934). *Theoretical physics.* London: Blackie & Son Ltd., 748 pp.

Joughin, I., W. Abdalati, and M. A. Fahnestock (2004). Large fluctuations in speed on Jakobshavn Isbræ glacier. *Nature* 432, 608–610.

Joughin, I. et al. (2008a). Ice-front variations and tidewater behavior on Helheim and Kangerdlugssuaq glaciers, Greenland. *Journ. Geoph. Res.* 113, F01004, doi: 10.1029/2007JF000837.

Joughin, I. et al. (2008b). Continued evolution of Jakobshavn Isbræ following its rapid speed up. *Journ Geoph. Res.* 113, F04006, doi: 10.1029/2008JF001023.

Joughin, I. et al. (2012). Seasonal to decadal scale variations in the surface velocity of Jakobshavn Isbræ, Greenland: Observation and model-based analysis. *Journ. Geoph. Res.* 117, F02030, doi: 10.1029/2011JF002110.

Kamb, B. (1970). Sliding motion of glaciers: Theory and observation. *Rev. Geoph. Space Phys.* 8, 673–728.

Kamb, B. (1972). Experimental recrystallization of ice under stress. In: *Flow and fracture of rocks* (eds. H. C. Heard, I. Y. Borg, N. L. Carter, and C. B. Raleigh). Washington, DC, American Geophysical Union, Geophysical Monograph 16, 211–241.

Kamb, B. (1986). Stress-gradient coupling in glacier flow: III. Exact longitudinal equilibrium equation. *Journ. Glaciol.* 32, 335–341.

Kamb, B. (1987). Glacier surge mechanism based on linked cavity configuration of the basal water conduit system. *Journ. Geoph. Res.* 92(B), 9083–9100.

Kamb, B. (1991). Rheological nonlinearity and flow instability in the deforming bed mechanism of ice stream motion. *Journ. Geoph. Res.* 96(B), 16,585–16,595.

Kamb, B., and K. A. Echelmeyer (1986a). Stress-gradient coupling in glacier flow: I. Longitudinal averaging of the influence of ice thickness and surface slope. *Journ. Glaciol.* 32, 267–284.

Kamb, B., and K. A. Echelmeyer (1986b). Stress-gradient coupling in glacier flow: IV. Effects of the "T" term. *Journ. Glaciol.* 32, 342–349.

Kamb, B. et al. (1985). Glacier surge mechanism: 1982–1983 surge of Variegated Glacier, Alaska. *Science* 227, 469–479.

Kavanaugh, J. L., and G. K. C. Clarke (2006). Discrimination of the flow law for subglacial sediment using in situ measurements and an interpretation model. *Journ. Geoph. Res.* 111, F01002, doi: L10.1029/2005JF000346.

Kehle, R. O. (1964). Deformation of the Ross Ice Shelf, Antarctica. *Geol. Soc. Amer. Bull.,* 75, 259–286.

Kennett, M., T. Lauman, and B. Kjølmoen (1997). Predicted response of the calving glacier Svarthisheibreen, Norway, and outbursts from it, to future changes in climate and lake level. *Ann. Glaciol.* 24, 16–20.

Koechlin, R. (1944). *Les glaciers et leur mécanisme.* Lausanne: F. Rouge & Cie S.A., 177 pp.

Konzelmann, T. et al. (1994). Parameterization of global and longwave incoming radiation for the Greenland Ice Sheet. *Global Planet. Change* 9, 143–164.

Krimmel, R. M. (1997). Documentation of the retreat of Columbia Glacier, Alaska. In: *Calving glaciers: Report of a workshop,* February 28–March 2 (ed. C. J. van der Veen). BPRC Report No. 15. Columbus, OH: Byrd Polar Research Center, The Ohio State University, 105–108.

Kuipers Munnike, P. et al. (2009). The role of radiation penetration in the energy budget of the snowpack at Summit, Greenland. *The Cryosphere* 3, 155–165.

Lambeck, K., and S. M. Nakiboglu (1981). Seamount loading and stress in the ocean lithosphere 2. Viscoelastic and elastic-viscoelastic models. *Journ. Geoph. Res.* 86(B), 6961–6984.

Lawson, D. E. (1993). *Glaciohydrologic and glaciohydraulic effect on runoff and sediment yield in glacierized basins.* CRREL Monograph 93–2. Hanover, NH: Cold Regions Research and Engineering Laboratory, 108 pp.

Lawson, W. (1997). Spatial, temporal, and kinematic characteristics of surges of Variegated Glacier, Alaska. *Ann. Glaciol.* 24, 95–101.

Lazzara, M. A. et al. (1999). On the recent calving of icebergs from the Ross Ice Shelf. *Polar Geog.* 23, 201–212.

Leng, W. et al. (2012a). A parallel high-order accurate finite element Stokes ice sheet model. *Journ. Geophys. Res.*, 117, F01001, doi: 10.1029/2011JF001962.

Leng, W. et al. (2012b). Manufactured solutions and the numerical verification of isothermal, nonlinear, three-dimensional Stokes ice-sheet models. *The Cryosphere Discuss* 6, 2689–2714.

Li, J., T. H. Jacka, and W. F. Budd (1996). Deformation rates in combined compression and shear for ice which is initially isotropic and after the development of strong anisotropy. *Annals Glaciol.* 23, 247–252.

Lighthill, M. J., and G. B. Whitham (1955a). On kinematic waves. I. Flood movement in long rivers. *Proc. Royal Soc. London Ser. A.* 229, 281–316.

Lighthill, M. J., and G. B. Whitham (1955b). On kinematic waves. II. A theory of traffic flow on long crowded roads. *Proc. Royal Soc. London Ser. A.* 229, 317–345.

Lile, R. C. (1978). The effect of anisotropy on the creep of polycrystalline ice. *Journ. Glaciol.* 21, 475–483.

Lingle, C. S., and T. J. Brown (1987). A subglacial aquifer bed model and water pressure dependent basal sliding relationship for a West Antarctic ice stream. In: *Dynamics of the West Antarctic Ice Sheet* (eds. C. J. van der Veen and J. Oerlemans). Dordrecht: Reidel, 249–285.

Lingle, C. S., and J. A. Clark (1985). A numerical model of interactions between a marine ice sheet and the solid earth: Application to a West Antarctic ice stream. *Journ. Geoph. Res.* 90(C), 1100–1114.

Lingle, C. S. et al. (1991). A flow band model of the Ross Ice Shelf, Antarctica: Response to a CO_2-induced climatic warming. *Journ. Geoph. Res.* 96(B), 6849–6871.

Lliboutry, L. (1958a). Frottement sur le lit et mouvement pas saccades d'un glacier. *Comptes Rendus Séances Acad. Sci.* 247, 228–230.

Lliboutry, L. (1958b). Contribution à la théorie du frottement du glacier sur son lit. *Comptes Rendus Séances Acad. Sci.* 247, 318–320.

Lliboutry, L. (1959). Une théorie du frottement du glacier sur son lit. *Ann. Géoph.* 15, 250–265.

Lliboutry, L. (1968). General theory of subglacial cavitation and sliding of temperate glaciers. *Journ. Glaciol.* 7, 21–58.

Lliboutry, L. (1979). Local friction laws for glaciers: A critical review and new openings. *Journ. Glaciol.* 23, 67–95.

Lliboutry, L. (1987a). Sliding of cold ice sheets. *Int. Assoc. Hydr. Sci. Publ.* 170, 131–143.

Lliboutry, L. (1987b). Realistic, yet simple bottom boundary conditions for glaciers and ice sheets. *Journ. Geoph. Res.* 92(B), 9101–9109.

Lliboutry, L. (1987c). *Very slow flow of solids.* Dordrecht: Martinus Nijhoff Publishers, 510 pp.

Lliboutry, L., and I. Andermann (1982). Extension de la loi de Norton-Hoff à un matériau orthotrope de révolution. *Comptes Rendus Acad. Sci. Paris* 294, 1309–1312.

MacAyeal, D. R. (1987). Ice-shelf backpressure: Form drag versus dynamic drag. In: *Dynamics of the West Antarctic Ice Sheet* (eds. C. J. van der Veen and J. Oerlemans). Dordrecht: Reidel, 141–160.

MacAyeal, D. R. (1989). Large-scale ice flow over a viscous basal sediment: Theory and application to Ice Stream B, Antarctica. *Journ. Geoph. Res.* 94(B), 4071–4087.

Male, D. H. (1980). The seasonal snowcover. In: *Dynamics of snow and ice masses* (ed. S. C. Colbeck). New York: Acad. Press, 305–395.

Man, C.-S., and Q.-X. Sun (1987). On the significance of normal stress effects in the flow of glaciers. *Journ. Glaciol.* 33, 268–273.

Marchi, L. de (1911). La propagation des ondes dans les glaciers. *Zeitsch. Gletscherk. Eiszeitforschung und Geschichte des Klimas* 5, 207–211.

Marshall, S. J., and K. M. Cuffey (2000). Peregrinations of the Greenland Ice Sheet divide in the last glacial cycle: Implications for central Greenland ice cores. *Earth Planet. Sci. Lett.* 179, 73–90.

Martin, C. et al. (2009). On the effects of anisotropic rheology on ice flow, internal structure, and the age-depth relationship at ice divides. *Journ. Geoph. Res.* 114, F04001, doi: 10.1029/2008JF001204.

McGee, W. J. (1894). Glacial cañons. *Journ. Geol.* 2, 350–364.

McGrath, D. et al. (2012). Basal crevasses and associated surface crevassing on the Larsen C Ice Shelf, Antarctica, and their role in ice-shelf stability. *Annals Glaciol.* 58(60), 10–18.

McIntyre, N. F. (1985). The dynamics of ice-sheet outlets. *Journ. Glaciol.* 31, 99–107.

McTigue, D. F., S. L. Passman, and S. J. Jones (1985). Normal stress effects in the creep of ice. *Journ. Glaciol.* 31, 120–126.

Means, W. D. (1976). *Stress and strain.* New York: Springer-Verlag, 339 pp.

Meier, M. F. (1958). The mechanics of crevasse formation. *Int. Assoc. Hydr. Sci. Publ.* 46, 500–508.

Meier, M. F. (1997). The iceberg discharge process: Observations and inferences drawn from the study of Columbia Glacier. In: *Calving glaciers: Report of a workshop,* February 28–March 2 (ed. C. J. van der Veen). BPRC Report No. 15. Columbus, OH: Byrd Polar Research Center, The Ohio State University, 109–114.

Meier, M. F., and A. Post (1969). What are glacier surges? *Can. Journ. Earth Sci.* 6, 807–817.

Meier, M. F., and A. Post (1987). Fast tidewater glaciers. *Journ. Geoph. Res.* 92(B), 9051–9058.

Meier, M. F. et al. (1974). Flow of Blue Glacier, Olympic Mountains, Washington, USA. *Journ. Glaciol.* 13, 187–212.

Mellor, M., and D. M. Cole (1982). Deformation and failure of ice under constant stress or constant strain-rate. *Cold Reg. Sci. Techn.* 5, 201–219.

Mercer, J. H. (1968). Antarctic ice and Sangamon sea level. *Int. Assoc. Hydr. Sci. Publ.* 79, 217–225.

Mercer, J. H. (1978). West Antarctic Ice Sheet and CO_2 greenhouse effect: A threat of disaster. *Nature* 271, 321–325.

Mitchell, A. R., and D. F. Griffiths (1980). *The finite difference method in partial differential equations.* Chichester: John Wiley & Sons, 272 pp.

Montagnat, M., and P. Duval (2004). The viscoplastic behavior of ice in polar ice sheets: Experimental results and modeling. *Comptes Rendues Physique*, 5, 667–671.

Morland, L. W. (1976). Glacier sliding down an inclined wavy bed with friction. *Journ. Glaciol.* 17, 463–477.

Morland, L. W. (1979). Constitutive laws for ice. *Cold Reg. Sci. Techn.* 1, 101–108.

Morland, L. W. (1987). Unconfined ice-shelf flow. In: *Dynamics of the West Antarctic Ice Sheet* (eds. C. J. van der Veen and J. Oerlemans). Dordrecht: Reidel, 99–116.

Morland, L. W., and R. Zainuddin (1987). Plane and radial ice-shelf flow with prescribed temperature profile. In: *Dynamics of the West Antarctic Ice Sheet* (eds. C. J. van der Veen and J. Oerlemans). Dordrecht: Reidel, 117–140.

Motyka, R. J. et al. (2011). Submarine melting of the 1985 Jakobshavn Isbræ floating tongue and the triggering of the current retreat. *Journ. Geoph. Res.* 116, F01007, doi: 10.1029/2009JF001632.

Munro, D. S. (1989). Surface roughness and bulk heat transfer on a glacier: Comparison with eddy correlation. *Journ. Glaciol.* 35, 343–348.

Murray, T. (1997). Assessing the paradigm shift: Deformable glacier beds. *Quat. Sci. Rev.* 16, 995–1016.

Murray, T. et al. (2003). Is there a single surge mechanism? Contrasts in dynamics between glacier surges in Svalbard and other regions. *Journ. Geoph. Res.* 108, 2237 doi: 10.1029/2002JB001906.

Nereson, N. A. et al. (1998). Migration of the Siple Dome ice divide, West Antarctica. *Journ. Glaciol.* 44, 643–652.

Newman, J. C. Jr., and I. S. Raju (1981). An empirical stress intensity factor equation for surface cracks. *Engineering Fracture Mechanics* 15, 185–192.

Nick, F. M., C. J. van der Veen, and J. Oerlemans (2007). Controls on advance of tidewater glaciers: Results from numerical modeling applied to Columbia Glacier. *Journ. Geoph. Res.* 112, F03S24, doi: 10.1029/2006JF000551.

Nick, F. M. et al. (2009). Large-scale changes in Greenland outlet glacier dynamics triggered at the terminus. *Nature Geoscience* 2, 110–114.

Nick, F. M. et al. (2010). A physically based calving model applied to marine outlet glaciers and implications for the glacier dynamics. *Journ. Glaciol.* 56, 781–794.

Nick, F. M. et al. (2012). The response of Petermann Glacier, Greenland, to large calving events, and its future stability in the context of atmospheric and oceanic warming. *Journ. Glaciol.* 58, 229–239.

Nicolas, A., and J. P. Poirier (1976). *Crystalline plasticity and solid state flow in metamorphic rock.* London: John Wiley, 444 pp.

Nye, J. F. (1951). The flow of glaciers and ice-sheets as a problem in plasticity. *Proc. Royal Soc. London Ser. A* 207, 554–572.

Nye, J. F. (1952a). The mechanics of glacier flow. *Journ. Glaciol.* 2, 82–93.

Nye, J. F. (1952b). A comparison between the theoretical and the measured long profile of the Unteraar Glacier. *Journ. Glaciol.* 2, 103–107.

Nye, J. F. (1953). The flow law of ice from measurements in glacier tunnels, laboratory experiments and the Jungfraufirn borehole experiments. *Proc. Royal Soc. London Ser. A* 219, 477–489.

Nye, J. F. (1955). Comments on Dr. Loewe's letter and notes on crevasses. *Journ. Glaciol.* 2, 512–514.

Nye, J. F. (1957). The distribution of stress and velocity in glaciers and ice sheets. *Proc. Royal Soc. London Ser. A* 239, 113–133.

Nye, J. F. (1958). A theory of wave formation in glaciers. *Int. Assoc. Hydr. Sci. Publ.* 47, 139–154.

Nye, J. F. (1959). A method for determining the strain-rate tensor at the surface of a glacier. *Journ. Glaciol.* 3, 409–419.

Nye, J. F. (1961). The influence of climatic variations on glaciers. *Int. Assoc. Hydr. Sci. Publ.* 54, 397–404.

Nye, J. F. (1963a). The response of a glacier to changes in the rate of nourishment and wastage. *Proc. Royal Soc. London Ser. A* 275, 87–112.

Nye, J. F. (1963b). On the theory of the advance and retreat of glaciers. *Geoph. Journ. Royal Astron. Soc.* 7, 431–456.

Nye, J. F. (1963c). Theory of glacier variations. In: *Ice and snow* (ed. W. D. Kingery). Cambridge, MA: MIT Press, 151–161.

Nye, J. F. (1965a). The frequency response of glaciers. *Journ. Glaciol.* 5, 567–587.

Nye, J. F. (1965b). A numerical method of inferring the budget history of a glacier from its advance and retreat. *Journ. Glaciol.* 5, 589–607.

Nye, J. F. (1965c). The flow of a glacier in a channel of rectangular, elliptic, or parabolic cross-section. *Journ. Glaciol.* 5, 661–690.

Nye, J. F. (1969a). A calculation on the sliding of ice over a way surface using a Newtonian viscous approximation. *Proc. Royal Soc. London Ser. A* 311, 445–467.

Nye, J. F. (1969b). The effect of longitudinal stress on the shear stress at the base of an ice sheet. *Journ. Glaciol.* 8, 207–213.

Nye, J. F. (1970). Glacier sliding without cavitation in a linear viscous approximation. *Proc. Royal Soc. London Ser. A* 315, 381–403.

Nye, J. F. (2000). A flow model for the polar caps of Mars. *Journ. Glaciol.* 46, 438–444.

Oerlemans, J. (1980). Model experiments on the 100,000-yr glacial cycle. *Nature* 287, 430–432.

Oerlemans, J. (1982). Response of the Antarctic Ice Sheet to a climatic warming: A model study. *Journ. Climat.* 2, 1–11.

Oerlemans, J., and C. J. van der Veen (1984). *Ice sheets and climate.* Dordrecht: Reidel Publ. Co., 217 pp.

O'Neel, S., K. A. Echelmeyer, and R. J. Motyka (2003). Short-term variations in calving of a tidewater glacier: LeConte Glacier, Alaska, U.S.A. *Journ. Glaciol.* 49, 587–598.

Oreskes, N., K. Shrader-Frechette, and K. Belitz (1994). Verification, validation, and confirmation of numerical models in the earth sciences. *Science* 263, 641–646.

Orowan, E. (1949). Remarks at joint meeting of the British Glaciological Society, the British Rheologists Club, and the Institute of Metals. *Journ. Glaciol.* 1, 231–236.

Paterson, W. S. B. (1972). Laurentide Ice Sheet: Estimated volumes during late Wisconsin. *Rev. Geoph. Space Phys.* 10, 885–917.

Paterson, W. S. B. (1994). *The physics of glaciers* (3rd. ed.). Oxford: Pergamon Press/Elsevier Ltd., 480 pp.

Paterson, W. S. B., and W. F. Budd (1982). Flow parameters for ice-sheet modeling. *Cold Regions Sci. Techn.* 6, 175–177.

Paterson, W. S. B., and E. D. Waddington (1984). Past precipitation rates derived from ice core measurements: Methods and data analysis. *Rev. Geoph. Space Phys.* 22, 123–130.

Pattyn, F. et al. (2008). Benchmark experiments for higher-order and full-Stokes ice sheet models (ISMIP-HOM). *The Cryosphere* 2, 95–108.

Pattyn, F. et al. (2012). Results of the Marine Ice Sheet Model Intercomparison Project, MISMIP. *The Cryosphere* 6, 573–588.

Payne, A. J. et al. (2004). Recent dramatic thinning of largest West Antarctic ice stream triggered by oceans. *Geoph. Res. Lettrs.* 31, L23401, doi: 10.1029/2004GL021284.

Pedlosky, J. (1982). *Geophysical fluid dynamics.* New York: Springer Verlag, 624 pp.

Peixoto, J. P., and A. H. Oort (1992). *Physics of climate.* New York: Amer. Inst. Physics, 520 pp.

Peltier, W. R. (1982). Dynamics of the ice-age earth. *Adv. Geoph.* 24, 1–146.

Peltier, W. R. (1985). New constraints on transient lower mantle rheology and internal mantle buoyancy from glacial rebound data. *Nature* 318, 614–617.

Peltier, W. R. (1987). Glacial isostasy and the ice age cycle. *Int. Assoc. Hydr. Sci. Publ.* 170, 247–260.

Peltier, W. R. (1988). Lithospheric thickness, Antarctic deglaciation history, and ocean basin discretization effects in a global model of postglacial sea level change: A summary of some sources of nonuniqueness. *Quat. Res.* 29, 93–112.

Peltier, W. R., and W. T. Hyde (1987). Glacial isostasy and the Ice Age cycle. *Int. Assoc. Hydr. Sci. Publ.* 170, 247–260.

Peltier, W. R. et al. (2000). Ice-age ice-sheet rheology: Constraints from the last glacial maximum form of the Laurentide ice sheet. *Ann. Glaciol.* 30, 163–176.

Pelto, M. S., and C. R. Warren (1991). Relationship between tidewater glacier calving velocity and water depth at the calving front. *Ann. Glaciol.* 15, 115–118.

Pettit, E. C., and E. D. Waddington (2003). Ice flow at low deviatoric stress. *Journ. Glaciol.* 49, 359–369.

Pettit, E. C. et al. (2007). The role of crystal fabric in flow near an ice divide. *Journ. Glaciol.* 53, 277–288.

Pfeffer, W. T. et al. (2000). In situ stress tensor measured in an Alaskan glacier. *Annals Glaciol.* 31, 229–235.

Phillips, T., H. Rajaram, and K. Steffen (2010). Cryo-hydrologic warming: A potential mechanism for rapid thermal response of ice sheets. *Geoph. Res. Ltrs.* 37, L20503, doi: 10.1029/2010GL044397.

Pimienta, P., P. Duval, and V. Ya Lipenkov (1987). Mechanical behaviour of anisotropic polar ice. *Int. Assoc. Hydr. Sci. Publ.* 170, 57–66.

Piotrowsky, J. A. et al. (2001). Were deforming subglacial beds beneath past ice sheets really widespread? *Quat. Int.* 86, 139–150.

Piotrowsky, J. A. et al. (2002). Reply to the comments by G. S. Boulton, K. E. Dobbie, S. Zatsepin on: Deforming beds under ice sheets: How extensive were they? *Quat. Int.* 97–98, 173–177.

Pitman, D., and B. Zuckerman (1967). Effect of thermal conductivity of snow at –99°, –27°, and –5°C. *Journ. Applied Physics* 38, 2698–2699.

Poirier, J. P. (1985). *The creep of crystals.* Cambridge UK: Cambridge University Press, 260 pp.

Pollard, D. (1984). Some ice-age aspects of a calving ice-sheet model. In: *Milankovitch and Climate, Part 2* (eds. A. L. Berger et al.). Dordrecht: Reidel, 541–564.

Popper, K. R. (1963). *Conjectures and refutations.* London: Routledge & Kegan Paul Ltd.

Pounder, E. R. (1965). *Physics of Ice.* London: Pergamon Press, 151 pp.

Prandtl, L. (1932). Meteorologische anwendungen der strömungslehre. *Beitr. Phys. Atmos.* 19, 188–202.

Press, W. H. et al. (1992). *Numerical Recipes. The art of scientific computing* (2nd. ed.). Cambridge: Cambridge University Press, 963 pp.

Price, R. J. (1970). Moraines at Fjallsjökull, Iceland. *Arctic Alpine Res.* 2, 27–42.

Price, S. F., E. D. Waddington, and H. Conway (2007). A full-stress, thermomechanical flow band model using the finite volume method. *Journ. Geophy. Res.* 112, F03020, doi: 10.1029/2006JF000724.

Raymond, C. F. (1969). *Flow in a transverse section of Athabasca Glacier, Alberta, Canada.* PhD Thesis. Pasadena, CA: California Institute of Technology,.

Raymond, C. F. (1971). Flow in a transverse section of Athabasca Glacier, Alberta, Canada. *Journ. Glaciol.* 10, 55–84.

Raymond, C. F. (1983). Deformation in the vicinity of ice divides. *Journ. Glaciol.* 29, 357–373.

Raymond, C. F. and W. D. Harrison (1988). Evolution of Variegated Glacier, Alaska, U.S.A., prior to its surge. *Journ. Glaciol.* 34, 154–169.

Raymond, M. J., and G. H. Gudmundsson (2005). On the relationship between surface and basal properties on glaciers, ice sheets, and ice streams. *Journ. Geoph. Res.* 110, B08411, doi: 10.1029/2005JB003681.

Readey, D. W., and W. D. Kingery (1964). Plastic deformation of single crystal ice. *Acta Metallica* 12, 171–178.

Reeh, N. (1982). A plasticity approach to the steady-state shape of a three-dimensional ice sheet. *Journ. Glaciol.* 28, 431–455.

Reeh, N., and N. S. Gundestrup (1985). Mass balance of the Greenland Ice Sheet at Dye 3. *Journ. Glaciol.* 31, 198–200.

Reeh, N., and W. S. B. Paterson (1988). Application of a flow model to the ice-divide region of Devon Island ice cap, Canada. *Journ. Glaciol.* 34, 55–63.

Reiner, M. (1945). A mathematical theory of dilatancy. *Amer. Journ. Math.* 67, 350–362.

Reynaud, L. (1973). Flow of a valley glacier with a solid friction law. *Journ. Glaciol.* 12, 251–258.

Riedel, H., and J. R. Rice (1980). Tensile cracks in creeping solids. In: *Proc. 12th Conf. on Fracture Mechanics.* Spec. Publ. No. 7000. Washington, DC: Amer. Soc. for Testing and Materials, 112–130.

Rignot, E. (1998). Hinge-line migration of Petermann Gletscher, north Greenland, detected using satellite-radar interferometry. *Journ. Glaciol.,* 44, 469–476.

Rignot, E. et al. (2004). Accelerated ice discharge from the Antarctic Peninsula following the collapse of the Larsen B Ice Shelf. *Geoph. Res. Ltrs.* 31, L18401, doi: 10.1029/2004GL020697.

Rignot, E. et al. (2008). Recent Antarctic ice mass loss from radar interferometry and regional climate modeling. *Nature Geoscience* 1, 106–110.

Rigsby, G. P. (1958). Effect of hydrostatic pressure on the velocity of shear deformation of single ice crystals. *Journ. Glaciol.* 3, 273–278.

Rist, M. A. et al. (1996). Experimental fracture and mechanical properties of Antarctic ice: Preliminary results. *Ann. Glaciol.*, 23, 284–292.

Robin, G. de Q. (1955). Ice movement and temperature distribution in glaciers and ice sheets. *Journ. Glaciol.* 2, 523–532.

Röthlisberger, H. (1972). Water pressure in intra- and subglacial channels. *Journ. Glaciol.* 11, 177–204.

Röthlisberger, H., and H. Lang (1987). Glacial hydrology. In: *Glacio-fluvial sediment transfer* (eds. A. M. Gurnell and M. J. Clark). New York: John Wiley, 207–284.

Russell-Head, D. S., and W. F. Budd (1979). Ice-sheet flow properties derived from bore-hole shear measurements combined with ice-core studies. *Journ. Glaciol.* 24, 117–130.

Rutt, I. C. et al. (2009). The Glimmer community ice sheet model. *Journ. Geophys. Res.* 114, F02004, doi: 10.1029/2008JF001015.

Sanderson, T. J. O. (1979). Equilibrium profile of ice shelves. *Journ. Glaciol.* 22, 435–460.

Sanderson, T. J. O. (1988). *Ice mechanics: Risks to offshore structures*. London: Graham & Trotman, 253 pp.

Sanderson, T. J. O., and C. S. M. Doake (1979). Is vertical shear in an ice shelf negligible? *Journ. Glaciol.* 22, 285–292.

Sargent, A., and J. L. Fastook (2010). Manufactured analytical solutions for isothermal full-Stokes ice sheet models. *The Cryosphere* 4, 285–311.

Scambos, T. et al. (2004). Glacier acceleration and thinning after ice shelf collapse in the Larsen B embayment, Antarctica. *Geoph. Res. Ltrs.* 31, L18402, doi: 10.1029/2004GL020670.

Scambos, T. A. et al. (2007). MODIS-based mosaic of Antarctica (MOA) data sets: Continent-wide surface morphology and snow grain size. *Remote Sens. Environ.* 111, 242–257.

Scheuchzer, J. J. (1723). *Helveticus, sive itinera per helvetiae alpines regions facta, annis 1702–1711*. Leyden: Lugduni Batavorum.

Schilling, D. H., and J. T. Hollin (1981). Numerical reconstructions of valley glaciers and small ice caps. In: *The Last Great Ice Sheets* (eds. G. H. Denton and T. J. Hughes). New York: John Wiley & Sons, 207–220.

Schlichting, H. (1968). *Boundary layer theory* (6th ed.). New York: McGraw Hill, 535 pp.

Schoof, C. (2005). The effect of cavitation on glacier sliding. *Proc. Royal Soc. London Ser. A* 461, 609–627.

Schoof, C. (2007a). Marine ice-sheet dynamics. Part 1. The case of rapid sliding. *Journ. Fluid Mech.* 573, 27–55.

Schoof, C. (2007b). Ice sheet grounding line dynamics: Steady states, stability, and hysteresis. *Journ. Geoph. Res.* 112, F03S28, doi: 10.1029/2006JF000664.

Schoof, C. (2010). Ice-sheet acceleration driven by melt supply variability. *Nature* 468, 803–806.

Schoof, C. (2011). Marine ice sheet dynamics. Part 2. A Stokes flow contact problem. *Journ. Fluid Mech.* 679, 122–155.

Schoof, C. (2012). Marine ice sheet stability. *Journ. Fluid Mech.* 698, 62–72.

Schowalter, W. R. (1978). *Mechanics of non-Newtonian fluids*. Oxford: Pergamon Press, 300 pp.

Schulson, E. M. (1999). The structure and mechanical behavior of ice. *JOM*, 51, 21–27.

Schulson, E. M., and P. Duval (2009). *Creep and fracture of ice*. Cambridge: Cambridge University Press, 401 pp.

Schweizer, J., and A. Iken (1992). The role of bed separation and friction in sliding over an undeformable bed. *Journ. Glaciol.* 38, 77–92.

Schwerdtfeger, W. (1963). Theoretical derivation of the thermal conductivity and diffusivity of snow. *Int. Assoc. Hydr. Sci. Publ.* 61, 75–81.

Schwerdtfeger, W. (1970). The climate of the Antarctic. In: *World survey of climatology, vol. 14* (ed. H. E. Landsberg). Amsterdam: Elsevier, 253–355.

Schytt, V. (1958). The inner structure of the ice shelf at Maudheim as shown by core drilling. In: *Norwegian-British-Swedish Antarctic Expedition, 1949–52, Scientific Results 4, Glaciology 2.* Oslo: Norsk Polarinstitutt, 115–151.

Seddon, J. A. (1900). River hydraulics. *Trans. Amer. Soc. Civ. Engrs* 43, 179–243.

Sellers, W. D. (1965). *Physical climatology.* Chicago: Univ. Chicago Press, 272 pp.

Sergienko, O. V. (2010). Elastic response of floating glacier ice to impact of long-period waves. *Journ. Geoph. Res.,* 115, F04028, doi: 10.1029/2010JF001721.

Shea, J. M., F. S. Anslow, and S. J. Marshall (2005). Hydrometeorological relationships on Haig Glacier, Alberta, Canada. *Ann. Glaciol.* 40, 52–60.

Shepherd, A., D. Wingham, and E. Rignot (2004). Warm ocean is eroding West Antarctic Ice Sheet. *Geoph. Res. Ltrs.* 31, L23402, doi: 10.1029/2004GL021106.

Shoemaker, E. M. (1986). Debris-influenced sliding laws and basal debris balance. *Journ. Glaciol.* 32, 224–231.

Shreve, R. L. (1972). Movement of water in glaciers. *Journ. Glaciol.* 11, 205–214.

Sih, G. C. (1973). *Handbook of stress intensity factors.* Bethlehem, PA: Institute of Fracture and Solid Mechanics, LeHigh University.

Sih, G. C. (1974). Strain energy density factor applied to mixed mode crack problems. *Int. Journ. Fracture,* 10, 305–322.

Sikonia, W. G. (1982). Finite-element glacier dynamics model applied to Columbia Glacier. *US Geological Survey Prof. Paper* 1258-B, 74 pp.

Smeets, C. J. P. P., P. G. Duynkerke, and H. F. Vugts (1999). Observed wind profiles and turbulence fluxes over an ice surface with changing surface roughness. *Bound. Layer Meteorol.,* 92, 101–123.

Smith, G. D. (1985). *Numerical solution of partial differential equations* (3rd. ed.). Oxford: Clarendon Press, 337 pp.

Smith, R. A. (1976). The application of fracture mechanics to the problem of crevasse penetration. *Journ. Glaciol.* 17, 223–228.

Smith, R. A. (1978). Iceberg cleaving and fracture mechanics: A preliminary survey. In: *Iceberg Utilization* (ed. A. A. Husseiny). New York: Pergamon Press, 176–190.

Squires, G. L. (2001). *Practical physics* (4th ed.). Cambridge: Cambridge University Press, 228 pp.

Stearns, C. R., and G. A. Weidner (1993). Sensible and latent heat flux estimates in Antarctica. In: *Antarctic Meteorology and climatology: Studies based on automatic weather stations* (eds. D. H. Bromwich and C. R. Stearns). Washington, DC: American Geophysical Union, Antarctic Research Series 61, 109–138.

Stearns, L. A. (2007). *Outlet glacier dynamics in east Greenland and east Antarctica.* PhD Thesis, University of Maine (Orono). 113 pp.

Steinemann, S. (1954). Results of preliminary experiments on the plasticity of ice crystals. *Journ. Glaciol.* 2, 404–412.

Steinemann, S. (1958). Experimentelle untersuchungen zur plastizität von eis. *Beitr. Geol. Schweiz. Hydrol.* 10, 72 pp.

Stern, T. A., and U. S. Ten Brink (1989). Flexural uplift of the Transantarctic Mountains. *Journ. Geoph. Res.* 94(B), 10315–10330.

Stokes, C. R., and C. D. Clark (2004). Evolution of late glacial ice-marginal lakes on the northwestern Canadian Shield and their influence on the location of the Dubawnt Lake palaeo-ice stream. *Palaeogeog. Clim. Ecol.* 215, 155–171.

Strahler, A. N. (1987). *Science and Earth history: The evolution-creation controversy.* Buffalo NY: Prometheus Books, 552 pp.

Stroeven, A., R. van de Wal, and J. Oerlemans (1989). Historic front variations of the Rhône Glacier: Simulation with an ice flow model. In: *Glacier fluctuations and climatic change* (ed. J. Oerlemans). Dordrecht: Kluwer, 391–405.

Sturm, M. et al. (1991). Non-climatic control of glacier-terminus fluctuations in the Wrangell and Chugach Mountains, Alaska, U.S.A. *Journ. Glaciol.* 37, 348–356.

Swithinbank, C. (1977). Glaciological research in the Antarctic Peninsula. *Philos. Trans. Royal Soc. London Ser. B*, 279, 161–184.

Szidarovsky, F., K. Hutter, and S. Yakowitz (1989). Computational ice-divide analysis of a cold plane ice sheet under steady conditions. *Ann. Glaciol.* 12, 170–177.

Tada, H., P. C. Paris, and G. R. Irwin (1973). *The stress analysis of cracks handbook.* Hellerton, PA: Del Research Corporation.

Tangborn, W. (1997). Using low-altitude meteorological observations to calculate the mass balance of Alaska's Columbia Glacier and relate it to calving speed. In: *Calving glaciers: Report of a workshop,* February 28–March 2 (ed. C. J. van der Veen). BPRC Report No. 15. Columbus, OH: Byrd Polar Research Center, The Ohio State University, 141–161.

Taylor, G. I. (1938). Plastic strain in metals. *Journ. Inst. Metals* LXII, 307–324.

Taylor, J. R. (1997). *An introduction to error analysis* (2nd ed.). Sausalito CA: University Science Book, 327 pp.

Tennekes, H., and J. L. Lumley (1972). *A first course in turbulence.* Cambridge, MA: MIT Press, 300 pp.

Thomas, R. H. (1973). The creep of ice shelves: Theory. *Journ. Glaciol.* 12, 45–53.

Thomas, R. H. (1977). Calving-bay dynamics and ice-sheet retreat up the St. Lawrence valley system. *Geogr. Phys. Quat.* 31, 347–356.

Thomas, R. H. (1979). The dynamics of marine ice sheets. *Journ. Glaciol.* 24, 167–177.

Thomas, R. H. (1985). Responses of the polar ice sheets to climatic warming. In: *Glaciers, ice sheets, and sea level: Effect of a CO_2-induced climatic change.* Washington, DC: U.S. Dept. of Energy Report DOE/EV/60235-1, 301–316.

Thomas, R. H., and C. R. Bentley (1978). A model for the Holocene retreat of the West Antarctic Ice Sheet. *Quat. Res.* 10, 150–170.

Thomas, R. H., and D. R. MacAyeal (1982). Derived characteristics of the Ross Ice Shelf, Antarctica. *Journ. Glaciol.* 28, 397–412.

Thomas, R. H., T. J. O. Sanderson, and K. E. Rose (1979). Effect of a climatic warming on the West Antarctic Ice Sheet. *Nature* 227, 355–358.

Thomas, R. H. et al. (1984). Glaciological studies on the Ross Ice Shelf, Antarctica, 1973–1978. In: *The Ross Ice Shelf: Glaciology and Geophysics.* Washington, DC: American Geophysical Union, Antarctic Research Series 42, 21–53.

Thorsteinsson, T. et al. (2003). Bed topography and lubrication inferred from surface measurements on fast-flowing ice streams. *Journ. Glaciol.* 49, 481–490.

Truesdell, C., and W. Noll (1965). The non-linear field theories of mechanics. In: *Handbuch der physik, Vol. III/3* (ed. S. Flügge). Berlin: Springer-Verlag, 1–579.

Truffer, M., and K. A. Echelmeyer (2003). Of isbræ and ice streams. *Ann. Glaciol.* 36, 66–72.

Turcotte, D. L., and G. Schubert (2002). *Geodynamics* (2nd ed.). Cambridge: Cambridge University Press, 456 pp.

Van de Wal, R. S. W., and J. Oerlemans (1995). Response of valley glaciers to climatic change and kinematic waves: A study with a numerical ice-flow model. *Journ. Glaciol.* 41, 142–152.

Van den Broeke, M. et al. (2008). Partitioning of melt energy and meltwater fluxes in the ablation zone of the west Greenland Ice Sheet. *The Cryosphere* 2, 179–189.

Van den Broeke, M. et al. (2009). Partitioning recent Greenland mass loss. *Science* 326, 984–986.

Van der Veen, C. J. (1983). *A note on the equilibrium profile of a free floating ice shelf.* IMOU Report v 83–15, State University Utrecht, 15 pp.

Van der Veen, C. J. (1985). Response of a marine ice sheet to changes at the grounding line. *Quat. Res.* 24, 257–267.

Van der Veen, C. J. (1986). Numerical modelling of ice shelves and ice tongues. *Ann. Geoph.* 4(B), 45–54.

Van der Veen, C. J. (1987). Longitudinal stresses and basal sliding: A comparative study. In: *Dynamics of the West Antarctic Ice Sheet* (eds. C. J. van der Veen and J. Oerlemans). Dordrecht: Reidel Publ. Co., 223–248.

Van der Veen, C. J. (1989). A numerical scheme for calculating stresses and strain rates in glaciers. *Math. Geol.* 21, 363–377.

Van der Veen, C. J. (1996). Tidewater calving. *Journ. Glaciol.* 42, 375–385.

Van der Veen, C. J. (1998a). Fracture mechanics approach to penetration of surface crevasses on glaciers. *Cold Regions Sci. Techn.* 27, 31–47.

Van der Veen, C. J. (1998b). Fracture mechanics approach to penetration of bottom crevasses on glaciers. *Cold Regions Sci. Techn.* 27, 213–223.

Van der Veen, C. J. (1999a). Crevasses on glaciers. *Polar Geog.* 23, 213–245.

Van der Veen, C. J. (1999b). *Fundamentals of glacier dynamics.* Rotterdam: A. A. Balkema, 462 pp.

Van der Veen, C. J. (1999c). Evaluating the performance of cryospheric models. *Polar Geog.* 23, 83–96.

Van der Veen, C. J. (2001). Greenland ice sheet response to external forcing. *Journ. Geoph. Res.* 106, 34047–34058.

Van der Veen, C. J. (2002a). Calving glaciers. *Prog. Phys. Geog.* 26, 96–122.

Van der Veen, C. J. (2002b). Polar ice sheets and global sea level: How well can we predict the future? *Global Planet. Change* 32, 165–194.

Van der Veen, C. J. (2007). Fracture propagation as means of rapidly transferring surface melt-water to the base of glaciers. *Geoph. Res. Ltrs.* 34, L01501, doi: 1029/2006GL028385.

Van der Veen, C. J., and K. C. Jezek (1993). Seasonal variations in brightness temperatures for Central Antarctica. *Ann. Glaciol.* 17, 300–306.

Van der Veen, C. J., K. C. Jezek, and L. A. Stearns (2007). Shear measurements across the northern margin of Whillans Ice Stream. *Journ. Glaciol.* 53, 17–29.

Van der Veen, C. J., and A. J. Payne (2003). Modelling land-ice dynamics. In: *Mass balance of the cryosphere: observations and modeling of contemporary and future changes.* (eds. J. L. Bamber and A. J. Payne). Cambridge: Cambridge University Press, 169–225.

Van der Veen, C. J., J. C. Plummer, and L. A. Stearns (2011). Controls on the recent speed-up of Jakobshavn Isbræ, West Greenland. *Journ. Glaciol.* 57, 770–782.

Van der Veen, C. J., and I. M. Whillans (1989a). Force budget: I. Theory and numerical methods. *Journ. Glaciol.* 35, 53–60.

Van der Veen, C. J., and I. M. Whillans (1989b). Force budget: II. Application to two-dimensional flow along Byrd Station Strain Network, Antarctica. *Journ. Glaciol.* 35, 61–67.

Van der Veen, C. J., and I. M. Whillans (1990). Flow laws for glacier ice: Comparison of numerical predictions and field measurements. *Journ. Glaciol.* 36, 324–339.

Van der Veen, C. J., and I. M. Whillans (1993). Location of mechanical controls on Columbia Glacier, Alaska, U.S.A., prior to its rapid retreat. *Arctic Alpine Res.* 25, 99–105.

Van der Veen, C. J., and I. M. Whillans (1994). Development of fabric in ice. *Cold Reg. Sci. Techn.* 22, 171–195.

Van der Veen, C. J., and I. M. Whillans (1996). Model experiments on the evolution and stability of ice streams. *Ann. Glaciol.* 23, 129–137.

Van der Veen, C. J. et al. (1998). Surface roughness on the Greenland Ice Sheet from airborne laser altimetry. *Geoph. Res. Lett* 25, 3887–3890.

Van der Veen, C. J. et al. (2009). Surface roughness over the northern half of the Greenland Ice Sheet from airborne laser altimetry. *Journ. Geoph. Res.*, 114, F01001, doi: 1029/2008JF001067.

Van Dusen, M. S. (1929). Thermal conductivity of non-metallic solids. In: *International critical tables of numerical data: Physics, chemistry, and technology, vol. 5* (ed. E. W. Washburn). New York: McGraw Hill, 216–217.

Vaughan, D. G. (1993). Relating the occurrence of crevasses to surface strain rates. *Journ. Glaciol.* 39, 255–266.

Vaughan, D. G., and C. S. M. Doake (1996). Recent atmospheric warming and retreat of ice shelves on the Antarctic Pensinsula. *Nature* 379, 328–331.

Vaughan, D. G. et al. (1999). Distortion of isochronous layers in ice revealed by ground-penetrating radar. *Nature* 398, 323–326.

Vialov, S. S. (1958). Regularities of glacial shields movement and the theory of plastic viscou[r]s flow. *Int. Assoc. Hydr. Sci. Publ.* 47, 266–275.

Vieli, A., M. Funk, and H. Blatter (2001). Flow dynamics of tidewater glaciers: A numerical modeling approach. *Journ. Glaciol.* 47, 595–606.

Vieli, A., and F. M. Nick (2011). Understanding and modeling rapid dynamic changes of tidewater outlet glaciers: Issues and implications. *Surv. Geophys.* 32, 437–458.

Vieli, A., and A. J. Payne (2005). Assessing the ability of numerical ice sheet models to simulate grounding line migration. *Journ. Geoph. Res.* 110, F01003, doi: 10.1029/2004JF000202.

Vieli, A. et al. (2006). Numerical modeling and data assimilation of the Larsen B Ice Shelf, Antarctic Pensinsula. *Phil. Trans. Royal Soc. London Series A* 364, 1815–1839.

Vieli, A. et al. (2007). Causes of pre-collapse changes of the Larsen B Ice Shelf: Numerical modeling and assimilation of satellite observations. *Earth Planet. Sci. Lett.* 259, 297–306.

Von Mises, R. (1928). Mechanik der plastischen formänderung von kristallen. *Zeitsch. Angew. Mathem. Mechanik* 8, 161–184.

Vornberger, P. L., and I. M. Whillans (1990). Crevasse deformation and examples from Ice Stream B, Antarctica. *Journ. Glaciol.*, 36, 3–10.

Vyalov, S. S. (1986). *Rheological fundamentals of soil mechanics.* Amsterdam: Elsevier, 564 pp.

Wakahama, G. (1968). The metamorphism of wet snow. *Int. Assoc. Hydr. Sci. Publ.* 79, 370–379.

Walcott, R. I. (1970). Flexural rigidity, thickness, and viscosity of the lithosphere. *Journ. Geoph. Res.* 75, 3941–3954.

Walder, J. S. (1986). Hydraulics of subglacial cavities. *Journ. Glaciol.* 32, 439–445.

Walder, J. S. (2010). Röthlisberger channel theory: Its origins and consequences. *Journ. Glaciol.* 56, 1079–1086.

Walker, J. C., and E. D. Waddington (1988). Early discoverers XXXV. Descent of glaciers: Some early speculation on glacier flow and ice physics. *Journ. Glaciol.* 34, 342–348.

Walter, F. et al. (2010). Iceberg calving during transition from grounded to floating ice: Columbia Glacier, Alaska. *Geoph. Res. Lett.* 37, L15501, doi: 10.1029/2010GL043201.

Warren, C., and M. Aniya (1999). The calving glaciers of South America. *Global Planet. Change* 22, 59–77.

Warren, C. R., D. Greene, and N. F. Glasser (1995). Glaciar Upsala, Patagonia: Rapid calving retreat in fresh water. *Ann. Glaciol.* 21, 311–316.

Weertman, J. (1957a). On the sliding of glaciers. *Journ. Glaciol.* 3, 33–38.

Weertman, J. (1957b). Deformation of floating ice shelves. *Journ. Glaciol.* 3, 38–42.

Weertman, J. (1958). Travelling waves on glaciers. *Int. Assoc. Hydr. Sci. Publ.* 47, 162–168.

Weertman, J. (1961b). Mechanism for the formation of inner moraines found near the edge of cold ice caps and ice sheets. *Journ. Glaciol.* 3, 965–978.

Weertman, J. (1961c). Equilibrium profile of ice caps. *Journ. Glaciol.* 3, 953–964.

Weertman, J. (1964). The theory of glacier sliding. *Journ. Glaciol.* 5, 287–303.

Weertman, J. (1968). Comparison between measured and theoretical temperature profiles of the Camp Century, Greenland, Borehole. *Journ. Geoph. Res.* 73, 2691–2700.

Weertman, J. (1972). General theory of water flow at the base of a glacier or ice sheet. *Rev. Geoph. Space Phys.* 10, 287–333.

Weertman, J. (1973a). Creep of ice. In: *Physics and chemistry of ice* (eds. S. J. Jones and L. W. Gold). Ottawa: Royal Soc. Canada, 320–337.

Weertman, J. (1973b). Can a water-filled crevasse reach the bottom surface of a glacier? *Int. Assoc. Hydr. Sci. Publ.* 95, 139–344.

Weertman, J. (1974) Stability of the junction of an ice sheet and an ice shelf. *Journ. Glaciol.* 13, 3–11.

Weertman, J. (1979). The unsolved general glacier sliding problem. *Journ. Glaciol.* 23, 97–115.

Weertman, J. (1980). Bottom crevasses. *Journ. Glaciol.*, 25, 185–188.

Weertman, J., and G. E. Birchfield (1982). Subglacial water flow under ice streams and West Antarctic ice-sheet stability. *Ann. Glaciol.* 3, 316–320.

Weiss, J. W. et al. (2002). Dome Concordia ice microstructure: Impurities effect on grain growth. *Ann. Glaciol.* 35, 552–558.

Whillans, I. M. (1987). Force budget of ice sheets. In: *Dynamics of the West Antarctic Ice Sheet* (eds. C. J. van der Veen and J. Oerlemans). Dordrecht: Reidel, 17–36.

Whillans, I. M., M. Jackson, and Y.-H. Tseng (1993). Velocity pattern in a transect across Ice Stream B, Antarctica. *Journ. Glaciol.* 39, 562–572.

Whillans, I. M., and K. C. Jezek (1987). Folding in the Greenland Ice Sheet. *Journ. Geoph. Res.* 92, 485–493.

Whillans, I. M., and S. J. Johnsen (1983). Longitudinal variations in glacial flow: Theory and test using data from the Byrd Station Strain Network, Antarctica. *Journ. Glaciol.* 29, 78–97.

Whillans, I. M., and C. J. van der Veen (1993). New and improved determinations of velocity of ice streams B and C, West Antarctica. *Journ. Glaciol.* 39, 483–490.

Whillans, I. M., and C. J. van der Veen (1997). The role of lateral drag in the dynamics of Ice Stream B, Antarctica. *Journ. Glaciol.* 43, 231–237.

Whillans, I. M., and C. J. van der Veen (2001). Transmission of stress between an ice stream and interstream ridge. *Journ. Glaciol.* 47, 433–440.

Whillans, I. M. et al. (1989). Force budget: III. Application to three-dimensional flow of Byrd Glacier, Antarctica. *Journ. Glaciol.* 35, 68–80.

Xu, Z. (1992). *Applied elasticity.* New York, NY: John Wiley & Sons, 373 pp.

Yen, Y.-C. (1981). *Review of thermal properties of snow, ice, and sea ice.* CRREL Report 81–10. Hanover, NH: Cold Regions Research and Engineering Laboratory, 27 pp.

Zumberge, J. H. et al. (1960). *Deformation of the Ross Ice Shelf near the Bay of Whales, Antarctica.* Columbus, OH: IGY Glaciological Report Series 3, 148 pp.

Index

Milton Keynes UK
Ingram Content Group UK Ltd.
UKHW030900141024
449569UK00025B/1305

9 781138 077218